Energy 2050

Energy 2050

Making the Transition to a Secure Low Carbon Energy System

Edited by
Jim Skea, Paul Ekins and Mark Winskel

First published by Earthscan in the UK and USA in 2011

For a full list of publications please contact:

Earthscan
2 Park Square, Milton Park, Abingdon, Oxfordshire OX14 4RN
711 Third Avenue, New York, NY 10017

First issued in paperback 2015

Earthscan is an imprint of the Taylor & Francis Group, an informa business

ISBN 13: 978-1-138-96879-0 (pbk)
ISBN 13: 978-1-84971-084-8 (hbk)

Typeset by Bookcraft Ltd, Stroud, Gloucestershire
Cover design by Rob Watts

A catalogue record for this book is available from the British Library

Library of Congress Cataloging-in-Publication Data

Energy 2050 : making the transition to a secure low carbon energy system / edited by Jim Skea, Paul Ekins, and Mark Winskel.
    p. cm.
Includes bibliographical references and index.
    ISBN 978-1-84971-084-8 (hardback)
    1. Energy policy–Great Britain. 2. Climatic changes–Government policy–Great Britain. 3. Environmental policy–Great Britain. 4. Energy security–Great Britain--Planning. I. Skea, Jim. II. Ekins, Paul. III. Winskel, Mark.
    HD9502.G72E58 2010
    333.790941–dc22

2010029332

Contents

List of Figures	*ix*
List of Tables	*xiii*
List of Contributors	*xv*
Acknowledgements	*xxi*
Acronyms and Abbreviations	*xxii*
Conversion Matrix	*xxv*

1 Introduction — **1**
Jim Skea, Paul Ekins and Mark Winskel
The long-term challenge of secure low carbon energy — 1
Thinking about energy futures — 4
How the book was written — 6
Structure of the book — 7

2 UK Energy in an Era of Globalization: Trends, Technologies and Environmental Impacts — **11**
Jim Skea, Xinxin Wang and Mark Winskel
Introduction — 11
Long-term energy trends — 12
Final energy demand — 14
How energy is used — 14
Future energy demand technologies — 17
Primary energy demand — 19
Trends in electricity generation — 21
Future electricity generation technologies — 23
Energy trade and self-sufficiency — 27
Energy infrastructure — 28
Environmental concerns — 31
Conclusions — 39

3   **UK Energy Policy and Institutions**   **41**

Introduction   41
Paul Ekins, Jim Skea and Mark Winskel
Ownership issues, late 1940s to mid-1990s   41
Managing and regulating the flow of electricity   43
Managing and regulating the flow of gas   44
New issues in energy policy   45
UK climate and energy policy development   47
UK policies for CO_2 emissions reduction, 2000–2010   49
Policies for energy security   57
Conclusion   61

4   **Energy Futures: The Challenges of Decarbonization and Security of Supply**   **67**

Jim Skea, Gabrial Anandarajah, Modassar Chaudry, Anser Shakoor, Neil Strachan, Xinxin Wang and Jeanette Whitaker
Introduction   67
Energy systems, decarbonization and resilience   69
The scenario framework   73
Scenario analysis and modelling tools   78
Key assumptions in the core scenarios   91
Reference scenario results   95
The gap between the Reference scenario and policy aspirations   101

5   **Pathways to a Low Carbon Economy**   **105**

Gabrial Anandarajah, Paul Ekins and Neil Strachan
Introduction   105
Scenario design   105
Scenario results   107
Insights and conclusions   133
Annex 5.1: Data for calculation of carbon tax implied by UK Climate Change Levy (CCL)   143

6   **A Resilient Energy System**   **145**

Jim Skea, Modassar Chaudry, Paul Ekins, Kannan Ramachandran, Anser Shakoor and Xinxin Wang
Introduction   145
What can go wrong: shocks to the energy system   146
Indicators of resilience   150
Quantifying resilience at the macro level   153
Resilience: implications for energy markets and technologies   155
Reliability in the network industries   172
Hypothetical system shocks   177
Mitigating the shocks   179
Adding up the costs of resilience   181
Policy implications   184

7 Accelerating the Development of Energy Supply Technologies:
The Role of Research and Innovation 187
Mark Winskel, Gabrial Anandarajah, Jim Skea and Brighid Jay
Introduction 187
Technological innovation and energy system change 188
The accelerated technology development scenarios 192
Scenarios, system modelling and the real world 196
Accelerated development scenarios and UK
   decarbonization pathways 198
Implications and challenges 208
Summary and conclusions 214

8 A Change of Scale? Prospects for Distributed Energy Resources 219
*Adam Hawkes, Noam Bergman, Chris Jardine, Iain Staffell,
Dan Brett, and Nigel Brandon*
Introduction 219
Challenges in the residential sector 219
Technology characteristics, performance and suitability 222
The human dimension: installers and householders 235
Policy challenges for distributed energy resources 245
Conclusions 252

9 The Way We Live From Now On: Lifestyle and Energy
Consumption 258
*Nick Eyre, Jillian Anable, Christian Brand, Russell Layberry
and Neil Strachan*
Introduction 258
Quantifying lifestyle 261
Lifestyle change at home 267
Lifestyle change in mobility and transport 271
Lifestyle change for a low carbon world 276
Public policy implications 281
Conclusions 287

10 Not Just Climate Change: Other Social and Environmental
Perspectives 294
*David Howard, Brighid Jay, Jeanette Whitaker, Joey Talbot,
Nick Hughes and Mark Winskel*
Introduction 294
Environmental pressures 295
Socio-environmental sensitivities 306
Conclusions 320

11 UK Energy in an Uncertain World   324
    Neil Strachan and Jim Skea
    Introduction   324
    Assumptions and scenarios   325
    Results   329
    Conclusions   337

12 Putting It All Together: Implications for Policy and Action   341
    Paul Ekins, Mark Winskel and Jim Skea
    Introduction   341
    Current energy system challenges   342
    The historic context   343
    Comparing the scenarios   344
    Multiple pathways to a low carbon economy   350
    Building in resilience   351
    Accelerating technological change   352
    Decarbonizing electricity supply   353
    Energy demand   357
    Lifestyle change   358
    Environmental concerns   360
    The international context   362
    In conclusion   362

Index   367

List of Figures

2.1   UK primary energy demand 1700–2008   12
2.2   UK GDP and primary energy demand 1948–2008   13
2.3   Final energy demand by sector 1970–2008   15
2.4   Final energy demand by fuel 1970–2008   15
2.5   Breakdown of household energy demand 1970–2008   16
2.6   Breakdown of transport energy demand 1970–2008   17
2.7   Breakdown of industrial energy demand 2007   18
2.8   Primary energy demand by fuel 1970–2008   20
2.9   Electricity generation mix 1970–2007   21
2.10   Renewable electricity generation 1990–2008   22
2.11   Changes in transmission-contracted generation capacity to 2015/16   23
2.12   Development of existing GB generating capacity   24
2.13   UK crude oil production and demand   27
2.14   UK natural gas production and demand   28
2.15   The electricity supply system in Great Britain   29
2.16   The gas supply system in Great Britain   31
2.17   UK CO_2 emissions 1970–2009   33
2.18   Sulphur dioxide emissions 1970–2007   34
2.19   Emissions of nitrogen oxides 1970–2007   35
2.20   Emissions of PM-10 particulates 1970–2007   36
2.21   Abstractions of water in England and Wales 1995–2007   37
3.1   Fuel poverty and real fuel prices 1996–2007   47
3.2   Development of the UK fridge freezer market by energy rating   57
3.3   Gas capacity versus demand and supply   59
4.1   Conceptualizing energy security   71
4.2   The UKERC core scenarios   74
4.3   Operation of the three models in tandem   81
4.4   Representation of MARKAL-MED supply–demand equilibrium   85
4.5   CGEN optimization model   88
4.6   Location of a CCGT plant   90
4.7   Network infrastructure expansion   90
4.8   Final energy demand by sector   96
4.9   Final energy demand by fuel   96
4.10   Transport fuel demand   97

4.11	Passenger cars distance travelled	97
4.12	Buses distance travelled	98
4.13	Installed electricity generation capacity mix	99
4.14	Electricity generation	99
4.15	Primary energy demand	100
4.16	CO_2 emissions by sector	100
5.1	CO_2 emissions under scenarios with different annual carbon constraints	108
5.2	CO_2 emissions under cumulative emissions scenarios	108
5.3	Sectoral CO_2 emissions during 2000–2050 in REF	109
5.4	Sectoral CO_2 emissions in years 2000, 2035, 2050: carbon ambition scenarios	110
5.5	Sectoral CO_2 emissions in years 2035, 2050: LC and cumulative emissions scenarios	111
5.6	Primary energy demand under different scenarios	112
5.7	Primary energy demand in selected years under different scenarios	113
5.8	Primary energy demand in selected years in LC and cumulative emissions scenarios	114
5.9	Share of biomass provided by imports	114
5.10	Final energy demand under different scenarios	115
5.11	Final energy demand by fuel under different scenarios	116
5.12	Final energy demand by fuel in LC, LC-EA, LC-LC and LC-SO	117
5.13	Electricity generation fuel mix in 2020 in REF	118
5.14	Electricity generation mix under different scenarios	119
5.15	Electricity generation mix in LC, LC-EA, LC-LC and LC-SO	119
5.16	Marginal price of CO_2 emissions under different scenarios	120
5.17	CO_2 emissions and marginal cost of CO_2 under different scenarios	121
5.18	CO_2 emissions and marginal cost of CO_2 under LC and cumulative emissions scenarios	121
5.19	Selected energy service demand reductions in 2050 under different scenarios	123
5.20	Change in social welfare under different scenarios	124
5.21	Sectoral energy demand under different scenarios	125
5.22	Sectoral energy demand under LC, LC-EA, LC-LC and LC-SO scenarios	126
5.23	Sectoral bio-fuel energy demand under different scenarios	127
5.24	Electricity demand for heat pumps under different scenarios	127
5.25	Transport sector energy demand by modes under different scenarios	128
5.26	Transport sector energy demand by modes in LC, LC-EA, LC-LC and LC-SO	129
5.27	Installed capacity in REF during 2000–2050	130
5.28	Installed capacity under different scenarios	132
5.29	Installed capacity under LC, LC-EA, LC-LC and LC-SO scenarios	132
5.30	Offshore wind installed capacity under different scenarios	133

6.1	Major world oil supply disruptions over the last 50 years	150
6.2	Primary energy supply in the core scenarios	157
6.3	Final energy demand in the core scenarios by fuel type	159
6.4	Final energy demand in the core scenarios by sector	160
6.5	Transport fuel demand in the core scenarios (2050)	161
6.6	Demand price response in the transport sector (2050)	161
6.7	Car fleets in the core scenarios (2050)	162
6.8	Residential energy system fuel demand in the core scenarios (2050)	163
6.9	Residential heating in the core scenarios by type (2050)	163
6.10	Electricity demand in the core scenarios by sector	165
6.11	Electricity generation mix in the core scenarios	166
6.12	Installed generation capacity in the core scenarios	167
6.13	CO_2 emissions in the core scenarios by sector	169
6.14	CO_2 intensity of electricity generation in the core scenarios	170
6.15	Marginal cost of CO_2 abatement	170
6.16	Welfare costs and changes in energy system costs in the core scenarios with respect to the Reference scenario	171
6.17	Capacity margin using different reliability approaches, Low Carbon scenario	174
6.18	Loss-of-load expectation using capacity margin approach, Low Carbon scenario	175
6.19	Gas supply/demand balance in the four core scenarios	176
7.1	UK Public spending on energy RD&D, 1974–2008	190
7.2	The innovation chain	192
7.3	Capital cost assumptions for the ATD solar PV scenarios	194
7.4	Marine energy installed capacity under accelerated and non-accelerated scenarios (selected data, smoothed)	202
7.5	Power generation portfolios under different ATD scenarios	203
7.6	Share of electricity generated, LC Acctech (late CCS) scenario	204
7.7	Use of biomass for power and residential heating, non-accelerated (LC-60) and accelerated (LC-60-Bio-energy) scenarios	205
7.8	Primary energy demand by fuel in 2050, LC and LC Acctech scenarios	206
7.9	Welfare increases associated with technology acceleration (2010–2050)	208
8.1	Change in mix of energy consumed in the UK residential sector between 1970 and 2008	220
8.2	Greenhouse gas contribution by source for the UK residential sector from 1970 to 2008	221
8.3	Diagram depicting the typical size and position of different micro-generation technologies within the home	223
8.4	A Sankey diagram showing the flows of energy required to heat and power a typical house for a year using a heat pump	226

8.5   A sankey diagram showing the flow of energy required to heat
       and power a typical house for a year using fuel cell based
       micro-CHP   230
8.6   Categories of adoption according to diffusion of innovation
       theory   238
8.7   A simplistic model of micro-generation diffusion in the
       residential sector   239
8.8   A model of micro-generation diffusion in a reoriented
       residential energy sector   244
8.9   Modelled CO_2 emissions reduction provided by a $1kW_e$ solid
       oxide fuel cell (SOFC) micro-CHP system versus a high
       temperature air source heat pump operating in a typical UK
       terrace house, with sensitivity to gas and electricity emissions
       rates   251
9.1   Outline of the modelling process   263
9.2   UK household energy use 1970–2007   268
9.3   Final energy use in the Reference and Lifestyle scenarios   278
10.1   The DPSIR (Driver-Pressure-State-Impact-Response) model as
       originally defined by EEA (1995)   296
10.2   The proportion of UK pollutant attributed to different sources 298
10.3   Total emissions of sulphur dioxide (SO_2) over time in the core
       scenarios   300
10.4   Total emissions of particulates (PM_{10}) in the core scenarios   301
10.5   Total emissions of carbon monoxide (CO) in the core scenarios   302
10.6   Total emissions of nitrous oxide (N_2O) in the core scenarios   303
10.7   Total radioactive releases in the core scenarios   304
10.8   Land-take for bio-energy in the core scenarios   305
10.9   Sectoral CO_2 emissions in 2050   314
10.10 Electricity generation in 2050   314
10.11 Installed capacity in 2050   315
10.12 Residential electricity demand reduction   316
10.13 Residential gas demand reduction   316
10.14 Marginal cost of CO_2 318
10.15 Societal welfare expressed as consumer and producer surplus   318
11.1   Final energy (petajoules, PJ)   331
11.2   UK sectoral CO_2 emissions (2050)   332
11.3   UK CO_2 emissions and use of emission purchases   332
11.4   Transport final energy demand (2035, 2050)   335
11.5   Welfare costs, £(2000) billion   336
11.6   Marginal CO_2 prices, £(2000)/tCO_2 337

List of Tables

2.1	UK radioactive waste materials inventory (cubic metres)	38
4.1	Events impacting on the energy system	72
4.2	List of scenarios and scenario variants	76
4.3	Key characteristics of the modelling tools	80
4.4	Key assumptions in the core scenarios	91
4.5	World energy price assumptions	92
4.6	Price elasticities of energy service demands	93
4.7	Intermittent generation capacity credit	94
4.8	Main car technologies included in MARKAL-MED	95
5.1	Carbon pathway scenarios	106
6.1	Major electricity blackouts and disturbances	147
6.2	Gas supply crises and accidents	148
6.3	Global crude oil production and peak supply loss in each disruption	149
6.4	Possible resilience indicators for primary energy supply	151
6.5	Possible resilience indicators for energy infrastructure	151
6.6	Possible resilience indicators for energy users	152
6.7	Macro-level resilience indicators	153
6.8	Final energy demand constraints	154
6.9	Demand reduction in the core scenarios with respect to the Reference scenario	172
6.10	Reliability indicators for gas and electricity	173
6.11	Additional capacity and system costs to ensure reliability	175
6.12	Gas infrastructure investments	176
6.13	Description of facilities lost in the hypothetical shocks	177
6.14	Impact of 40-day shocks in the four core scenarios	178
6.15	Impact of the loss of Easington for different periods	179
6.16	Gas infrastructure projects	179
6.17	Impact of mitigating investments: Bacton out for 40 days	181
6.18	Estimated costs associated with different aspects of resilience in 2025	182
7.1	Accelerated technology development (ATD) scenario set	195
7.2	ATD scenarios: summary of assumptions and impacts	200
7.3	Innovation themes and research priorities for ATD scenarios	210

8.1	A comparison of metrics for micro-generation technologies	233
9.1	The lifestyle scenarios related to the core scenarios	262
9.2	Summary results of the Lifestyle scenario (LS-REF)	277
11.1	Fossil fuel import prices ($, £ 2009)	326
11.2	International CO_2 emissions credits availability and costs	327
11.3	Scenarios undertaken on global energy uncertainties	328
11.4	Primary energy (petajoules, PJ)	330
11.5	Electricity generation (TWh) and capacity (GW) in 2050	334
12.1	Comparative table of Energy 2050 scenarios: carbon and energy	346
12.2	Comparative table of Energy 2050 scenarios: economic variables	348

List of Contributors

Dr Jillian Anable is a Senior Lecturer at the Centre for Transport Research, University of Aberdeen. She is an expert in travel behaviour, climate change and energy policy with particular emphasis on the potential for demand-side solutions. She has developed methods of combining social psychological theories of behaviour and market segmentation techniques to assess the potential to influence travel and car choice. Recent research has specialized in the monitoring and evaluation of travel behaviour in response to 'soft measures' or 'smarter choice' interventions as well as the likely consumer response to electric vehicle technology.

Dr Gabrial Anandarajah is a Senior Research Associate at the UCL Energy Institute. He gained his PhD from University College Dublin (UCD) in 2006, Master's degree from Asian Institute of Technology (AIT) in the field of Energy Economics and Planning. He has over seven years' research experience in energy, environment and climate change mitigation modelling for UK and for Asian developing countries. Gabrial is an energy system modeller. He is involved in UK MARKAL modelling work identifying long-term energy, technology and policy options for climate change mitigation. Presently, he is leading the TIAM-UCL global model development for the UKERC Phase II project. Gabrial is the author of over ten peer reviewed journal papers.

Dr Noam Bergman is a Research Analyst in the Lower Carbon Futures group in the Environmental Change Institute (ECI) at the University of Oxford. He is currently working on micro-generation and small-scale renewables and their part in a shift to a low carbon future. Noam joined the ECI from the Tyndall Centre at the University of East Anglia, where he worked on the European MATISSE project, focusing on transitions to sustainable development, and developing modelling tools for policy-makers to assess sustainability implications of proposed policies, and leading on a case study of transitions to sustainable housing in the UK. Noam completed a PhD in Earth System Sciences at the School of Environmental Sciences of UEA in Norwich in 2003, where he developed COPSE, a geological time-scale bio-geochemical model.

Dr Christian Brand is Senior Researcher in Transport and Energy at the University of Oxford. He has over 15 years' research experience in both academic and consultancy environments focusing on integrated analysis of the interface between transport, energy and the environment. He is co-investigator of the EPSRC-funded iConnect study, measuring and evaluating the travel and carbon effects of physical interventions for walking and cycling. He led the ESRC-funded project Counting your carbon: the development and deployment of integrated travel emissions profiles, covering annual travel activity across all modes of transport and associated carbon emissions at the individual, household and local levels. He is a transport modeller within the Energy Demand theme of the UK Energy Research Centre, focusing on strategic modelling of low carbon transport policies, including pricing, electrification and 'smarter choices'.

Prof. Nigel Brandon FREng is Director of the Energy Futures Lab at Imperial College London. His research interests focus on electrochemical power systems, with a particular interest in fuel cell science, engineering and technology. He was awarded the 2007 Silver Medal from the Royal Academy of Engineering for his contribution to engineering leading to commercial exploitation.

Dr Dan Brett has a PhD in Electrochemistry from Imperial College London and is a Lecturer in Energy at UCL (Department of Chemical Engineering). He is the Co-director of the Centre for CO_2 Technology where he leads the Electrochemical Energy Storage and Conversion Group. Dan specializes in developing novel diagnostic techniques for the study of fuel cells and is also active in modelling, engineering design, device fabrication, materials development and techno-economics of electrochemical energy conversion technologies. He is interested in micro-generation technologies and particularly the application of fuel cells in this area.

Dr Modassar Chaudry is a Research Fellow in the School of Engineering at Cardiff University. Modassar's previous appointments were as Research Associate at the University of Manchester and Analyst at an energy consultancy. Modassar's research spans a range of energy topics and in particular the modelling of the gas and electricity supply sector. This includes knowledge of the whole energy supply chain from power generation, fuel supply, transmission networks and final end demand. He contributed extensively to the UKERC Energy 2050 project, and is currently extending this work within the UKERC Energy Supply Theme.

Prof. Paul Ekins has a PhD in economics from the University of London and is a Co-Director of the UK Energy Research Centre, heading its Energy Systems theme. He is Professor of Energy and Environment Policy at the UCL Energy Institute, University College London. His academic work focuses on the conditions and policies for achieving an environmentally sustainable economy, with a special focus on energy technologies and policy, innovation, the role of economic instruments, sustainability assessment and environment and trade.

Dr Nick Eyre is Programme Leader of the Lower Carbon Futures group in the Environmental Change Institute at the University of Oxford, and a Senior Research Fellow at Oriel College, Oxford. He is a Co-Director of the UK Energy Research Centre, leading its work on energy demand. Nick previously worked at the Energy Saving Trust as Director of Strategy and, on second-ment, in Cabinet Office, where he was a co-author of the Government's 2002 Review of Energy Policy. Nick has been a researcher, consultant, policy analyst and programme manager on energy and environment issues for 25 years. His current research interests focus on energy policy, especially with respect to energy demand and small-scale generation technologies. Nick has published extensively on energy and climate issues, including being a co-author of a recent book on carbon markets. He is a lead author in the ongoing Global Energy Assessment.

Dr Adam Hawkes graduated a decade ago with a First in Mechanical Engineering from the University of Western Australia. Initially he pursued consultancy, working in a large energy trading house and on infrastructure projects. He began energy policy and technology research with the Department of Premier and Cabinet in Australia, considering liberalization of energy markets and sustainable transport futures. In 2003 he joined Imperial College London as a Research Associate, focusing on decentralized energy systems, supporting policy, and energy market issues. He was appointed as a Research Fellow in 2008. Adam is currently working with AEA Technology leading their modelling activities.

Dr David Howard leads Lancaster Environment Centre's Centre for Sustainable Energy. He has worked for the Centre for Ecology and Hydrology (and its predecessor the Institute of Terrestrial Ecology) for nearly a quarter of a century studying land use and energy. He takes a landscape ecology approach to the capacity of the environment to provide energy and the impacts of its use. Current projects include an investigation of the relationships between energy-scapes and ecosystem services. David was a Co-Director of the UK Energy Research Centre, leading the theme on Environmental Sustainability. He is a member of the Board of Governors of the Joule Centre.

Nick Hughes has undertaken research on low carbon energy technology and policy, in posts at the Policy Studies Institute and at King's College London, for the UK Energy Research Centre, and as part of the SUPERGEN-funded UK Sustainable Hydrogen Energy Consortium. He has also worked as a policy advisor at the Department of Energy and Climate Change, and has co-authored book chapters on future low carbon electricity networks, and on the socio-economics of hydrogen energy. He is currently engaged in doctoral research on future electricity networks at the Imperial College Centre for Energy Policy and Technology.

Dr Chris Jardine is a Senior Research Analyst in the Lower Carbon Futures group in the Environmental Change Institute (ECI) at the University of Oxford. His research focus is on studying technologies for greenhouse gas emission reductions. Chris is currently leading the ECI's work on the SUPERGEN consortium on Highly Distributed Power Systems. This project is examining how the electricity network might function in future, if there is a very high penetration of small-scale renewables and micro-generation on the grid. Chris has completed a number of other initiatives at the ECI, most notably the PV-Compare project, which was a technical comparison of different photo-voltaic technologies under UK and Mallorcan climatic conditions. Chris has a degree in Chemistry and a PhD in Computational Inorganic Chemistry, both from the University of Oxford.

Brighid Moran Jay is a Research Associate at the University of Edinburgh's Institute for Energy Systems. She has worked on energy issues at the university and small NGOs, as well as on large national programmes in the United States where she managed and delivered energy trainings and programmes for organizations such as the US Department of Energy. As a member of the Institute's Innovation & Policy Group, she has been involved in projects on innovation theory, marine energy roadmaps, accelerated development of energy technologies and the impact of public acceptance on the deployment of energy technologies with the UK Energy Research Centre, the EU-ORECCA Project and others.

Dr Russell Layberry is a Senior Researcher in the Lower Carbon Futures group of the Environmental Change Institute in the University of Oxford. He has been responsible for development of models of energy use in buildings, including the UK Domestic Carbon Model, within the programme of the UK Energy Research Centre. He is a computational physicist specializing in modelling, remote sensing and its environmental applications, and software development. He holds a doctorate from the University of Aston and has previously been a post-doctoral researcher at the universities of Bristol and Sussex.

Dr Kannan Ramachandran is a leading international energy expert in MARKAL/TIMES energy system modelling, energy management, life cycle costing, cash-flow analysis, techno-economic feasibility study, environmental life cycle assessment, energy economics, climate change and related issues with more than ten years of experience from the UK, India, Singapore, Thailand, Germany, Switzerland and many other Asian countries. He has an interdisciplinary academic background in energy science, energy engineering and project management. He has authored a number of international energy-environmental-economics journal papers, book chapters, project documents and conference papers. He has been referee for a number of journals. He has also reviewed the International Energy Agency's *World Energy Outlook 2009*.

Dr Anser Shakoor is a Research Fellow at Imperial College London and a senior staff member of the UK Centre for Distributed Generation and Sustainable Electrical Energy. He specializes in the evaluation of security and economic performance of electricity systems. His recent research has focused on technical, market and regulatory requirements associated with the strategic development of energy systems with large-scale integration of renewable energy and emerging demand sources. His earlier work has contributed to UK Energy White Papers (2003–2007). He was among the leading researchers in several high profile European projects commissioned by the European Commission, and in UKERC Phase I. He has been working on collaborative projects with the World Bank, IEA, IAEA and as a consultant for leading energy utilities and consulting companies.

Prof Jim Skea's research interests are in energy, climate change and technological innovation. He has been Research Director of the UK Energy Research Centre since 2004, and a Professor of Sustainable Energy at Imperial College London since 2009. He previously directed the Policy Studies Institute and the Economic and Social Research Council's Global Environmental Change Programme. He is a founding member of the UK's Committee on Climate Change and a Vice-Chair of the Intergovernmental Panel on Climate Change Working Group III.

Dr Iain Staffell graduated from the University of Birmingham with degrees in Physics (BSc) and the Physics and Technology of Nuclear Reactors (MSc), before moving to the Fuel Cells Group in Chemical Engineering. There he studied for an interdisciplinary PhD reviewing the economic and environmental potential of fuel cells for domestic heat and power generation. After graduating in 2010, he now manages the University's fleet of hybrid fuel cell electric vehicles, and continues his cross-disciplinary research into micro-generation.

Dr Neil Strachan is an interdisciplinary energy economist. He is a Reader in Energy Economics and Modelling at University College London. He received his PhD in Engineering and Public Policy from Carnegie Mellon University in 2000. At the UCL Energy Institute, Neil's research interests revolve around energy-environment-economic modelling, the quantification of scenarios and transitions pathways, and interdisciplinary issues in new energy technology diffusion. He is a lead author of the Energy Systems chapter of the IPCC's 5th Assessment Report. He is the author of over 20 peer reviewed journal papers and book chapters.

Joey Talbot is currently reading for a PhD at the University of Leeds, on relationships between biodiversity and carbon dynamics in tropical forests. While working as an environmental energy researcher at CEH Lancaster, he investigated the impacts of future energy scenarios on the emissions of a range of common pollutants. He has gained a First in BSc Geography from Swansea University, a Distinction in MRes Science of the Environment from Lancaster University, and a BTEC in Tropical Habitat Conservation. Voluntary work undertaken has included the production of village land-use maps for the Ujamaa Community Resource Trust, Arusha, Tanzania.

Dr Xinxin Wang has been a Research Associate at the UK Energy Research Centre since late 2006. Her research interests include energy technology, energy and environmental policy and climate change. Previously, she completed her PhD in the Department of Electrical and Electronic Engineering at Imperial College London. Xinxin is also a graduate in Electrical Engineering from Xi'an Jiaotong University in China.

Dr Jeanette Whitaker is an ecologist at the Centre for Ecology and Hydrology, Lancaster. She has 12 years' research experience in the impacts of human activity on terrestrial ecosystems, with special emphasis on soil–plant–microbe interactions. Over the last 12 years she has investigated the impacts of a range of pollutants on soil communities (metals, organics, radionuclides, N deposition). In the last four years her research has focused on the environmental impacts of changing the UK energy system to mitigate climate change, with a particular emphasis on bio-energy and carbon sequestration. She has published 19 journal papers (more than 500 citations).

Dr Mark Winskel is Research Co-ordinator of the UK Energy Research Centre, and Senior Research Fellow in the Institute for Energy Systems, University of Edinburgh. He has held a number of research grant awards on innovation in energy systems, and acted as an advisor for organizations such as the Carbon Trust and the International Energy Agency. He has an interdisciplinary education across the natural, environmental and social sciences, and his PhD addressed technological change in the UK electricity supply sector.

Acknowledgements

The research on which this book is based was funded by the UK Natural Environment Research Council (NERC) under award NE/C513169/1. NERC made the award on behalf of two other research Councils, the Engineering and Physical Sciences Research Council (EPSRC) and the Economic and Social Research Council (ESRC). UKERC plays a central role in the UK Research Councils Energy Programme and continues to be funded by these three research councils. Work on the UKERC Energy 2050 project was conducted by members of the UK Energy Research Centre (UKERC) during its Phase I activities (2004-09).

The editors thank all of the chapter authors for their timely and considered contributions to the book and for so willingly fitting in with the over-arching framework under which the research and the book were conceived. During the course of the work, many other members of UKERC contributed to discussions and debates which helped to sharpen the analysis and conclusions. We would like to acknowledge these contributions. A number of external reviewers commented on earlier working papers, helping to improve both the presentation and content. Our particular thanks go to Jeannie Cruickshank (DECC), Chris Dent (University of Durham), Nikos Hatziargyriou (National Technical University of Athens), David Joffe (Committee on Climate Change) and Denise Van Regemorter (Center for Economic Studies, Catholic University of Leuven). Our thanks are also extended to the many academics, policymakers and others – too numerous to mention individually – who attended a workshop held in London in April 2009, when the initial results from the project were presented. A number of other workshops and briefings were also held between 2007 and 2009 to allow for more focused feedback on particular parts of the research.

Staff from the UKERC headquarters and the Meeting Place function helped to facilitate UKERC internal meetings during the course of the project. Our thanks to Angie Knight, Jen Otoadese and Karyn John. The UKERC communications team – Lex Young and Patricia Luna – helped to bring the work to the attention of stakeholders and the media. Special thanks to UKERC Research Associate Xinxin Wang. Without her admirable ability to manipulate documents, files and graphics, this book would not have been completed on time. Finally, special thanks to Julie Laws at Bookcraft and Emma Barnes, Michael Fell, Claire Lamont, Gina Mance, Anna Rice and Lee Rourke at Earthscan for all their work and help in the production of this book.

Acronyms and Abbreviations

AGT	auxiliary gas turbines
ATD	accelerated technology development
bbl	oil barrel
bcm	billion cubic metres
BERR	Department for Business, Enterprise and Regulatory Reform
BETTA	British Electricity Transmission and Trading Arrangements
BEV	battery electric vehicles
BIEE	British Institute of Energy Economics
BREDEM	Building Research Establishment Domestic Energy Model
BWEA	British Wind Energy Association
CAA	Civil Aviation Authority
CCC	Committee on Climate Change
CCGT	combined cycle gas turbine
CCS	carbon capture and storage
CDM	Clean Development Mechanism
CEC	Commission of the European Communities
CEGB	Central Electricity Generating Board
CERT	Carbon Emission Reduction Target
CFL	compact fluorescent lamp
CGEN	Combined Gas and Electricity Networks (as in CGEN model)
CH_4	methane
CHP	combined heat and power
CLG	Communities and Local Government
CO	carbon monoxide
CO_2	carbon dioxide
CRC	Carbon Reduction Commitment
CSA	Chief Scientific Adviser
CSR	Corporate Social Responsibility
DECC	Department of Energy and Climate Change
DEFRA	Department for Environment, Food and Rural Affairs
DETR	Department of the Environment, Transport and the Regions
DfT	Department for Transport
DIUS	Department for Innovation, Universities and Skills
DNO	District Network Operator

DTI	Department of Trade and Industry
DUKES	Digest of UK Energy Statistics
EEC	Energy Efficiency Commitment
EHCS	English House Condition Survey
ENA	Energy Networks Association
ESD	energy service demand
EST	Energy Saving Trust
ETI	Energy Technologies Institute
ETSAP	Energy Technology and Systems Analysis Program
EU	European Union
EU ETS	EU Emissions Trading Scheme
EUA	EU allowance unit
EWP	Energy White Paper
FGD	flue gas desulphurization
GDP	gross domestic product
GHG	greenhouse gases
GW	gigawatts
GWh	gigawatt hour
HEV	hybrid electric vehicle
HFC	hydrogen fuel cell
HGV	heavy goods vehicle
HHV	higher heating value
HPR	heat to power ratio
IAEA	International Atomic Energy Authority
ICT	information and communication technology
IEA	International Energy Agency
km pppa	kilometres travelled per person per annum
kW	kilowatt
kWth	kilowatt (thermal)
LCBP	Low Carbon Buildings Programme
LED	light emitting diode
LGV	light goods vehicle
LNG	liquefied natural gas
LOLE	loss-of-load expectation
LOLP	loss-of-load probability
MARKAL	MARKet ALlocation (as in MARKAL model)
mcm	million cubic metres
MEA	Millennium Ecosystem Assessment
$MtCO_2$	million tonnes of carbon dioxide
$MtCO_2e$	million tonnes of carbon dioxide equivalent
MW	megawatts
MWh	megawatt hours
N_2O	nitrous oxide
NO_x	nitrogen oxides
NTM	National Transport Model
OFGEM	Office of the Gas and Electricity Markets

OFT	Office of Fair Trading
ONS	Office for National Statistics
OPEC	Organisation of Petroleum Exporting Countries
PEM	polymer electrolyte membrane (fuel cell)
PHEV	plug-in hybrid electric vehicle
PJ	petajoules
PM	particulate matter
PSA	Public Service Agreement
PTE	Passenger Transport Executive
PV	photovoltaics
RCEP	Royal Commission on Environmental Pollution
RD&D	research, development and demonstration
REC	Regional Electricity Company
RO	Renewables Obligation
ROC	Renewables Obligation Certificate
RTFO	Renewable Transport Fuel Obligation
SAP	Standard Assessment Procedure
SMMT	Society of Motor Manufacturers and Traders
SO_2	sulphur dioxide
TNO	Transmission Network Operator
TRL	technology readiness level
TSB	Technology Strategy Board
TWh	terawatt hour
UKDCM	UK Domestic Carbon Model
UKERC	UK Energy Research Centre
UKTCM	UK Transport Carbon Model
UNEP	UN Environment Programme
UNFCC	UN Framework Convention on Climate Change
VAT	Value Added Tax
VOLL	value of lost load
WASP	Wien Automatic System Planning (as in WASP model)

Conversion Matrix

To:	Million tonnes of oil	Petajoules (PJ)	TWh	bcm	Million therms
From:	Multiply by				
Million tonnes of oil	1	41.868	11.630	1.06264	396.83
Petajoules (PJ)	0.023885	1	0.27778	0.025381	9.4778
Terawatt-hours (TWh)	0.085985	3.6	1	0.091371	34.121
Billion cubic metres (bcm)	0.94105	39.4	10.94444	1	373.42532
Million therms	0.00252	0.10551	0.029307	0.0026779	1

1
Introduction

Jim Skea, Paul Ekins and Mark Winskel

The long-term challenge of secure low carbon energy

Not for three decades has interest in the energy sector, and the role that energy plays in the economy and society more widely, been at such a high level. Although the world has changed hugely during that time, the anxieties that have driven energy to the top of the policy agenda are remarkably unchanged. Concerns about 'peak oil' in the early 21st century echo concerns about oil depletion prompted by the 1970s oil crises. In both periods, concerns have given rise in some quarters to prophecies of crisis, if not catastrophe, such that the functioning of modern society is seen as being fundamentally challenged. The concentration of energy resources in a limited number of countries has been seen as a problem in itself. The 1970s oil crises partly reflected the capacity of OPEC countries to act in a concerted manner. In Europe at least, the concentration of gas supplies in Russia and the limited number of transit routes have caused a similar anxiety.

There are also similarities in the environmental domain. In the 1970s, modern environmentalism was still in a nascent phase, but already the energy sector was under pressure. Tighter controls on air pollution were emerging in Japan, and in the US through the 1970 Clean Air Act. Global climate change poses far more fundamental challenges to the current energy system. Basic technical fixes and fuel switching relieved the problems associated with air pollution. The measures needed to deal with climate change go to the heart of the energy system and have fundamental implications for the economy also.

Although the economic implications of energy and of changes to the energy system go far beyond cost, cost remains a major concern for governments. The affordability of energy for private consumers and the impact of energy costs on business competitiveness are major issues. But the fact that the energy system is so intimately tied to other aspects of critical infrastructure, in terms of transport, information and/or communications, and provides an essential input into so many economic processes, gives energy a strategic role far greater than that suggested by the roughly 10 per cent of UK GDP accounted for by energy expenditures.

It has been common for some time to characterize the policy imperatives surrounding energy as a 'triangle' of concerns relating to economy, environment and security (McGowan et al, 1993). This can be invoked in the EU context as a *Lisbon-Copenhagen-Moscow* triangle, reflecting the Lisbon Treaty which came into force in 2009, the UNFCCC Copenhagen climate conference that took place in December of that year and discussion between the EU and Russia on access to energy supplies The three sets of forces may reinforce or act in tension with each other. This presents a particular challenge in 2010 and the years following as the security and environmental imperatives are both strong, but must be delivered against the background of difficult economic conditions following the banking crisis and subsequent recession.

While the current renewed interest in energy issues may evoke a certain degree of *déjà vu*, much has changed. Since the early 1990s, the world has experienced an unparalleled growth in economic activity and trade. The process of globalization has created much stronger trade links between countries. Emerging economies such as China and India have been leading economic growth. This growth has been associated with rising energy demand and growing levels of emissions of carbon dioxide (CO_2) and other greenhouse gases (GHGs) that give rise to climate change. These pressures on energy supplies and the environment have enhanced the sense of crisis surrounding the energy sector.

This theme of greater integration also applies at the European level. The very first treaties bringing the European Union countries together related to energy: the coal and steel treaty in 1951 and the Euratom Treaty in 1957. But when the UK joined the then European Economic Community in 1973, energy overall was very much a reserved matter for member states. A series of treaty revisions have increased the role of the EU, first with respect to environmental and subsequently energy issues. The Lisbon Treaty which came into force in 2009 identifies the EU role as being to 'ensure the functioning of the energy market; ensure security of energy supply in the Union; promote energy efficiency and energy saving and the development of new and renewable forms of energy; and promote the interconnection of energy networks' (OJEU, 2007, p88). The EU can implement new energy measures on the basis of weighted majority decision-making, unless the measures are of a fiscal nature in which cases a unanimous decision of all the member states is required. There is a general acknowledgement of member states' sovereignty over their energy resources, however, as 'measures shall not affect a member state's right to determine the conditions for exploiting its energy resources, its choice between different energy sources and the general structure of its energy supply' (OJEU, 2007, p88).

The other key development of the 1990s and early 21st century was the liberalization of energy markets, notably the network industries electricity and gas. Lower energy prices in that period offered the opportunity to see energy as 'just another commodity' which did not justify public ownership, special protection for key players or exclusive market arrangements. The UK led the way in privatizing state monopolies, breaking up vertically integrated structures and introducing competition at all levels from bulk electricity generation through to retail supply to households. The EU has been on the same path as the UK, though resistance from some member states has resulted in slower progress. Within the EU and the UK, this freeing up of structures has helped to make capital increasingly mobile and there are clear signs that pan-European utilities are emerging.

The net result is that, in 2010, the capacity of any country, especially an EU member state, to exercise sovereignty over its energy system is severely limited. Trade ties, shared decision-making and a greatly enhanced role for the private sector constrain a government's ability to act decisively at all times. A UK government, for example, can certainly introduce new policies and regulations, but must do so in the full knowledge that the measures that are introduced will influence the attractiveness of the country as a site for the new investment needed to lead the drive towards a secure, low carbon energy system. Nor does the UK government any more have a direct hand on the levers that are used to implement policy. Increasingly, policy delivery has been devolved to companies operating under a variety of regulatory requirements and obligations. Against this background, a key objective must be to create a co-operative partnership between government and industry, which generates in the private sector a willingness to invest. Government needs to facilitate investment by holding to a stable and predictable policy framework, establishing transparent and cost-effective regulatory and planning procedures, and putting in place (within the constraints of EU competition law) suitable incentives and reinforcing research and development support.

Faced with resurgent environmental and security challenges, there is a strong pull towards 're-regulating' energy markets and swinging back towards a planned, as opposed to a market-based, approach to energy policy. But any attempt to wind back the clock would simply fail. The world cannot return to the 1970s when governments had greater capacity to direct developments in the energy sector. Globalization and the sharing of power with the private sector have created a complex web of mutual commitments and dependencies which cannot be untangled but must be managed. The challenge going forward for governments is how to reconcile the pursuit of strategic goals in an interconnected world where the private sector plays a much more pivotal role. Governments have a limited capacity to direct or to play the role of economic actors in their own right. They must now learn to facilitate and co-ordinate action involving business and citizens – arguably an altogether more difficult task.

The UK provides a fascinating case study for this new energy governance challenge. As far as energy is concerned, the UK has had a long history of considerable self-reliance, through first coal and later oil and gas. The rundown of its coal industry, and later of supplies of North Sea oil and gas, means that the UK will become a substantial importer of energy from this point forwards. The UK has also set an ambitious long-term target of reducing GHG emissions (including those from international aviation and shipping) by 80 per cent by 2050 compared to 1990 levels. This target is enshrined as a formal legal obligation in the 2008 Climate Change Act. This book is essentially about how the UK can meet these twin challenges – of achieving energy security, while drastically reducing carbon emissions – over the next 40 years, while taking into account wider global, social, technological and environmental uncertainties.

Thinking about energy futures

It is impossible to make reliable predictions 40 years into the future. The advantage of reflecting on how the world has changed in the past is that it provides a healthy reminder of how much can change – and what may not. Going back 40 years takes us to a world where the oil crises had never happened, UK coal miners had not struck for 50 years, the Iron Curtain between East and West still stood, 9/11 was unimaginable and global warming was the forgotten theory of obscure 19th century physicists. Almost certainly, the next 40 years will throw up a comparable set of surprises. The challenge for the UK, as for other countries, is how to pursue clearly defined long-term energy security and environmental goals against deeply uncertain background conditions.

Any good dictionary of quotations will yield a wealth of remarks about prediction and dealing with the future. One common theme is that prediction simply cannot be done ('you can only predict things after they have happened', Ionesco), while another is that the future can be actively managed rather than passively anticipated ('the best way to predict the future is to invent it', Alan Kay, Apple Computers).

Just because long-term energy futures are not predictable does not mean that the various possibilities for these futures do not need to be thought through carefully. Indeed, uncertainty places a premium on generating a variety of options with which to respond to new or unexpected circumstances, and to keep them open for as long as the potential benefits exceed the costs of doing so. The management of energy uncertainties is made more difficult by the fact that many energy investments have lifetimes measured in decades, and infrastructure may leave its mark even longer. Emissions associated with energy use, notably CO_2, persist in the atmosphere for as long as a hundred years. Thinking about energy futures is therefore dominated by an approach known as 'scenario analysis', which is explicitly based on the premise that the future cannot be predicted. A scenario can be thought of as an 'imagined future' in which the drivers and developments are both plausible and internally consistent. It has been found that exploring a diverse range of scenarios provides an aid to thinking about present actions and their possible consequences. This is the spirit in which the scenarios in this book are presented.

Many UK energy scenarios have been constructed and a number of reviews of such scenarios exist (e.g. Hughes et al, 2009; Energy Research Partnership, 2010). These exercises involving thinking about energy futures can be divided into three broad categories: extrapolations or 'business-as-usual' exercises; exploratory scenarios; and back-casting scenarios (McDowall and Eames, 2006). Each has a qualitative, story-telling element and a quantitative element usually backed up by formal modelling.

Most scenario exercises start by considering how conditions might change if current trends and relationships between different variables (e.g. GDP, prices and energy demand) will continue into the future. Often called *business-as-usual* scenarios, in stable times these may be regarded essentially as forecasts or predictions. In less stable times, their lack of predictive power causes their scenario nature to be stressed, and they can simply be termed *Reference* scenarios, as in this book, to emphasize that 'business' is rather unlikely to be 'as usual', and that they therefore serve as comparators for other scenarios that may be equally, or more, plausible. The *Reference* scenario in such a case is simply the first of a wider scenario set and may well have a minimal qualitative dimension, being essentially driven by quantitative modelling assumptions. The *Reference* scenarios produced as part of the International Energy Agency's annual World Energy Outlook exercise (IEA, 2009) are perhaps the most high-profile examples of this type of approach. The energy projections of the UK Department of Energy and Climate Change (DECC) (e.g. DECC, 2009) also fall into this category, though uncertainties about economic growth and global energy prices are systematically explored by starting with a *central case* and then altering key variables.

Exploratory scenarios can be designed to shake complacency and unsettle fixed expectations about what the future might bring. The most celebrated sets of exploratory scenarios in the energy domain are perhaps those produced by Shell's Group Planning function. These started in the 1970s and early exercises were said to have anticipated the first oil crisis in 1973. Exploratory scenario exercises generally begin with a narrative element which sketches out, in a qualitative and imaginative way, a possible future which is dissimilar from the past. Once the basic concept has been fleshed out, the details can be filled in by assigning quantitative values to key variables and undertaking formal modelling to ensure internal consistency. In general, a number of contrasting scenarios will be developed to emphasize that no specific scenario is a prediction, and it is usually also emphasized that no single scenario is either more likely than or preferable to any other. However, the latest set of Shell scenarios (Shell, 2008) departed from this convention. The *Blueprints* scenario, which envisaged a world in which coalitions of interest combined to address shared goals, was seen to be preferable to the world described in the *Scramble* scenario in which individual countries focus on national energy security in a more competitive way.

The energy futures scenarios developed for the UK Foresight Programme in the early 21st century (Energy Futures Task Force, 2001) also fall into the exploratory category. These looked at four worlds defined by a 2 × 2 matrix with the two axes reflecting: (a) a world in which either individualistic or more communitarian values come to dominate; and (b) a world in which actions are well co-ordinated internationally versus one in which individual countries strive to further their own interest. The Foresight approach to scenarios has been termed 'deductive' as a result of the formal matrix structure while the Shell scenarios can be seen as 'inductive' (Wilkinson, 2008).

Finally, *back-casting* scenarios, as the name implies, take a normative approach by defining a set of desirable conditions for the future and constructing or modelling the pathways that are required to arrive at the final goals. As with *business-as-usual* scenarios, the qualitative aspects are often minimal or non-existent and such scenarios can be highly quantitative and model driven. Examples include the Royal Commission on Environmental Pollution scenarios aimed at establishing how the UK could achieve a 60 per cent reduction in CO_2 emissions by 2050 (RCEP, 2000) or the scenarios in the IEA's Energy Technology Perspectives exercise (IEA, 2008) which addressed how the world could cut global emissions by 50 per cent.

This book takes an eclectic approach to scenario building. A *Reference* scenario has been constructed that reflects *business-as-usual* trends – where these trends derive from energy system realities and the state of UK policy at the time of the 2007 Energy White Paper. A *back-casting* dimension has then been added by defining objectives which the UK aspires to achieve. The first objective is the current goal of cutting GHG emissions by 80 per cent by 2050. This has been interpreted as an 80 per cent reduction goal for energy-related CO_2 emissions by 2050. The second set of objectives relates to energy security, or energy resilience, which has been translated into a quantified set of objectives as described in Chapter 4. Together, these two objectives plus the *Reference* scenario define, using the deductive 2 × 2 matrix approach, a four-scenario core set which we designate as *Reference*, *Low Carbon*, *Resilient* and *Low Carbon Resilient*. Up to this point the narrative, qualitative aspects of the scenarios have been minimal.

Finally, to take account of the deep background uncertainties against which these goals must be pursued, elements of the exploratory approach are introduced. The business-as-usual assumptions are relaxed by considering a range of issues that will impact on the development of the energy sector. These include social attitudes, lifestyles and behaviours in respect of energy demand, the social acceptability of supply technologies based on their environmental characteristics, support for technological innovation, and uncertainties about global developments over which the UK will have no control.

The technical details of scenario construction are described in Chapters 4–11. The following section describes the broader approach taken.

How the book was written

This book is neither an edited book with disparate invited contributions, nor a single-authored monograph. It describes the fruits of a closely integrated, collaborative project that engaged members of the UK Energy Research Centre (UKERC) over the period 2006–2009. Part of UKERC's mission is to undertake interdisciplinary, 'whole systems' research that covers all aspects of the UK energy system – physical, engineering, environmental, economic and social. It engages people from a wide range of disciplines and institutions in order to realise that goal (UKERC, 2010). The work described in this book was part of the UKERC Energy 2050 Project which became the flagship project of UKERC's initial phase.

The project fulfilled two broad goals. First, as already noted, UKERC's membership encompasses a wide range of disciplines and perspectives that offer a uniquely wide insight into all aspects of the UK's energy circumstances and how they might develop. Therefore, the first goal was systematically to integrate this knowledge into a unified framework, in order to help inform debate about UK energy futures. The second goal, deriving from the fact that UKERC was a new collaboration set up by the UK Research Councils, was to initiate a collaborative activity that would bring the new team together and forge interdisciplinary alliances.

The scenario approach outlined above provided the initial mechanism for creating a shared vision of what the project could achieve. The desired balance between the business-as-usual, exploratory and back-casting approaches was debated before the more eclectic approach was finally adopted. This was seen as an approach with which a wider range of disciplines and perspectives would be comfortable.

The second element was to establish a set of workstreams, each of which would take responsibility for one of the topics, such as carbon pathways or socio-environmental sensitivities, addressed in Chapters 5–10 of the book. Workstream members comprised both those with underlying scientific and policy knowledge of the topic in question and modellers who would undertake the quantitative analysis needed to assess the system-level implications of the scientists' insights. Each workstream sought to combine off-model expertise and modelling-derived insights, in order to ensure that both contributed to the scenario specifications, the interpretation of modelling results and their relationship to input assumptions. The relationship between modellers with a systems perspective, policy and innovation analysts and those with fundamental knowledge of the underlying science was key to project dynamics.

The benefits of this structure were threefold. First, it provided a way of allowing basic scientific insights to have full play. Second, it allowed a systems-level perspective to emerge. Third, the use of a common scenario set allowed comparability across workstreams and topics, and a mechanism for integrating the project as a whole.

Structure of the book

Chapters 2 and 3 describe the current UK energy system. Chapter 2 focuses on current and past trends in energy supply and demand, energy trade, links with economic growth, CO_2 emissions and other environmental releases, and energy infrastructure. It also addresses emerging energy technologies which could help change patterns of supply and demand. Chapter 3 moves on to look at UK policies and institutions for governing the energy sector. It takes a longer-term perspective on the development of institutions and policies but focuses mainly on the post-1990 period when the energy sector went through major reforms. It also sets out the current policy framework through which policy-makers are seeking to realize the UK's ambitious policy goals.

Chapter 4 explains the analytical framework for thinking about UK energy futures. It identifies low carbon objectives and energy system 'resilience' as the key goals of UK energy policy. It describes the basic framework of four core scenarios and explains the range of issues and scenarios addressed by the book as a whole. The concept of energy system resilience is defined, by analogy with resilience in other systems, and the types of event to which an energy system might be required to be resilient are identified. The chapter then describes the key features of the main modelling tools used to explore the scenarios and finishes by describing, in some depth, the *Reference* scenario for the development of the UK energy system. As explained above, the *Reference* scenario is intended to form an analytical baseline and does not achieve either the UK's CO_2 targets or enhanced energy system resilience.

Chapter 5 considers in detail one of the book's main themes – the ambition and timing of the UK's CO_2 reductions. As well as the core *Low Carbon* scenario which entails an 80 per cent reduction by 2050, it considers both weaker and more ambitious levels of reduction and the implications of early action to reduce emissions. It also considers two scenarios which involve capping cumulative emissions over the period 2010–2050, while meeting the 2050 80 per cent reduction target. These two scenarios involve discounting the future at different rates – one with a low social discount rate and the other with a higher investor discount rate which takes account of risk and other factors. In all cases, the impact on demand and supply patterns, technology choice and economic welfare are considered.

Chapter 6 looks at the resilience of the UK energy system to external shocks in the context of CO_2 ambitions. It starts with a literature review of incidents in gas, electricity and oil markets that have impacted energy systems in the past. It then moves on to define and quantify indicators of resilience that can be used to define formally two core scenarios – *Resilient* and *Low Carbon Resilient*. The resilience indicators cover macro-level indicators relating to supply diversity and demand levels as well as more detailed indicators relating to system reliability and infrastructure provision. The chapter then systematically explores the outcomes of the *Reference*, *Low Carbon*, *Resilient* and *Low Carbon Resilient* scenarios and the degree to which there are synergies or tensions between the low carbon and resilience objectives. Finally, the chapter postulates a set of severe incidents, or 'shocks', affecting gas supply and assesses the capacity of the UK energy system to respond under each of the four scenarios. The value of a set of alternative mitigating investments – for example, in gas storage – is investigated.

Chapter 7 assesses the potential benefits of accelerating innovation in key energy technologies through enhanced R&D and other measures. It shows that technology acceleration brings substantial long-term economic benefits that are especially pronounced with more ambitious CO_2 emission reduction targets. It also shows that there is a case for continuing to support a wide range of energy technologies.

The capacity of distributed energy technologies to bring about more radical change in energy provision is the subject of Chapter 8. Focusing on the residential sector, a range of energy conversion and micro-generation technologies is considered, including biomass boilers, heat pumps, fuel cell combined heat and power (CHP), solar thermal, photovoltaics and micro-wind.

Chapter 9 postulates a set of lifestyle scenarios in which people voluntarily, as part of a wider social and cultural shift, adopt less energy-intensive patterns of living. This involves changes in transport patterns, comfort expectations in the home and the propensity to purchase energy-efficient technology. The assumptions are ambitious but not extreme. The result is that the costs of attaining stringent CO_2 emission reduction targets are greatly reduced.

Chapter 10 is concerned with the wider environmental implications of energy system development, going beyond CO_2 and other GHG emissions. It explores the implications of the four core scenarios in terms of air pollution and, more qualitatively, in terms of land and water use. It goes on to consider the implications of stylized scenarios reflecting three different types of environmentalism that preclude the development of various supply technologies. The NIMBY scenario reflects objections that relate to a sense of place; the ECO scenario reflects objections based on impacts on land use and associated ecosystem services; and the DREAD scenario reflects objections to unfamiliar technologies. In each case, the cost of reaching CO_2 goals increases and patterns of energy supply and demand are altered, radically so under the ECO scenario.

Chapter 11 considers the implications of a range of global uncertainties over which UK policy-makers have no control. These include higher world market fossil fuel prices, the availability of international CO_2 credits as an alternative to emissions abatement in the UK, and restrictions on global biomass availability. Again, there are important implications in terms of economic welfare and energy supply and/or demand patterns.

Finally, Chapter 12 systematically compares the outcomes of the various scenarios described throughout the book, pulls together the key themes that emerge from the work and draws, at a fairly high level, some conclusions about policy and action needs.

References

DECC (Department of Energy and Climate Change) (2009) *UK Low Carbon Transition Plan Emissions Projections*, DECC, London

Energy Futures Task Force (2001) *Energy for Tomorrow: Powering the 21st Century*, Foresight Programme, London, www.foresight.gov.uk/Energy/Energy_For_Tomorrow_Sep_2001.pdf

Energy Research Partnership (2010) *Energy innovation milestones to 2050*, Energy Research Partnership, London

Hughes, N., Mers, J. and N. Strachan (2009) Review and Analysis of UK and International Low Carbon Energy Scenarios, A Joint Working Paper of the UKERC and the e.on UK/EPSRC Transition Pathways Project. UKERC, London, www.lowcarbonpathways.org.uk/lowcarbon/publications/

IEA (International Energy Agency) (2008) *Energy Technology Perspectives 2008: Scenarios and Strategies to 2050*, IEA, Paris

IEA (International Energy Agency) (2009) *World Energy Outlook 2009*, IEA, Paris, www.worldenergyoutlook.org/

McDowall, W. and Eames, M. (2006) 'Forecasts, scenarios, visions, backcasts and roadmaps to the hydrogen economy: A review of the hydrogen futures literature', *Energy Policy*, vol 34, pp1236–1250

McGowan, F., Mitchell, C. and J. Skea (eds) (1993) *Energy policy: an agenda for the 1990s*, SPRU, University of Sussex, Brighton

OJEU (Official Journal of the European Union) (2007) *Treaty of Lisbon*, 2007/C 306, Office for Official Publications of the European Communities, Luxembourg

RCEP (Royal Commission on Environmental Pollution) (2000) 22nd report, *Energy: the Changing Climate*, Royal Commission on Environmental Pollution, London, www.rcep.org.uk/reports/22-energy/22-energyreport.pdf

Shell (2008) *Shell energy scenarios to 2050*, Shell International BV, The Hague, www-static.shell.com/static/public/downloads/brochures/corporate_pkg/scenarios/shell_energy_scenarios_2050.pdf

UK Energy Research Centre (2010) *Outline of UKERC Research Activities*, UKERC, London, www.ukerc.ac.uk/support/Research

Wilkinson, A. (2008) 'Approaches to Scenario Building: Scenarios and models – a marriage of effectiveness or dialogue for the deaf?', UKERC Annual Energy Modelling Conference: Scenario definition, quantification and modelling, Oxford, 30–31 January

2
UK Energy in an Era of Globalization: Trends, Technologies and Environmental Impacts

Jim Skea, Xinxin Wang and Mark Winskel

Introduction

Three factors define the role that energy currently plays within the UK economy:

1   the lack of any significant growth trend in primary and final energy demand;
2   a long-term decline in carbon dioxide (CO_2) emissions since a high point in 1970; and
3   the very recent emergence of the UK as a net energy importer following decades of self-sufficiency.

Against a background of concern about climate change and energy security, these factors help to define the possibility of transforming the UK energy system through to the mid-21st century. The purpose of this chapter is to examine the current state of the UK energy system, and key underlying trends, in some detail.

The chapter starts with a very long-term perspective on energy demand. It then looks at patterns of final energy demand, that is the energy used by households, business, the public sector and in transport. It covers fuels, sectors and the uses to which energy is put. The chapter then moves on to look at primary energy demand, which reflects the nation's total use of energy taking account of conversion losses in power stations and refineries, and losses in transmission and distribution.

The electricity generation sector, which has a critical role to play in the transformation of the UK energy system, is then covered in more depth. In contrast to the lack of growth in fossil fuel demand, electricity demand is growing. In addition, the electricity sector is a key arena for competition between different energy sources, making it critical in working towards climate change and energy policy goals. The next section addresses UK trade in energy, highlighting the current move away from self-sufficiency in oil and gas towards a situation of import dependence.

After this, there is a description of the current energy infrastructure in the oil and gas sectors, illustrating the substantial investments that have been made in the past. Finally, the chapter concludes with trends in environmental emissions from the energy sector, focusing on greenhouse gas (GHG) emissions and conventional pollutants such as sulphur dioxide (SO_2) and nitrogen oxides (NO_X).

Long-term energy trends

Figure 2.1 shows long-term trends in UK primary energy demand going back to 1700. The UK was one of the first economies to industrialize and consequently energy demand built up rapidly during the 19th century, almost entirely fuelled by indigenous coal. Primary energy demand grew by 2.8 per cent pa between 1800 and 1850 and then by 2.1 per cent pa between 1850 and 1900. However, the First World War, the depressed inter-war period and the Second World War caused a lengthy dip in demand. Primary energy demand did not recover to

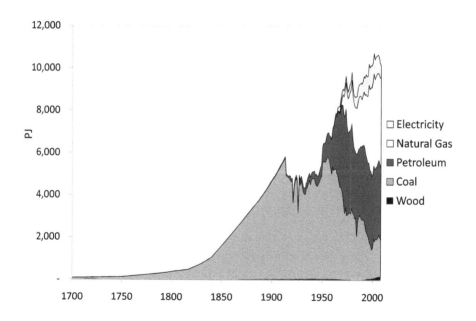

Source: CDIAC (2010)

Figure 2.1 *UK primary energy demand 1700–2008*

1913 levels until 1950. Growth then resumed at a more modest 0.9 per cent pa until the 1970s, when the oil crises of 1973 and 1979 induced further dips in demand that were partly market-driven and partly policy-induced. As oil prices fell in the mid-1980s demand grew again, but at a more modest rate of 0.4 per cent pa. Demand fell by 6 per cent between 2001 and 2007 (DECC, 2009a).

The primary energy mix has changed considerably over time. Coal had little competition until after the Second World War but was substituted firstly by imported oil and then, in the 1970s, by indigenous supplies of natural gas. Primary electricity – hydro, nuclear and subsequently other forms of renewables – have played a material role since the 1950s but expanded more rapidly in the 1970s with investment in nuclear power. Nuclear and renewables, which have been dominated by large-scale hydro until recently, still play a limited role in the overall energy picture.

The relationship between energy demand and GDP, and whether these can be decoupled, is a key policy concern. Figure 2.2 plots primary energy demand against UK GDP over the period 1948–2008. The dates of key turning points are marked. Over that period, GDP grew by a factor of four while energy demand grew by a factor of just under two. There was a strong positive correlation between energy demand and GDP until the first oil crisis. This has not been re-established since. There was a weak positive correlation between energy and

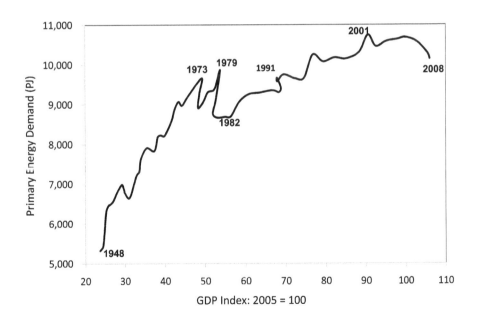

Source: Based on CDIAC (2010), DECC (2010) and HM Treasury (2010)

Figure 2.2 *UK GDP and primary energy demand 1948–2008*

GDP in the periods 1975–1979 and 1982–2001. However, since the turn of the century, primary energy demand has trended downwards in spite of rising GDP.

The strong message from this is that the *relationships between economic activity, energy and, as will be seen later, atmospheric emissions are not set in stone and can be profoundly shifted by changes in the global and domestic economy and purposeful policy action.* Engineering further such shifts is among the key challenges for UK energy policy.

Final energy demand

Although the absolute level of UK final energy demand has not changed much between 1970 and today, the sectoral mix has changed considerably (Figure 2.3). The only declining sector is industry, where demand has halved, with its share of final energy demand falling from 43 per cent to 20 per cent. There has been a long steady decline since the early 1980s but this was preceded by two precipitous declines following the 1970s oil crises as energy intensive industries contracted. Industrial decline is mirrored by growth in the transport sector, whose final energy use more than doubled between 1970 and 2008. Its share of final energy demand grew from 19 per cent in 1970 to 38 per cent in 2008. Household energy demand is 24 per cent higher than in 1970 but most of that increase took place in the period up to 1986. Its share of final energy demand now stands at 29 per cent. There has been little overall change in final energy demand in the commercial and public sectors since 1970.

The fuel mix in final energy demand has also changed considerably since 1970 (Figure 2.4). Coal use has been virtually eliminated from the picture as industry, and particularly energy intensive industry, has declined. Coal now accounts for only 2 per cent of final energy demand, whereas it had a 32 per cent share in 1970. Oil use, with a 44 per cent market share, is virtually unchanged since 1970 although there was a dip following the oil crisis which bottomed out in 1983. Oil use then grew until 2000, largely driven by expanding transport energy demand, before stabilizing. Natural gas use more than tripled between 1970 and the late 1980s, displacing coal in the household sector and oil in industry. Its share of final energy demand was only 10 per cent in 1970 but has hovered at 30–35 per cent since the early 1980s. In 2008, gas accounted for 33 per cent of final energy demand. Lastly, electricity's share of final energy demand has grown from 11 per cent in 1970 to 19 per cent in 2008. The expansion of electricity's market share has been much more gradual than is the case for natural gas. It market share has locked in at 18–19 per cent since 2001. Renewables and heat traded as a commodity are beginning to play a small role in final energy demand, with a 2 per cent overall share in 2008.

How energy is used

It is critical to understand the ways in which energy is used in order to assess the opportunities for change. In this section, we examine the use of final energy in the household, transport and industry sectors.

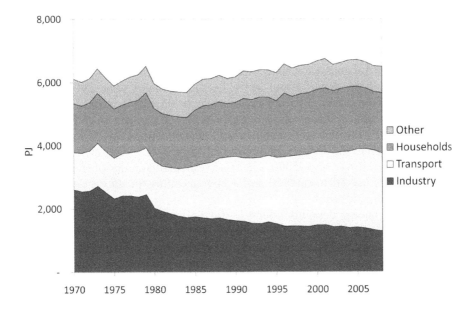

Source: DECC, 2009a

Figure 2.3 *Final energy demand by sector 1970–2008*

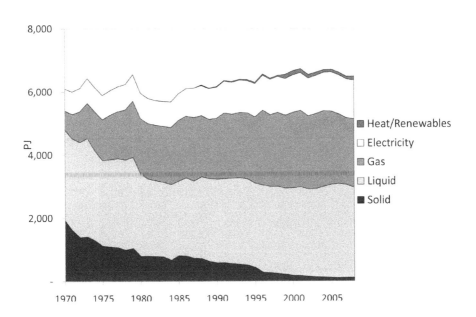

Source: DECC, 2009a

Figure 2.4 *Final energy demand by fuel 1970–2008*

Figure 2.5 shows that household energy demand is dominated by space heating (56 per cent) and water heating (26 per cent). Total household energy demand has shown a general upward trend since 1970 and this very much reflects increased demand for space heating. However, there has been a turn-down in both total demand and space heating demand since 2003 which suggests that some energy efficiency policies may be starting to work. Electricity demand for lighting and appliances has grown by 150 per cent since 1970 and accounted for 15 per cent of final household energy demand in 2008. Perhaps surprisingly, given the increasing uptake of consumer electronics, demand growth has been lower since 2000. The relatively small amount of energy used for cooking has halved since 1970 and it now accounts for only 3 per cent of household energy demand.

Transport energy demand, which is still dominated by oil products, has doubled since 1970. Figure 2.6 shows that road transport still accounts for 72 per cent of the total. Overall transport energy demand and demand for road fuels grew more rapidly during 1970–1990, with lower levels of growth being experienced since then. The most rapid growth has been in aviation. Figure 2.6 includes aviation fuels used for international flights. Demand has more than tripled since 1970 and aviation now accounts for 23 per cent of transport energy demand compared to 14 per cent in 1970.

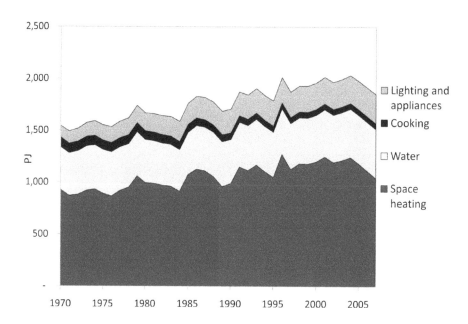

Source: DECC, 2009a

Figure 2.5 *Breakdown of household energy demand 1970–2008*

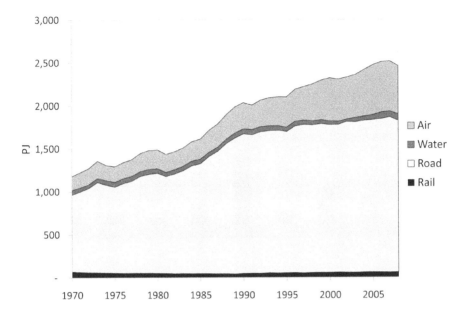

Source: DECC, 2009a

Figure 2.6 *Breakdown of transport energy demand 1970–2008*

Figure 2.7 provides a snapshot of industrial energy demand in 2007. The dominant use is process heat which accounts for more than half of industrial energy demand. Process heating needs, and the contribution of individual fuels, are very varied across industry, as Figure 2.7 shows.

Future energy demand technologies

The two main sectors in which significant changes in energy-using technologies are possible are residential buildings and road transport. In residential buildings, the main alternatives to conventional gas-fired condensing boilers are heat pumps, biomass boilers and combined heat and power units based either on Stirling engines or fuel cells. These technologies are described in Chapter 8 on distributed energy resources. All these alternatives would reduce dependence on natural gas either by 'electrifying' heat supply, or by substituting biomass for natural gas. Similar electrification or bio-energy conversions are also possible in the transport sector.

Road transport currently relies almost exclusively on the internal combustion engine fuelled by refined petroleum products, either petrol (known as gasoline in the US) or diesel. However, a range of alternative fuels and technologies are becoming available that could radically transform

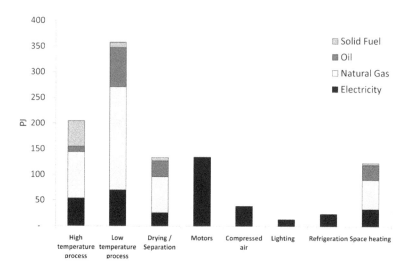

Source: DECC, 2009b

Figure 2.7 *Breakdown of industrial energy demand 2007*

road transport over the coming decades. The main motivations are to reduce CO_2 emissions and lessen dependence on oil. There are three main options available: the use of bio-fuels derived either from food crops (e.g. corn, sugar cane) or dedicated energy crops; electric or hybrid (e.g. petrol-electric) drives; and the use of alternative fuels, of which the most discussed is hydrogen. The scenarios and modelling set out in the book allow for all three options and all are taken up to some degree across different scenarios. Also, the three options are not mutually exclusive: for example bio-fuels can be used in hybrid vehicles.

With respect to bio-fuels, the scenarios allow the possibility of 'E85' private cars and light goods vehicles (E85 refers to vehicles that are capable of running on a blended fuel containing up to 85 per cent bio-ethanol). This fuel is now available in the US and parts of Europe and many existing vehicle models are capable of running on it. In Brazil, bio-ethanol derived from sugar cane is a major transport fuel and 'flex' vehicles that can run on pure bio-ethanol or 'E25', a blend containing 25 per cent bio-ethanol, are available. Bio-ethanol derived from food crops may have a limited impact on CO_2 emissions when a full life cycle assessment is conducted (Whitaker et al, 2010). Biodiesel can be manufactured from food crops or dedicated 'second generation' energy crops such as miscanthus, willow or poplar. In the book, it is assumed that some bio-fuels can be produced in the UK, subject to land-use constraints, and that the UK has access to traded bio-fuels on international markets. The availability of bio-fuels that meet sustainability criteria relating to life cycle CO_2 emissions and biodiversity impacts is deeply uncertain. The advantage of bio-fuels is that the existing fuelling infrastructure associated with petrol/diesel can be used.

Pure electric vehicles are already established in small niche markets. Recent improvements in battery technology leading to longer vehicle ranges mean that battery electric vehicles potentially have a much wider application. There are potential markets for private cars, light goods vehicles and buses. The 'electrification' of transport goes beyond pure battery electric vehicles. Hybrid petrol-electric vehicles are now beginning to enter the mainstream. The electric motor is used at lower speeds and to boost acceleration, while a conventional petrol engine is used at higher speeds. The car's battery is recharged through regenerative braking. 'Plug-in' hybrid vehicles are also under consideration. These rely to a greater degree on electricity than do basic hybrid vehicles as their batteries can be recharged while stationary, as is the case with pure battery electric vehicles. There is no reason why diesel could not be used in hybrid vehicles although most existing applications are based on petrol. Hybrid technology could be used across all market segments: private cars, goods vehicles and buses. The large-scale adoption of battery electric or plug-in hybrids would require the development of a recharging infrastructure. This would not be a simple exercise as the impact on local electricity distribution networks could be considerable. With battery electric and plug-in hybrid vehicles, the impact on CO_2 emissions is crucially dependent on the extent to which grid electricity is decarbonized.

There has been considerable interest in hydrogen as a transport fuel (e.g. Ekins, 2010). The most likely option is that hydrogen vehicles would rely on fuel cells that generate electricity through a chemical reaction. However, it is possible to burn hydrogen in an internal combustion engine. One of the biggest challenges in developing hydrogen-fuel cell vehicles is the on-board storage of hydrogen, which is much bulkier than petrol or diesel. As with electric vehicles, it would also require the development of a new refuelling infrastructure. The final challenge is that hydrogen is a fuel rather than a source of energy. It must be generated either by reforming fossil fuels, which would not lead to any CO_2 benefits, or through electrolysis. In the longer term, high temperature heat, for example from a nuclear reactor, could in principle be used to generate hydrogen. The carbon footprint of electrolytic hydrogen depends entirely on the carbon intensity of the electricity used to generate it. The use of hydrogen in vehicles on a large scale would need to be associated with a substantial move to renewable or nuclear energy.

There is deep uncertainty as to which of the three broad options – bio-fuels, electricity or hydrogen – might replace fossil fuel-based transport in the longer term. Much depends on the extent to which technology can be developed, the sustainability of bio-energy and wider energy sector developments, notably the pace of decarbonization in the electricity sector. Succeeding chapters of this book highlight the trade-offs involved.

Primary energy demand

Primary energy demand includes the energy input for electricity generation and petroleum refining. Overall changes in primary energy demand will broadly reflect changes in final energy demand. However, both the efficiency with which electricity is generated from fossil fuels, and the degree to which it is generated from non-fossil sources, affect the ratio between primary and final demand. The conventions used to calculate primary energy demand associated with electricity generation are critical. In the UK, for fossil fuels, nuclear power and biomass energy, primary energy is defined to be the heat input to the power station. For other renewables, primary energy is defined in terms of electricity output. Other conventions are used in other countries and by other organizations.

Figure 2.8 shows primary energy demand by fuel since 1970. A key feature is the decline in demand for coal which had only a 17 per cent market share in 2008, mainly for electricity generation. Oil demand is also lower than it was in 1970 but most of the decline in oil demand took place in the 1970s as it was withdrawn from electricity generation and industrial heat markets. Oil consumption is now mainly associated with transport. Natural gas demand has increased by a factor of eight since 1970. In 2008, gas had the largest share of primary energy demand (41 per cent). Initially, the increase in gas demand was associated with its use for final demand in buildings and industry. But there was a second expansion of gas use in the 1990s – the 'dash for gas' – as it displaced coal for electricity generation. The next section focuses specifically on electricity.

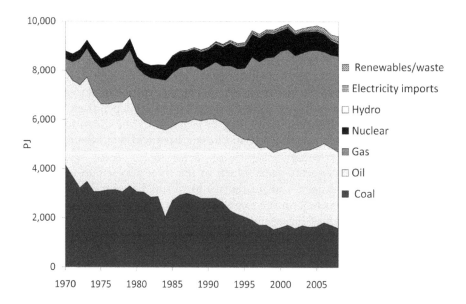

Source: DECC, 2009a

Figure 2.8 *Primary energy demand by fuel 1970–2008*

Trends in electricity generation

Figure 2.9 shows the development of the UK electricity generation mix between 1970 and 2008. Between 1982 and 2003, the quantity of electricity generated grew at 2 per cent per year reaching a peak of 380TWh in 2003. Since 2003, the volume of electricity generated has declined. There had been previous declines in electricity generated following the oil crises of 1973 and 1979. Up to 1990 the system was dominated by conventional power generation using coal- and gas-fired steam condensing sets. However, nuclear had steadily expanded its market share from 10 per cent in 1970 to 20 per cent in 1990. Nuclear's share peaked at 26 per cent in 1998. Since then plant closures of first UK-designed Magnox reactors and then subsequently Advanced Gas-cooled Reactors (AGRs) led to a reduction in the nuclear share to below 20 per cent.

The most striking feature of the 1990s was the rapid expansion of gas-fired power generation. Investment in efficient Combined Cycle Gas Turbine (CCGT) plant was precipitated by the liberalization of electricity markets and greater aversion to investment risk which favoured low capital cost/higher running cost plant. Starting from zero in 1990, CCGTs had a 35 per cent share of generation by the year 2000. In 2008, CCGTs overtook conventional coal plant in terms of generation share for the first time:

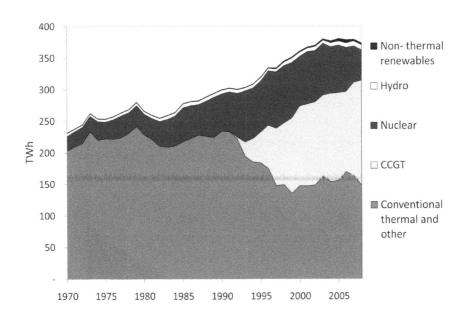

Source: DECC, 2009a

Figure 2.9 *Electricity generation mix 1970–2007*

CCGTs had 45 per cent of the market compared to 40 per cent for conventional coal plant.

Renewable energy still plays a very small role in UK generation markets in spite of the significant policy ambitions, described in Chapter 3. Figure 2.9 separates out hydro-electricity, much of which is based in Scotland and has been developed for some time, and non-thermal renewables (e.g. wind). Figure 2.10 looks in more detail at renewable electricity generation in the UK. In 1990, the 6TWh of electricity generated by renewables was almost completely dominated by large-scale hydro. Since then total renewable generation has increased almost four-fold at a rate approaching 9 per cent per year. Now both wind (7.1TWh) and landfill gas (4.8TWh) exceed large-scale hydro. Other important sources of renewable electricity are biomass co-firing in coal-fired power stations (1.6TWh) and the combustion of municipal solid waste (1.2TWh). Small-scale hydro, and the combustion of sewage sludge, animal wastes and plant biomass each contribute around 0.6TWh.

Due to the need to apply for grid access and planning permission, it is possible to foresee reasonably well how plant capacity may change over the coming few years. The National Grid Company's 2009 Seven-Year Statement (see Figure 2.11) identifies 35GW of 'transmission-contracted' generation capacity due to come online by 2015/16. Not all of this capacity will necessarily proceed to completion, and hence Figure 2.11 is not a forecast. Of this, roughly 15GW each is for renewables (mainly wind) and CCGTs. About 5GW

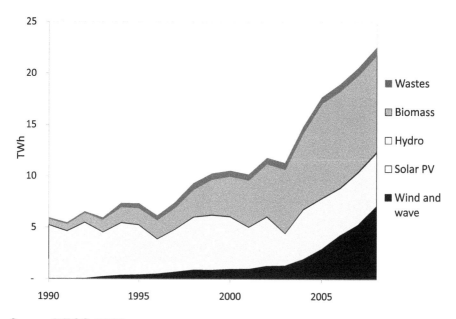

Source: DECC, 2009a

Figure 2.10 *Renewable electricity generation 1990–2008*

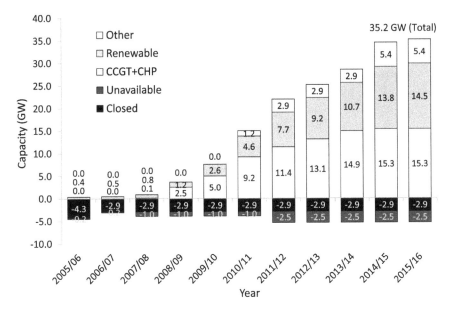

Source: National Grid, seven-year statement, 2009

Figure 2.11 *Changes in transmission-contracted generation capacity to 2015/16*

of existing plant is expected to be unavailable or closed. This signals the clear direction of generation mix over the next decade or so: an acceleration of the expansion of renewable energy with CCGTs acting as the default plant to meet overall system needs.

New capacity is needed because, by the end of 2015, 8GW of coal-fired plant which was opted-out of the requirements of the EU Large Combustion Plant Directive (LCPD) will be required to close (Figure 2.12). In addition, 7.4GW of older nuclear capacity is expected to close between 2010 and 2020, leaving only one operational nuclear plant by 2023, unless new build takes place.

Future electricity generation technologies

The UK's electricity system may accommodate a more diverse set of generation technologies in the future. As Chapter 3 will show, there are considerable policy ambitions to expand the use of renewable energy, re-invest in nuclear power and open up the possibility of low carbon fossil fuel power through carbon capture and storage technology. This section briefly summarizes the key characteristics of the most important alternatives: wind, marine renewables, solar PV, bio-energy, nuclear and carbon capture and storage for fossil plant; for more details, see Winskel et al, 2009.

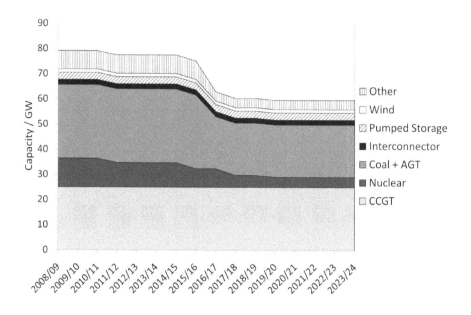

Source: DECC, Energy Markets Outlook 2009b

Figure 2.12 *Development of existing GB generating capacity*

Wind

Onshore wind energy technology systems are mature, with well understood cost components. While there is some scope for increased turbine efficiency, much of the scope for increasing the contribution of onshore wind relates to non-technical institutional and wider societal issues, such as the planning system and grid access reforms.

Offshore wind technology is much less established, and the costs of installation and access for maintenance are highly uncertain. Offshore wind technology is likely to evolve in distinctive directions compared to onshore, given the lower importance of noise and visual impact, the relatively higher cost of installation and grid connection – providing an imperative towards larger unit size – and the generally more severe operating environment.

The UK has a huge offshore wind resource, but it is still largely untapped. Given this, and despite a relative lack of experience compared to onshore, offshore wind is seen as a key technology for meeting UK and international policy targets for renewable deployment and energy system decarbonization, especially over the short and medium terms.

Marine renewables

Marine renewables comprise wave and tidal current technology. Compared to wind power, marine renewables are at a relatively early stage of development.

There are a number of engineering concepts for capturing wave energy, including oscillating water columns, overtopping devices, point absorbers, terminators, attenuators and flexible structures. Tidal current energy exhibits less variety, with most prototype designs based on horizontal axis turbines.

Development of a full-scale device prototype is time consuming and expensive. The established route for device development starts by testing at model scale in tank facilities, developing hydrodynamic models to design larger-scale models to be tested in larger tanks or offshore, and using results from these tests to verify the modelling before going to a full-scale design.

Unlike many other areas of low carbon innovation, the UK has a leading position in the emerging marine sector, with a significant resource and research base, related skills in offshore engineering, and a relatively strong funding and policy support framework. A significant proportion of all marine energy developer companies and support facilities are based in the UK. Marine renewables could potentially play a significant role in the UK energy mix.

Bio-energy

Bio-energy is a highly flexible option for decarbonizing the economy, spanning a wide range of fuel feedstocks, conversion technologies and end-use applications. This makes for multiple development possibilities. Bio-energy feedstocks include wood, 'first generation' food crops such as sugar cane, maize and wheat, and also 'second generation' dedicated energy crops such as willow, miscanthus and other woody shrubs, trees and grasses. Other feedstocks include waste products such as sewage and agricultural residues like straw and wood. Conversion techniques include simple combustion, fast pyrolysis, gasification and fermentation; and end uses include bio heat, power and fuels for transport.

Controversies surround direct and indirect land-use changes, life cycle carbon reduction potential and other environmental impacts of bio-energy technologies. There are particular concerns about the use of first generation food crops for bio-energy due to the potential impact on food prices and competition for land, including deforestation. A great deal of research is seeking to develop second generation dedicated energy crops that aim not to compete with food crops or negatively shift patterns of land use.

Solar PV

Solar photovoltaic (PV) technology – the direct conversion of sunlight into electricity – could provide a significant part of the world's future energy supply. The international PV market has shown sustained growth of over 30 per cent per annum in recent years. Research efforts have led to a reduction in the cost of PV cells by a factor of more than 20 over the last two decades. The main drivers for this development have been government incentives – despite the substantial price reductions, PV generation costs are still too high to make PV a commercially attractive investment for the private sector. Solar PV has the potential to make a substantial contribution to meeting UK energy needs in the medium and long term.

First generation PV devices are based on crystalline silicon, drawing heavily on the knowledge of that material that developed out of the electronics industry. The initial second generation thin film device was based on amorphous silicon, but a range of alternative thin film cells have since been developed. Efficiencies for commercial thin film modules can now be up to 12 per cent, compared to up to 18 per cent for commercial mono-crystalline silicon modules. Recent research has opened up the possibility of third generation technologies, including low cost–moderate efficiency molecular based cells (dye sensitized and organic semiconductor devices), high efficiency–high cost devices, and other novel concepts and nanotechnologies.

Nuclear power

Nuclear fission has been part of the UK energy mix since the 1950s, and contributed 18 per cent of all electricity generation in 2009. However, no new capacity has been ordered in the last two decades, and all of the UK's existing nuclear power stations are scheduled to close over the next few decades.

Since 2000 there has been renewed interest internationally in nuclear power development and deployment, and the technology is expected to play a significant role in many countries' future energy mixes. Fission power plants are being constructed in France, Finland and the US, with fleet build programmes in China, India and Brazil. The new UK coalition government intends to continue promoting nuclear power though there are fundamental differences of view between the two governing parties.

Reactors currently operating in the UK are based on 'Generation II' technologies. The Pressurised Water Reactor (PWR), such as Sizewell B in the UK, has become the *de facto* global standard. 'Generation III' technologies such as the Westinghouse AP1000 or Areva's European Pressurised Water Reactor (EPR) are evolutionary developments of PWRs and are ready to be deployed. These products are regarded as standard globalized systems, and there is little desire by the vendors to modify such systems.

Generation IV systems could be deployed by the 2030s. These systems attempt to address further improvements in reactor safety, as well as more sustainable use of resources. Some of the technologies, such as the sodium-cooled fast reactor, have already been demonstrated. France has committed to building a demonstration sodium-cooled fast reactor by 2020, and this project is now being considered as a central theme of the EU's EURATOM research programmes.

Carbon capture and storage

Carbon Capture and Storage (CCS) involves abating emissions from power plants and other industries by chemically capturing – or separating out – the CO_2, transporting it to a storage site, and injecting it for long-term storage in sealed geological formations. Depending on the capture technology, the CO_2 is captured from flue gases, from the fuel before combustion, or directly from the combustion process.

CCS technology is generally understood as consisting of three major steps: capture, transport and storage. Although each step can be realized with technologies proven in other applications, these technologies need to be adapted for use in the CCS application. Depending on the configuration of the power plant and capture technology, CCS can be a more or less tightly integrated part of the overall system. Post-combustion capture can be added on to existing power plants. The other capture technologies, pre-combustion and oxyfuel, are to a greater extent integral parts of power plant designs. Moreover, all three steps of CCS need to be integrated together with a power plant, to form a functioning overall system, which also entails technical and other challenges.

The UK government has initiated a competition for funding a full-scale, integrated CCS demonstration plant. The aim is first to demonstrate that CCS is technically feasible and then, in a subsequent phase, to demonstrate its economic feasibility. The UK's Committee on Climate Change has recommended that no new coal- or gas-fired power stations should operate without CCS after the early 2020s.

Energy trade and self-sufficiency

Other than for oil, the UK has traditionally been self-sufficient in energy. Even for oil, from 1980 until recently, the UK was a net oil exporter. This is no longer the case and the UK is now a net importer of all fossil fuels. Figure 2.13 compares the production of North Sea oil with oil demand. Production built up very rapidly from 1975 but reached a first peak in 1985, then fell back following the *Piper Alpha* offshore accident and falling global prices. A

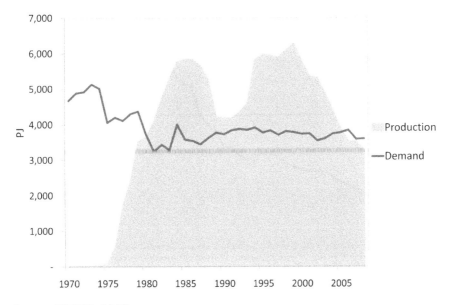

Source: DECC, 2009a

Figure 2.13 *UK crude oil production and demand*

second wave of investment led to another peak in 2000, since when production has declined rapidly. Even during the production dip in the early 1990s, the UK remained a net oil exporter. Production fell below demand in 2005 and dependence on imports is expected to increase as production falls further.

Figure 2.14 shows that the use of natural gas in the UK was strongly associated with the development of offshore gas production. There was a period when the UK was a net importer during the 1980s but throughout the 1990s the UK remained self-sufficient in gas. Production has been falling since 1998.

Energy infrastructure

Until recently, the adequacy of the infrastructure that brings energy to UK consumers has given little cause for concern. As Chapter 3 will discuss, the main priority during the 1990s and early part of the 21st century has been to drive costs out of the regulated gas and electricity sectors and 'sweat' existing assets. The prospect of radical changes in the pattern of electricity generation and the decline of UK gas production in the North Sea is opening up new challenges.

Figure 2.15 shows the shape of the current electricity supply system in Great Britain. Northern Ireland forms part of a separate, integrated Irish network. The greatest demand for electricity arises in southern England, while coal-fired generation in the north of England has dominated the generation side. A robust infrastructure of high voltage (400kV) transmission lines connecting

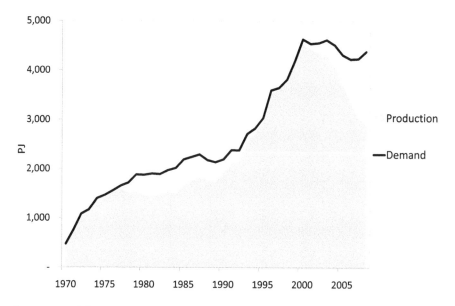

Source: DECC, 2009a

Figure 2.14 *UK natural gas production and demand*

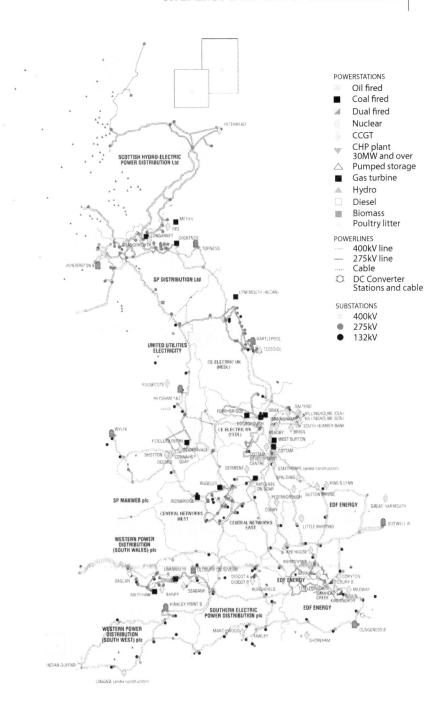

POWERSTATIONS
▨ Oil fired
■ Coal fired
◢ Dual fired
◉ Nuclear
✦ CCGT
▽ CHP plant
  30MW and over
△ Pumped storage
■ Gas turbine
▲ Hydro
□ Diesel
▨ Biomass
  Poultry litter

POWERLINES
— 400kV line
— 275kV line
···· Cable
◌ DC Converter
  Stations and cable

SUBSTATIONS
◌ 400kV
● 275kV
● 132kV

Source: DECC, 2009c

Figure 2.15 *The electricity supply system in Great Britain*

North and South allowed the system to meet demand. In the 1990s, the move from coal- to gas-fired generation, which is not so dependent on being close to a mine or a port, helped to ease further transmission constraints. There have been only two interconnectors linking Great Britain to other parts of Europe. These are the 2,000MW interconnector to France and a 500MW connection with Northern Ireland. A new 1,000MW interconnector to the Netherlands is under construction.

The expectation that renewable energy, especially wind, will play a bigger role in the electricity system sets new challenges. Much of the onshore wind capacity could be concentrated in Scotland, while a very considerable expansion in offshore wind is anticipated. The capacity of the transmission system to bring power from Scotland and the most northern parts of England is limited, as Figure 2.15 shows. There is therefore a pressing need to upgrade the onshore transmission system. A substantial step forward was taken in 2010 with the approval of the 2,500MW Beauly–Denny line which will connect northern and central Scotland. However, planning permission was first applied for in 2005. Speeding up planning approval for major infrastructure projects is essential if the electricity generating system is to be transformed.

The offshore infrastructure will need to be developed from scratch. This is a considerable engineering challenge, but a new regulatory regime involving competitive tenders for Offshore Transmission Owner licences was agreed in 2009. It is likely that the offshore planning process will be much faster than onshore. There is also the possibility of constructing undersea high voltage DC (HVDC) cables to bring power directly from Scotland to England. This would be more expensive than reinforcing the onshore network but could be commissioned more quickly.

As more intermittent renewables come on to the system, the benefits of further interconnection with continental Europe will increase. This is discussed in more depth in Chapter 6. Further interconnections with France and Belgium are being discussed. A Norwegian interconnector is also a possibility as pumped storage capacity in Norway would complement intermittent wind capacity in the UK.

Figure 2.16 shows that the gas supply system in Great Britain has been reliant on five terminals on the east coast and two on the west coast to bring in natural gas. The St Fergus and Easington terminals bring in supply from Norway while Bacton is connected to the continental gas system via Zeebrugge. With UK gas supplies declining, the challenges are now to build up liquefied natural gas (LNG) import facilities and further gas storage. Three LNG import facilities have been completed recently, two in South Wales and one on the Isle of Grain in Kent. Further investment is taking place, on a commercial basis, in gas storage facilities. In Chapter 6, the case for further strategic investment is considered.

A further consideration is that, with the prospect of a higher penetration of intermittent wind on the electricity system, gas-fired CCGT plant will be called on to supply large amounts of power at relatively short notice. The physical capacity of the gas system to supply the required quantities of gas may be limited (Qadrdan et al, 2010). For that reason, the possibility of requiring the storage of back-up distillate fuel oil at CCGT plant has been mooted.

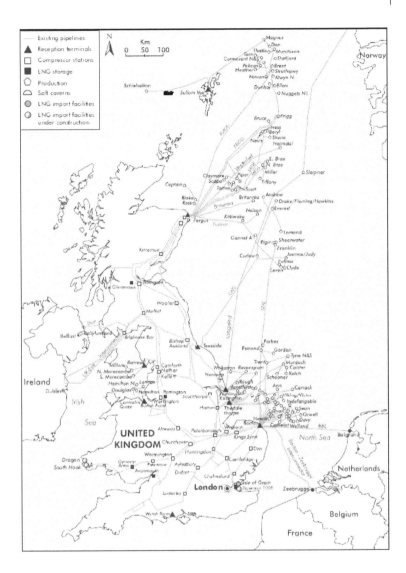

Source: DECC, 2009c

Figure 2.16 *The gas supply system in Great Britain*

Environmental concerns

Although greenhouse gas emissions (especially CO_2 emissions) are by far the most high profile and difficult environmental challenge facing the energy sector, there are other environmental concerns, notably air pollution associated with fossil fuel combustion. Coal use creates solid waste which must be re-used, recycled or disposed of. Electricity generation based on a steam cycle – whether nuclear, coal, oil or combined cycle gas turbine – requires considerable quantities of water. Nuclear power gives rise to a particular set of challenges, relating to the storage and disposal of radioactive waste. This problem has yet to be resolved in the UK, in the sense that no final storage site for high level wastes has been decided on. The development of renewable energy brings its own challenges which reflect the diversity of renewable sources. Among the most critical are the visual amenity consequences of onshore wind energy development. This particular environmental impact is strongly subjective, but has been key in preventing the development of a number of projects. The large-scale development of biomass crops would raise a number of issues relating to landscape, biodiversity and water availability.

Chapter 10 explores the wider environmental implications of our scenarios. In this section, we review past and recent trends in key environmental areas.

CO₂ emissions

UK CO_2 emissions account for around 85 per cent of the UK's total greenhouse gas emissions. Of the 15 per cent of total GHG emissions not arising from CO_2, almost 90 per cent are from methane and nitrous oxide emissions, mostly associated with agriculture and landfill waste. Non-CO_2 GHG emissions have declined substantially since 1990 – and an overall decline of 50 per cent is projected from 1990 to 2020 (CCC, 2008). The rest of this section considers trends in UK CO_2 emissions – around 95 per cent of which arise from energy production and consumption (the remainder arise from land use, land-use change and forestry (LULUCF)).

As discussed in Chapter 3, UK policy targets for CO_2 and GHG emission reductions are benchmarked against 1990 emission levels. UK CO_2 emissions trends between 1990 and 2009 reflect a number of factors, with a gradual decline in emissions during the 1990s, a levelling-off for several years from the late 1990s but then a very sharp fall between 2008 and 2009. Net UK CO_2 emissions in 2009 were 19 per cent below 1990 levels, at $481MtCO_2$. Excluding the net contribution of LULUCF, CO_2 emissions were 18 per cent below 1990 levels (provisional figures from DECC, 2010). Over the same period, UK GDP rose by almost 50 per cent, so there has been a sustained reduction in the carbon intensity of the UK economy over the last two decades, of around 3 per cent pa.

Emissions fell by $52MtCO_2$ in 2009 compared to 2008 levels, an almost 10 per cent reduction, and twice as large as any previous annual fall in the period since 1990. This was driven by fuel switching from coal to nuclear for electricity generation, estimated at $23MtCO_2$ reduction, and also demand reductions from sectors such as industry and transport as the economy contracted, estimated at $20MtCO_2$ reduction (provisional figures from DECC, 2010).

In terms of emissions by source, the most significant contributor to reduced CO_2 emissions over time has been the power sector, where emissions have fallen by 26.5 per cent between 1990 and 2009 (including a 13 per cent fall 2008–2009), even though electricity consumption rose by 16 per cent over the same period. Over half of this reduction since 1990 was driven by the rise of gas-fired generation and the decline of coal-fired generation, but improved generation efficiency has also been important. Emissions from transport rose gradually between 1990 and 2007, but have since declined back to 1990 levels, despite significant increases in transport-related energy demand since 1990. Industrial emissions declined significantly, by around 27.5 per cent, between 1990 and 2009. Emissions from the domestic (household) sector declined slightly, despite increases in domestic energy consumption over the same period.

By fuel type, CO_2 emissions from coal and other solid fuels fell by 54 per cent between 1990 and 2009, largely driven by reduced coal-fired electricity generation; emissions from oil were 19 per cent lower, while emissions from gas rose by 67 per cent (DECC, 2010).

It is important to note that UK official CO_2 emission figures are based on a methodology agreed by the Intergovernmental Panel on Climate Change (IPCC), which excludes emissions from international aviation and shipping and, also, emissions from the production of goods and services imported by the UK.

Air pollution

Figure 2.18 shows that sulphur dioxide (SO_2) emissions have fallen by 90 per cent since 1990. Both SO_2 and nitrogen oxides (NO_X) are associated with acid rain which was the subject of intense public debate in the 1970s and 1980s. The decline partly reflects the fact that technical fixes are generally available, at a cost, to deal with 'conventional' air pollutants. Up till 1990, the declines

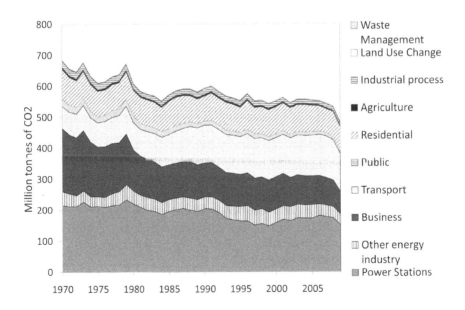

Source: National Atmospheric Emissions Inventory

Figure 2.17 *UK CO_2 emissions 1970–2009*

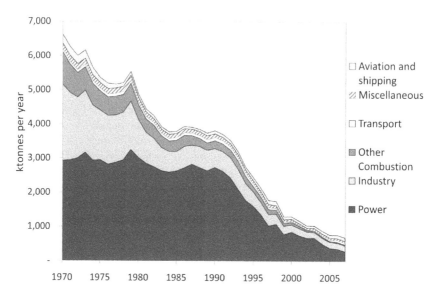

Source: National Atmospheric Emissions Inventory

Figure 2.18 *Sulphur dioxide emissions 1970–2007*

in SO_2 emissions largely took place in industry and 'other combustion' which includes the residential sector. In both cases, the substitution of natural gas, which is virtually sulphur-free, for coal and oil was the key driving force. The long-term effect of the 1956 Clean Air Act also played a role.

The large decline in power station emissions from 1990 onwards can be attributed to changes in both legislation and markets. The EU Large Combustion Plant Directive (LCPD) introduced in 1988 put in place a progressively declining set of caps on SO_2 emissions. It was originally anticipated that a large programme of investment in flue gas desulphurization (FGD) equipment would be required. However, in the event, the 1990s 'dash to gas' achieved much of the reduction required, and a more modest FGD programme in the UK was the result. The LCPD has been revised a number of times. In the most recent version, coal-fired power plants were required to fit FGD equipment or else be 'opted out'. From 2007 onwards, opted-out plants are allowed only a certain number of hours' operation, and must close by 2015. Between 2010 and 2015, 8,000MW of opted-out coal plant is expected to close in the UK for this reason, as was shown in Figure 2.12.

Figure 2.19 presents sources of NO_X emissions since 1970. Unlike SO_2, transport makes a bigger contribution to NO_X emissions than stationary sources. Emissions from industry have fallen very significantly since 1970 and power station emissions fell in the 1990s. In both cases the decline was due to a switch from coal to gas. Transport emissions rose from 1970 to 1990 as road traffic grew but have since fallen as a result of regulatory requirements for lean

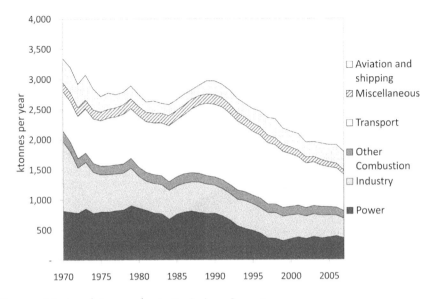

Source: National Atmospheric Emissions Inventory

Figure 2.19 *Emissions of nitrogen oxides 1970–2007*

burn engines and catalytic convertors. The LCPD currently requires relatively low cost combustion modifications to meet NO_X standards at power plants. A successor directive, the Industrial Emissions Directive (IED), may require the installation of more expensive 'selective catalytic reduction' (SCR) equipment in the early 2020s. There has been speculation that this would prompt a further wave of coal-fired power plant closures.

Figure 2.20 shows that there has been a 70 per cent decline in emissions of small particulate matter (PM-10s) since 1970. PM-10s are particles less than 10 microns in size which are easily inhaled and have significant health implications. In 1970, most small particulate emissions were associated with residential coal use. These emissions have been virtually eliminated. Electrostatic precipitation equipment is highly effective in removing particulate matter from flue gases at large stationary sources such as power stations. The largest source of particulate matter is now transport and, more specifically, older diesel-engined vehicles. Increasingly stringent vehicle standards are helping to drive emissions down, but this has been partly reversed by a switch from petrol to diesel engine vehicles. Under standards that have applied since 2009 ('Euro-5'), diesel engines give rise to no more emissions than do petrol engines. Over time, therefore, PM-10 emissions from transport will fall further.

In some of the scenarios we present in this book (see Chapter 10), the expanded use of biomass for residential heating could lead to particulate emissions rising once more. In practice, technical measures to abate emissions are available, but these would add to the cost of biomass utilization.

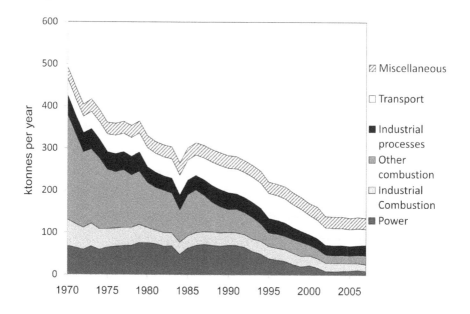

Source: National Atmospheric Emissions Inventory

Figure 2.20 *Emissions of PM-10 particulates 1970–2007*

Water

The power sector is one of the largest users of water in Britain, accounting for more than half of the abstractions from ground- and surface-water sources in England and Wales (Figure 2.21). This makes it a bigger user than public water supply. This dependence on water has become an issue in some other countries, though not yet the UK. In France, the output of nuclear power stations, many of which are located on rivers, has had to be curtailed on occasions because of the non-availability of cooling water under hot and dry conditions. Climate change could exacerbate these conditions.

In practice, a large number of the UK's power stations, and all of its nuclear stations, are located on the coast rather than on rivers. To the extent that large steam cycle plant plays a major role in the UK, it is likely that this will continue. But in scenarios where large capacities of nuclear or coal-fired CCS plant are installed, there may be a need to exploit sites next to rivers. Under certain scenarios, this could put pressure on water resources and, also, threaten the availability of generation capacity at times of water shortage.

Solid waste

The amount of solid waste generated from energy production and use is modest in the context of total arisings of waste from industrial and commercial processes. Most energy-related solid waste is associated with combustion processes (particularly the use of coal). In 2002/03, the most recent year for

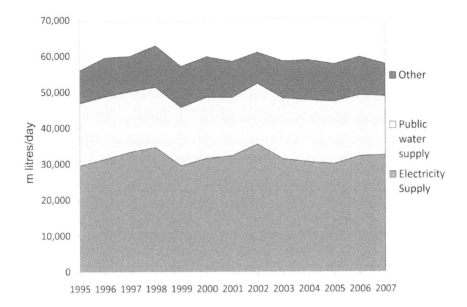

Source: Defra, 2010a

Figure 2.21 *Abstractions of water in England and Wales 1995–2007*

which data is available, 10 million tonnes of solid waste was generated from combustion processes of which 6 million tonnes came from the coke, oil, gas, electricity and water sectors combined (Defra, 2010b). Of this 10 million tonnes, 15 per cent was re-used (for example in aggregates for road construction), 43 per cent was recycled and 41 per cent went to landfill. This needs to be seen in the context of 67 million tonnes of waste generated from industrial and commercial processes overall plus a further 28–29 million tonnes of municipal waste. However, any future reversion to coal-firing for power generation could increase solid waste generation from combustion.

Radioactive waste

The storage and disposal of nuclear waste is now seen as the key environmental issue associated with the development of nuclear power. Concerns about catastrophic accidents, which were in the fore following the Three Mile Island and Chernobyl events of 1979 and 1986 respectively appear to have receded. With modern plants, the risks associated with low level releases of radiation have also diminished.

In 1976, an influential report from the Royal Commission on Environmental Pollution (the Flowers Report, RCEP (1976)) argued that it was necessary to demonstrate 'beyond reasonable doubt that a method exists to ensure the safe containment of long-lived highly radioactive waste for the indefinite future' when developing nuclear power. The 2006 report of the Committee on Radioactive Waste Management produced a firm recommendation that deep geological disposal was feasible and was the preferred option for the UK (CoRWM, 2006). CoRWM's recommendations were based on considerations of both the science and social acceptability. Nevertheless, a specific site for deep storage has yet to be identified.

In practice, most of the UK's need for radioactive waste disposal is attributable to legacy wastes associated with the nuclear weapons programme, experimental reactors and existing nuclear power stations reaching the end of their operating lives. Table 2.1 identifies the volume of packaged legacy waste which cannot be disposed of in the existing low level waste repository at Drigg in Cumbria. It also identifies the volume of waste arising from a programme of ten new nuclear power stations, assuming that spent fuel is not re-processed. A new programme would add less than 10 per cent to the volume of waste disposal to which we the UK is already committed. However, it would increase the volume of spent fuel to be disposed of by 400 per cent. It is argued that the waste arising from new build would not change the nature of the existing disposal challenge in any significant way (DECC, 2009d).

However, the timescale for identifying and developing a single depository for all of the wastes identified in Table 2.1 is considerable. The Nuclear Decommissioning Authority estimates that it will take until 2040 to begin disposing of legacy intermediate level waste. Disposing of legacy high level waste and spent fuel could begin around 2075. Disposal of all legacy waste could be complete by around 2130 and only then would disposal of waste from nuclear new build begin. In the meantime, interim storage might be required at power stations for around 160 years after the start of their operation to allow discharged fuel to cool sufficiently (DECC, 2009d). These timescales are similar to those involved in climate change projections and the challenge of long-term interim storage at power station sites will clearly need to be addressed as plans for specific nuclear power stations are developed.

Table 2.1 *UK radioactive waste materials inventory (cubic metres)*

	Base inventory	Ten plant nuclear programme
High level waste	1,290	–
Intermediate level waste	353,000	9,000
Low level waste (non-Drigg)	37,200	–
Plutonium	3,270	–
Uranium	74,950	–
Spent nuclear fuel	8,150	31,900
Total	477,860	40,900

Notes: assumes no re-processing of spent nuclear fuel; new build assumed to be Westinghouse AP1000 reactors

Source: CoRWM (2006)

Conclusions

This chapter has presented an overview of the UK energy sector in terms of its physical characteristics, supply and demand trends and environmental impacts, and how these have changed over time. The chapter started by taking a very long-term view, which showed shifts in the UK energy economy in terms of changing emphasis on coal, oil and gas. Despite the increasing attentions and expectations on new energy technologies, fossil fuels are still the backbone of the UK energy economy, and a substantial shift away from them, as is implied by recently introduced policy objectives, will present many challenges.

This said, a review of more recent historical timescales suggests the possibility of change. Encouraging trends here, given climate change and security concerns, are the decoupling of UK economic growth from both energy demand and carbon emissions – though it should be added that these official figures overlook the increased global interdependencies of UK production and consumption, such that demands and emissions which support the UK economy are unrepresented.

The overall picture that emerges from this overview is of accumulating trends building slowly over time (such as the growing importance of electricity, the rise of transport as an important source of demand and CO_2 emissions, or the gradual loss of energy self-sufficiency in the UK), punctuated by sometimes abrupt changes (such as the dash for gas following electricity industry privatization, or the sudden changes in energy demand and emissions 2008–2009). There is some predictability about likely changes over the next two decades: the further rise of gas-fired electricity generation over the next decade, as well as a more important role for renewables. However, history also suggests that many of the changes affecting energy supply and demand are sudden and unexpected, and energy policy history is littered with unrealized projections.

A final point that emerges from this overview is the increasing complexity of the energy system, both in terms of the many different possible ways of supplying and using energy in the future, and in terms of interdependencies between different parts of the system. Thus, changing one aspect may have unexpected consequences elsewhere. An example is the impact of electric vehicles on energy demand patterns. Looking ahead, it is clear that the energy sector faces fundamental changes as the UK faces up to the challenges of important dependence and radical decarbonization. The next chapter turns to the policy framework that will be needed to facilitate these changes.

References

CCC (Committee on Climate Change) (2008) *Building a low-carbon economy – the UK's contribution to tackling climate change*, Committee on Climate Change, TSO, London

CDIAC (Carbon Dioxide Information Analysis Centre) (2010) *National CO₂ Emissions from Fossil-Fuel Burning, Cement Manufacture, and Gas Flaring: 1751-2006*, Oak Ridge National Laboratory, TN. doi 10.3334/ CDIAC/00001, http://cdiac.ornl.gov/ftp/trends/emissions/uki.dat

CoRWM (Committee on Radioactive Waste Management) (2006) *Annex 3: Inventory, Managing our Radioactive Waste Safely: CoRWM's recommendations to Government*, Committee on Radioactive Waste Management, London

DECC (Department of Energy and Climate Change) (2009a) *Digest of UK Energy Statistics: Long-term Trends*, Department of Energy and Climate Change, London

DECC (2009b) *Energy Markets Outlook*, Department of Energy and Climate Change, London

DECC (2009c) *Digest of UK Energy Statistics*, Department of Energy and Climate Change, London

DECC (2009d) Draft National Policy Statement for Nuclear Power Generation (EN-6), London, November

DECC (2010) *Energy Trends – March 2010*, Department of Energy and Climate Change, London

DEFRA (Department for Environment, Food and Rural Affairs) (2010a) *Estimated abstractions from all surface and groundwaters by purpose and Environment Agency region: 1995–2007*, Department for Environment, Food and Rural Affairs, London

DEFRA (2010b) *Industrial and commercial waste arisings by waste type and waste management method, 2002*, DEFRA statistics, Department for Environment, Food and Rural Affairs, London

Ekins, P. (ed.) (2010) *Hydrogen Energy: Economic and Social Challenges*, Earthscan, London

HM Treasury (2010) *Gross Domestic Product (GDP) deflators: a user's guide*, www.hm-treasury.gov.uk/data_gdp_index.htm

National Grid (2009) *GB Seven Year Statement 2009*, National Grid, London, www.nationalgrid.com/uk/Electricity/SYS/archive/sys09, accessed 21 June 2010

Qadrdan, M., Chaudry, M., Wu J., Jenkins, N. and Ekanayake, J. (2010) 'Impact of a large penetration of wind generation on the GB gas network' *Energy Policy*, vol 38 (10), pp5684–5695

RCEP (1976) *Sixth Report: Nuclear Power and the Environment*, Royal Commission on Environmental Pollution, HMSO, London

Whitaker, J., Ludley, K., Rowe, R., Taylor, G. and Howard, D. (2010) 'Sources of Variability in Greenhouse Gas and Energy Balances for Biofuel Production: a Systematic Review', *Global Change Biology Bioenergy*, vol 2 (3), pp99–112, doi 10.1111/j.1757-1707.2010.01047.x

Winskel, M., Markusson, N., Moran, B., Jeffrey, H., Anandarajah, G., Hughes, N., Candelise, C., Clarke, D., Taylor, G., Chalmers, H., Dutton, G., Howarth, P., Jablonski, S., Kalyvas, C. and Ward, D. (2009) *Decarbonising the UK Energy System: Accelerated Development of Low-carbon Energy Supply Technologies*, UKERC Energy 2050 Research Report No. 2, March 2009, UK Energy Research Centre, London

3
UK Energy Policy and Institutions

Paul Ekins, Jim Skea and Mark Winskel

Introduction

The developments in the supply and demand of energy in the UK reviewed in Chapter 2, and their further evolution in the future, have been and will be largely influenced by interactions between two kinds of institutions: markets, which may be global, European, national or local; and public institutions, which make and administer public policy.

This chapter briefly reviews the changing nature of these institutions in the UK since the Second World War, and recent policies that have emanated from or acted on them. However, this is not the place for a detailed review of the post-war evolution of the UK energy system. From 1979 to 2002 that evolution has already been admirably described by Helm (2003), and for the purposes of this book, the period before 1979 may be sketched in the lightest outline. The bulk of this chapter will therefore concentrate on developments in energy (and related climate) policy and institutions since the turn of this century, focusing in particular on current energy policies and institutional arrangements. Later chapters will discuss, in the context of the different scenarios to be explored, to what extent these arrangements are likely to be appropriate, or will need to be changed, if the goals and aspirations of current energy policies are to be achieved.

Ownership issues, late 1940s to mid-1990s

Following the Second World War many of the substantial private undertakings in the energy industry were nationalized, creating a range of national corporations or other public bodies, including: the National Coal Board (1947); the Central Electricity Authority (1948), followed by the Central Electricity Generating Board, Area Boards and the Electricity Council (1957); the Gas Council and Area Boards (1948), followed by the British Gas Corporation (1972), to oversee the distribution and sale of natural gas from the North

Sea; the United Kingdom Atomic Energy Authority (1954) and British Nuclear Fuels (1971). For oil, the Government had a significant stake in BP, and the British National Oil Company was created in 1977 following the large-scale exploitation of North Sea oil (see Helm, 2003, p18).

In the years after 1982 these industries were successively privatized, oil first (1982–1987), then gas (1986), non-nuclear electricity (1990–1991), coal (1995) and nuclear power (1996). For gas and electricity 'arm's-length' regulators were created, which were merged by the Utilities Act (2000) to form a single regulator, the Office of Gas and Electricity Markets (Ofgem), governed by the Gas and Electricity Markets Authority, in order to promote competition in the industries and regulate more directly those aspects of them, transmission and distribution, which operated as natural monopolies.

Electricity as an energy carrier is generated from fossil fuel, nuclear or renewable sources, flows through the high-voltage transmission system into the lower-voltage distribution systems and thence through to end users, which may be industry, commercial or other service buildings or households. Because opportunities for electricity storage are limited, the demand for and supply of electricity must be balanced on a minute-by-minute basis. The wholesale electricity market has a complicated set of arrangements to achieve this.

On privatization in 1990, the electricity industry consisted of four distinct sets of entities: generators, of which the two largest were National Power and Powergen; the transmission operators; the distribution network operators (DNOs); and the electricity suppliers (to final users). Initially the DNOs and suppliers were one and the same, but were separated with the opening of retail competition later in the decade. Over time a great deal of entry of foreign companies into the UK market, cross-ownership and vertical integration among these entities has developed, such that in 2010 the four sets of entities comprised:

- 'The Big 6' Generators: EDF, E.ON, RWE, Iberdrola (which owns Scottish Power), Centrica, Scottish and Southern Energy (SSE), plus some much smaller companies;
- The major electricity suppliers, who are now owned by 'the Big 6' generators (shown in brackets): British Gas (Centrica), npower (RWE), London Energy (EDF), E.ON UK (formerly Powergen), Scottish and Southern (SSE), plus some smaller companies;
- The transmission network operators (TNOs): National Grid (which is also the overall electricity System Operator – see below), Scottish and Southern (owned by SSE), Scottish Power (owned by Iberdrola) and Northern Ireland Electricity (NIE);
- Fourteen DNOs, three owned by EDF, two owned by Scottish Power, two owned by SSE, two owned by E.ON, four owned by American companies, two each by PPL Corporation and MidAmerican Energy Holdings, and one owned by United Utilities. NIE is the DNO for Northern Ireland.

As noted above, the gas industry was privatized in the 1980s. The National Transmission System (NTS) for gas is owned and operated by National Grid (NG), which sells the gas to power stations, to some very large users, and to eight gas distribution network (GDN) owners: four of these are owned by NG, SSE owns two, United Utilities owns one, and one is owned by the global Macquarie Group. The gas suppliers take the gas from the GDNs and sell it on to households and businesses. As retail competition in the utilities sector developed, multi-utility suppliers emerged, such that the main gas suppliers are now the same as the main electricity suppliers.

Managing and regulating the flow of electricity

In Northern Ireland there is still a single electricity market, managed by NIE. The rest of this chapter therefore focuses on the rather more complicated arrangements in Great Britain (GB).

The four sets of entities described above together manage the flow of electricity in GB through a complex market structure. The generators sell their electricity to National Grid (NG) and directly to large users through the British Electricity Trading and Transmission Arrangements (BETTA), which is managed by a company called Elexon. The DNOs buy their electricity from the generators through BETTA, control its flow through their distribution system, and sell it on to the suppliers, who in turn sell it to households and businesses. NG, as the overall System Operator, and working in collaboration with the other TNOs, the DNOs, the generators and Elexon, has the responsibility to ensure overall minute-by-minute balancing of the electricity system.

The sale of electricity by suppliers to end users now has no price regulation and is subject to open competition, although suppliers have obligations through their licences and in other areas, some of which are described below. Ofgem, the regulator, has the job of ensuring that the competition is effective, and conducts periodic enquiries, when it believes this might not be the case, followed by interventions when it perceives these to be necessary. For example, following an Energy Supply Probe in 2008, it changed licence conditions in September 2009 to ensure that tariff differentials in the supply of electricity only reflected cost differentials, and banned 'undue' discrimination against customers (Ofgem, 2009).

The regulatory treatment of the DNOs and TNOs is different, as the markets in their case are perceived to be natural monopolies, and they are therefore subject by Ofgem to periodic price control reviews, which determine what they can charge for connection to their systems, what investments they can make in their systems (the cost of which needs to be passed on to customers in their charges) and what standards and conditions of service they must deliver.

The UK electricity system is currently very centralized, in that generation is dominated by large fossil-fuel and nuclear stations, located in the main fairly close to centres of electricity demand, and electricity flows through the transmission and distribution systems in a manner that leaves little scope for involvement from the demand side. One of the ways in which the electricity system might develop (discussed in much detail in Chapter 8) is for more generation to be delivered locally on a small scale, and for consumers to become more involved in managing electricity loads (known as 'demand side management'). As will be seen below, the Government has put in place some policies to encourage this. Should this occur on a substantial scale, there will need to be major modifications to the extent and management of distribution networks to ensure that the flows of electricity and demand side information are properly handled, such that balance in the electricity system overall is maintained. Modification of the transmission system (mainly by building new transmission lines where none currently exist) will also be required if there is to be a major expansion of renewables in remote areas (mainly wind, both onshore and offshore), far from centres of demand. Again, there have been recent changes in this area, notably the Scottish Government's approval in 2010 of a major transmission line upgrade in northern Scotland which will allow power to flow from North to South.[1]

Managing and regulating the flow of gas

The gas system is similar in structure to that for electricity, but less complex, not least because gas (unlike electricity) can be stored in large quantities so that the generation of and demand for gas does not need to be balanced on a minute-by-minute basis – fluctuations in the amount of gas stored in transmission pipelines ('linepack') can accommodate daily fluctuations in demand while dedicated storage facilities can be used to deal with some longer-term fluctuations. The principal mechanism used to balance supply with seasonal fluctuations in demand so far has been the rate of off-take of gas supplies from the North Sea.

Natural gas flows into the UK through pipelines into seven coastal terminals in Great Britain: one in Scotland, two in the northwest, one in the northeast, two in the East Midlands and one in the east of England. Three of these pipelines import foreign gas into Britain. An interconnector takes natural gas from Scotland to Northern Ireland. The gas coming into Britain enters the high-pressure National Transmission System (NTS).

The NTS is also fed by a variety of storage facilities. The largest of these is the Rough depleted gas field, which accounts for about 75 per cent of current UK storage. The UK has relatively little storage by European standards, but there are plans to increase it, both underground, and in the form of liquefied natural gas (LNG) storage facilities. In 2009 there were four LNG terminals, some only recently completed, at Teesside, the Isle of Grain and two in Milford Haven in Wales.

The gas network distribution owners take the gas from the NTS and sell it to the suppliers. As with electricity the gas suppliers operate in a competitive market, with Ofgem overseeing the competition, but the GDNs and NTS, being natural monopolies, are directly regulated through price controls in a similar way to the electricity TNOs and DNOs.

New issues in energy policy

In the 1990s, following privatization, UK energy policy mainly consisted of promoting competition in the retail markets through Ofgas (for gas) and Offer (for electricity) and, from 2000, Ofgem for both, and regulating the distribution networks and transmission systems, as briefly described above. It was assumed that the price of electricity would be set at a level high enough to incentivize the building of new power stations, and that seasonal variations in the price of gas would justify the building of new import and storage infrastructure, when they were needed. Up until now such assumptions have proved to be broadly justified, with very considerable private investment being forthcoming in all kinds of energy infrastructure, such as new Combined Cycle Gas Turbine (CCGT) power stations in the 1990s, and new gas import and storage capabilities in the 2000s.

However, there have been two major developments which have now called these assumptions into question. The first has been the depletion of North Sea oil and gas, as discussed in Chapter 2, which means that the UK is now a net importer of both fossil fuels (as also of coal), and this import dependence is set to increase markedly over the next decade. This has raised new concerns about energy security. The second development has been the high policy priority that has come to be accorded to climate change over the last decade, reflecting the hardening scientific judgement (IPCC, 2007), that it is very likely that global warming in recent decades has been induced by anthropogenic GHG emissions, which at current or increased rates will cause further warming and induce many changes in the global climate system during the 21st century with serious negative consequences for human societies.

These developments have resulted in a plethora of new legislation on energy, as the Government has sought to take the powers it perceived necessary to address these new concerns, including the Utilities Act (2000), the Sustainable Energy Act (2003), the Energy Act (2004), the Climate Change and Sustainable Energy Act (2006), the Energy Act (2008), the Climate Change Act (2008), and most recently the Energy Act (2010). The age in which it was perceived that energy provision was basically a market-led issue that required the lightest touch from government, which until 2000 was largely the post-privatization view, was clearly at an end.

The new salience of issues related to energy security and GHG emissions has brought into question not only the perceived primary role of markets in achieving the policy objectives relating to energy, but more specifically the role of the regulator, Ofgem (SDC, 2007). From its inception, Ofgem was supposed to take account of 'social and environmental guidance' from the relevant government department (which has changed from DTI to BERR to DECC[2] over the last ten years), but it was never clear what weight it should give to this guidance compared to its pursuit of its primary duties to promote competition and, thence, protect the interests of consumers. The Energy Act (2004) made it a duty for Ofgem to contribute to sustainable development, and the Energy Act (2008) put this duty on a par with Ofgem's other secondary duties – to meet reasonable energy demands and finance authorized activities – and made it clear that 'consumers' should include those of future generations as well as those alive today. The Energy Act (2010) specified explicitly that Ofgem should include the reduction of carbon emissions and the delivery of secure energy supplies in its assessment of the interests of consumers, as well as proactively protecting their other interests and considering longer-term actions to promote competition.

Ofgem has also been given key administrative tasks in respect of important government policies such as the Renewables Obligation and the Carbon Emissions Reduction Target (discussed below). Ofgem has responded to the new situation by restructuring to separate its core obligations from its support and delivery functions, and creating a new Sustainable Development Division, focusing on five themes: 'managing the transition to a low carbon economy; eradicating fuel poverty and protecting vulnerable customers; promoting energy saving; ensuring a secure and reliable gas and electricity supply; and supporting improvement in all aspects of the environment'.[3]

Whether these changes will adequately address the inherent tensions in Ofgem's new complex remit, in relation to both energy security and environmental and social matters, remains to be seen, and this issue will be revisited in the final chapter of this book, in the context of the changes in the UK energy system that the scenarios discussed in subsequent chapters suggest may be necessary if the UK's policy goals in these areas are to be achieved. The future role of Ofgem within the UK's wider regulatory and policy institutions is uncertain, given the likely restructuring of public bodies by the new UK Government following the General Election of May 2010.

Finally the principal social issue connected to energy, which is also one of Ofgem's five sustainable development themes, needs to be mentioned: fuel poverty, defined as a condition in which a household would need to spend 10 per cent or more of its disposable income to attain a minimum defined level of warmth and other energy services. The 2003 Energy White Paper had as one of its four objectives 'to ensure that every home is adequately and affordably heated' (DTI, 2003, p11). Encouraged by reductions in the number of households in fuel poverty in the 1990s (which turned out to be largely due to reductions in energy prices), the Government then committed itself to abolish fuel poverty for vulnerable households by 2010 and for all households by 2016. Since then, energy prices have risen substantially, as have the numbers of households in fuel poverty (see Figure 3.1), and it is clear that neither of these targets will be met.

Much of the rest of this book is about how the UK energy system might develop in order to respond to the two principal concerns identified above, with energy security, and the related issue of the resilience of the energy system, being covered especially in Chapters 4 and 6, and all other chapters giving a central role to the reduction of greenhouse gas (GHG) emissions from the use of energy in the UK. Given that most UK energy policy over the last decade has been motivated by climate concerns, much of the rest of this chapter focuses on the UK policy context in respect of energy and climate change as of 2010, to set the scene for the very large reductions in GHG emissions and, more

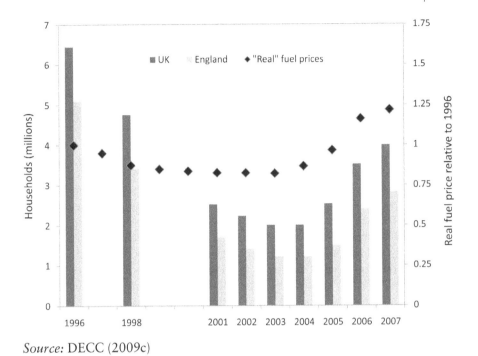

Source: DECC (2009c)

Figure 3.1 *Fuel poverty and real fuel prices 1996–2007*

particularly in respect of the energy system, emissions of carbon dioxide (CO_2, the main GHG) that are now envisaged by UK policy, the implications of which are explored in later chapters. A final section of the chapter sets out UK policy thinking on energy security in this period, in order to prepare for the more detailed treatment of this topic in Chapters 4 and 6.

UK climate and energy policy development

Climate change is a global environmental issue, and if the extent of climate change is to be mitigated below what will arise from business-as-usual (BAU) developments, it will only be through international agreement. Realization of this led to the signing at the Rio Earth Summit in 1992 of the UN Framework Convention on Climate Change (UNFCCC) and of the Convention's Kyoto Protocol in 1997, which for the first time sought to place mandatory limits on the emissions of GHGs from industrial countries. The Kyoto Protocol expires in 2012. It had been hoped that the Conference of the Parties to the Convention in Copenhagen in 2009 would have agreed at least the framework of a successor protocol, but in the event what emerged was the Copenhagen Accord (UNFCCC, 2009), which failed to agree on emissions reduction targets or timetables. The climate agreement that will be put in place at the global level beyond 2012, if any, is therefore still very unclear.

Notwithstanding the uncertainties in the global policy process, the European Union in 2008 proposed a 'climate and energy package' (EC, 2008), subsequently agreed by EU leaders in December of that year. This package committed EU member states to reducing their GHG emissions overall by at least 20 per cent from 1990 levels (to be raised to 30 per cent following an acceptable global agreement on GHG reduction, which failed to materialize in Copenhagen in 2009), to produce at least 20 per cent of their final energy demand from renewable sources, and to reduce their primary energy use by at least 20 per cent below its projected BAU level, all by 2020.

The main policy mandated and largely implemented by the EU to achieve these targets is the EU Emissions Trading System (EU ETS), which covers about half the energy-related CO_2 emissions in the EU (those from power generation and large combustion installations) and envisages that over three Phases (Phase 1 2005–2007; Phase 2 2008–2012; Phase 3 2013–2020) the emissions from these sources will be reduced by 21 per cent below their 2005 level (EU, 2009). This reduction requires that emissions from the covered installations reduce by an average of 1.74 per cent per year to 2020. Countries in which this rate of reduction is exceeded will be able to sell emissions permits to those countries which fall short of it. The EU has passed a number of other policy measures to achieve its targets, including the Directives on the energy performance of buildings and energy services[4] and the regulation on carbon emissions from light vehicles.[5] Apart from those delivered by the EU ETS and such specific regulations, the great majority of the emission reductions envisaged by the European targets, and any entailed by any additional national commitments, will have to be achieved through national policy.

The UK adopted its first programme of climate change policy in 2000 (DETR, 2000), which set a target of reducing UK CO_2 emissions by 20 per cent below 1990 levels by 2010, a significantly more stringent target than that for the UK under the Kyoto Protocol (which was an average annual 12½ per cent GHG emissions reduction from 1990 over 2008–2012). The 2000 Climate Change Programme put forward many policies, which were implemented in succeeding years, and was updated in 2006 (HMG, 2006) but CO_2 emissions have proved hard to reduce, such that by 2008 energy-related CO_2 emissions were still only 9 per cent lower than 1990 levels, and had only fallen by 3 per cent since 1995 (AEA Technology, 2010) (the other 6 per cent of the reduction having been due to the large-scale substitution in 1990–1995 of gas for coal in power generation). However, helped by the fact that non-CO_2 GHG emissions fell by 67 per cent over 1990–2006 (CCC, 2008, p356), the UK is expected to meet its Kyoto target comfortably; and the 2010 CO_2 target, which had seemed likely to be missed by a large margin for a number of years, now looks to be within reach, largely because of a 9.8 per cent reduction in CO_2 emissions in 2009 because of the recession.[6]

Climate change policy-making over this period connected closely with the rebirth in the UK of energy policy, reflected in an unprecedented spate of policy documents from central government, including two energy reviews (PIU, 2003; DTI, 2006), two Energy White Papers (DTI 2003, 2007a), Nuclear White Paper

(BERR, 2008), culminating in *The UK Low-Carbon Transition Plan: National strategy for climate and energy* (HMG, 2009b), as well as numerous associated consultation and more detailed policy documents and strategies, some of which are mentioned below. This interaction between energy and climate resulted in the 2007 Energy White Paper articulating the two over-arching challenges of UK energy policy as:

- tackling climate change by reducing carbon dioxide emissions both within the UK and abroad; and
- ensuring secure, clean and affordable energy as we become increasingly dependent on imported fuel.

(DTI, 2007a, p6)

These objectives still guide policy today.

UK policies for CO_2 emissions reduction, 2000–2010

The current context of UK climate change policy is the Climate Change Act 2008 (HMG, 2008), which sets a target for 2050 that UK GHG emissions must be 80 per cent below 1990's level, and follow a trajectory to be adopted following the recommendation of five-yearly 'carbon budgets' by the Committee on Climate Change (CCC). The first three such budgets to 2022 were adopted with the fiscal budget in 2009, following the recommendations in CCC, 2008. Energy systems modelling of the kind discussed in later chapters played a key underpinning role in assessing the costs, trade-offs and pathways related to achieving such long-term targets (Strachan et al, 2009)

In terms of existing and future policy instruments to meet such targets, and other policy objectives such as energy security, one typology groups instruments under four generic headings (see Jordan et al, 2003), although in practice a policy instrument may be a hybrid in that it may have the characteristics of more than one type. These four are:

1 *Market/incentive-based (also called economic) instruments* (see EEA, 2006, for a recent review of European experience). These instruments include 'emissions trading, environmental taxes and charges, deposit-refund systems, subsidies (including the removal of environmentally-harmful subsidies), green purchasing, and liability and compensation' (EEA, 2006, p13). Except for green purchasing, these instruments change the relative prices and costs of inputs or processes in favour of those with less environmental impact.
2 *Classic regulation instruments*, which seek to define legal standards in relation to technologies, and environmental performance, pressures or outcomes. Kemp (1997) has documented how such standards may bring about innovation. Regulation can also include the imposition of obligations on economic actors, such as the renewable and energy efficiency obligations that have been imposed on energy suppliers in the UK (see below).

These instruments impose penalties on actors who fail to meet the standards or obligation. Where the obligation is tradeable, the instrument is a hybrid regulation-economic instrument.

3 *Voluntary (also called negotiated) agreements* between governments and producing organizations, which amount to agreed self-regulation (see ten Brink, 2002, for a comprehensive discussion). These change the investment:return ratio either by forestalling the introduction of market-based instruments or regulation (i.e. they are less costly than the counter-factual, which is perceived to involve more stringent government intervention, rather than necessarily the status quo). They can also lead to greater awareness of technological possibilities for eco-innovation that increase profitability as well as improving environmental performance (see Ekins and Etheridge, 2006, for a discussion of this in relation to the UK Climate Change Agreements).

4 *Information and/or education-based instruments* (the main example of which given by Jordan et al. (2003) is eco-labels, but there are others, such as the UK Government's 'Act on CO_2' advertising campaign, or corporate reports of environmental impacts or CO_2 emissions), which may be mandatory or voluntary. These promote awareness of the relevant issue, or of more eco-efficient products. They can also improve corporate image and reputation.

It has been increasingly common in more recent times to seek to deploy these instruments in so-called 'policy packages', which combine them in order to enhance their overall effectiveness across the three (economic, social and environmental) dimensions of sustainable development. Instrument packages have been implemented in the UK for both the demand side in end-use sectors (industry, households, commerce, agriculture, government and transport) and the supply-side, including key energy supply chains (notably electricity, biomass, and hydrogen).

The Stern Review (Stern, 2007, p349) considered that a policy framework for carbon reduction should have three elements: *carbon pricing* (for example, through carbon taxes or emission trading); *technology policy* (to promote the development and dissemination of both low carbon energy sources and high-efficiency end-use appliances and/or buildings); and the *removal of barriers to behaviour change* (to promote the take-up of new technologies and high-efficiency end-use options, and low-energy and/or low carbon behaviours).

In general, a broad policy document such as a White Paper would be followed up by more detailed policy strategies in particular areas, which would articulate in some detail the actual policies to be adopted to achieve the desired objectives. In energy, such policies may relate to energy supply or the various sectors of energy demand, as distinguished in Chapter 2.

Policies for reducing carbon emissions from energy supply

Because of the relative difficulty of reducing GHG emissions from agriculture and aviation, achieving the carbon reduction targets which the Government has set for 2050 will require the almost complete decarbonization (in terms of its emissions to the atmosphere) of energy supply, including electricity: through the use of renewables, nuclear power and carbon capture and storage (CCS); heat: through the use of low carbon biomass or low carbon electricity; transport fuels: through the use of low carbon bio-fuels, low carbon electricity, and low carbon hydrogen; and the increased efficiency of energy generation for power, heat and mobility: through CHP, heat pumps, power generation and more efficient engines of all kinds. An exploration of how this might be achieved is a major topic in subsequent chapters.

A key requirement of policies in this area is their ability to mobilize very large investments from the private sector, given that the investments required are well outside the level which can be financed by governments alone. For example, IEA (2008a, pp41–43) estimates that, in its low carbon scenario, the extra (global) investment requirements (i.e. over and above the investment in the global energy system that would be necessary if carbon were of no concern) are USD 7.4 trillion in buildings and appliances, USD 3.6 trillion for the power sector, USD 33 trillion in the transport sector and USD 2.5 trillion in industry. These are enormous numbers, which make climate change mitigation easily the largest civil public policy thrust ever attempted, in terms of its direct economic impacts.

With regard to energy supply, the most comprehensive strategy to have been issued by the UK Government to date is the Renewable Energy Strategy (DECC, 2009a). The purpose of this is to set out how the UK can meet its now statutory obligation to source 15 per cent of its final energy demand from renewable energy by 2020, to contribute to the achievement of the EU's overall target of 20 per cent by the same date, adopted in 2008 as part of the EU climate policy package. The 'lead scenario' in this strategy suggests that by 2020 the UK will need to derive 30 per cent of its electricity from renewables (up from about 6 per cent in 2009), 12 per cent of its heat (up from about 1 per cent in 2009) and 10 per cent of its transport energy (up from 2.6 per cent in 2009).

The policy instruments that are available to government to achieve the objective of decarbonizing energy supply are carbon pricing (e.g. carbon tax, emissions trading); price support for low carbon technologies (for example, feed-in tariffs, obligations and/or quotas with tradeable certificates); investment support, such as through capital grants, Enhanced Capital Allowances or tax credits); the removal of barriers to the deployment of low carbon technologies, such as ensuring access to infrastructure (e.g. transmission, grid connection); and timely planning, regulation and licensing procedures; availability of skills; administrative requirements; and public funding or co-funding of research, development and demonstration of the whole range of low carbon technologies.

In its analysis of policies for deploying renewables, the IEA (2008b, p23) identified a number of principles for successful policies for renewables support, namely: removal of non-economic barriers (relating to administrative hurdles, planning, grid access, skills, social acceptance); a predictable, transparent policy framework to support investment; technology-specific incentives based on technological maturity; transition incentives to foster innovation and move technologies towards competitiveness; and due attention to system effects (e.g. penetration of intermittent renewables). In particular, for effective deployment each technology that was not yet competitive on the energy market needed to receive a minimum level of remuneration, which varied with the technology, through the policy framework (for onshore wind and biomass electricity, which are among the renewables closest to market, this was USD 0.07–0.08/kWh [IEA, 2008b, pp100, 109]).

Economic instruments that have been (or soon will be) applied in the UK to reduce supply-side carbon emissions include:

- Emissions trading, including the UK Emissions Trading Scheme (ETS), which operated from 2002–2006;[7] the EU ETS for energy-intensive industry, as discussed above;
- Feed-in Tariffs for small-scale electricity generation from renewables such as solar PV and micro-wind, which were introduced from April 2010;[8]
- A Renewable Heat Incentive, which was the subject of consultation in 2010, and will begin operation from April 2011 (DECC, 2010a). It is intended that this will encourage the installation of such technologies as biomass boilers, heat pumps and solar thermal panels;
- Subsidies for a number of large-scale demonstration CCS schemes.

The main regulatory instrument applied to decarbonize energy supply has been the Renewables Obligation (RO), which obliges the major electricity suppliers to purchase each year a certain proportion of the energy they supply from renewable generators, in return for tradeable Renewables Obligation Certificates (ROCs) to be presented to Ofgem, or pay a penalty 'buy-out' price, the revenues from which are distributed *pro rata* to those companies which present ROCs. The target for the RO increases over time and has been adjusted a number of times. It is now subject to a rather complicated calculation each year, which is designed to keep 'headroom' between the fixed target and the actual obligation required, in order to maintain the incentive to increase renewable generation. The target is currently around 10 per cent for 2010, rising to around 15 per cent by 2015, with a current maximum obligation of around 20 per cent through to 2037.

Renewable generation is currently somewhat behind its current target, the RO buy-out price having been set at a level insufficient to stimulate the required investment. To increase uptake of renewable technologies further from the market, technology 'banding' (the allocation of different numbers of ROCs to different technologies) was introduced in 2009 to allow for different levels of RO support to be given to technologies at different levels of maturity. Even so, it seems likely that the RO will need to be increased further, or some other subsidy mechanism introduced (perhaps the extension of feed-in tariffs), if the UK is to get to 30 per cent renewable generation as envisaged in the 'lead scenario' of the Renewable Energy Strategy. IEA (2008b, p17) found that for onshore wind, the RO had proved substantially more expensive per unit of generation deployed, and been significantly less successful in deploying capacity, than the feed-in tariffs in a number of other countries, indicating the importance in the UK of non-economic barriers to deployment.

The desire to limit the cost of carbon reduction has meant that, in addition to the RO not being as effective as hoped, the various capital grants schemes (addressing various low carbon generation technologies, buildings, energy crop planting grants, bio-energy plants) have been so limited that they have also not succeeded in bringing about the widespread implementation and deployment of the technologies that they have sought to encourage. For example, over £40m the bulk of £50m Marine Renewables Deployment Fund set up by the DTI in 2004 was completely unused.

Building on the experience of the Renewables Obligation for electricity, in 2008 the Government introduced a Renewable Transport Fuel Obligation, with the target that 5 per cent of transport fuels should come from renewable sources by 2014. This may be increased to 10 per cent by 2020, provided adequate safeguards are in place to ensure that the bio-fuels come from sustainable sources.

In respect of nuclear power, the Government is keen to ensure that there is a favourable policy framework for a new generation of nuclear power stations. The Nuclear White Paper (BERR, 2008) confirmed the Government's previously expressed support for a new generation of nuclear power stations, on the grounds that it was an affordable, dependable, safe and low carbon energy technology, that would also contribute to energy security; the Nuclear White Paper envisaged the start of construction in 2013 and power output from 2018 (BERR, 2008, p136).

The issues set out in the Nuclear White Paper as needing to be addressed include planning, site assessment, assessment of potential health impacts, design assessment and licensing, and review of the regulatory regime in general. It also stated that the Government would undertake a Strategic Siting Assessment and Strategic Environmental Assessment, and further develop the planning and regulatory processes for new nuclear power so that private investors would have the confidence to come forward with proposals to build new nuclear power stations. Since then the consultation for the six energy National Policy Statements (NPSs, comprising an over-arching energy NPS, and NPSs on fossil-fuel generating infrastructure, renewable energy infrastructure, gas supply infrastructure and gas and oil pipelines, electricity networks infrastructure, as well as nuclear power) has taken place;[9] the Infrastructure Planning Commission (IPC) has been set up with the objective of facilitating planning for large-scale infrastructure (including energy) projects;[10] a number of sites have been identified for the next generation of nuclear power stations; and a number of companies, including EDF, E.ON and Centrica have expressed an interest in building or intention to build new nuclear power stations.

On the economics of nuclear power, and therefore the potential need for public subsidy of new nuclear build, the Government has had a remarkable change of mind since 2003. The Energy White Paper of 2003 said unequivocally 'the current economics of nuclear power make it an unattractive option for new generating capacity' (DTI, 2003, p61), but the next Energy White Paper in 2007 stated in contrast 'Based on this conservative analysis of the economics of nuclear power, the Government believes that nuclear power stations would yield economic benefits to the UK in terms of reduced carbon emissions and security of supply benefits' (DTI, 2007a, p191), although 'it would be for the private sector to fund, develop, and build new nuclear power stations in the UK, including meeting the full costs of decommissioning and their full share of waste management costs' (DTI, 2007a, p17). This implies that new nuclear build would neither need nor receive public subsidy. The wording in DTI 2007b (p59) is slightly different: 'As for any type of power station, energy companies would decide whether to propose, develop, construct and fund any new nuclear power stations. Private sector financing would also need to cover the full costs of decommissioning and full share of waste management costs', and does not imply so strongly that there will be no public subsidy. Indeed, later on BERR 2008 (p154) says:

> It is not intended that incentives will be provided through the fiscal regime to invest in nuclear power generation in preference to other types of electricity generation. The Treasury and HMRC are, however, exploring the possibility that the timing of nuclear decommissioning could create a potential tax disadvantage for nuclear operators and, if so, whether it may be appropriate to take action to ensure a level fiscal playing field between nuclear power and other forms of electricity generation.

This may open the door to some public subsidy of decommissioning costs, at least. However, against this it should be said that one of the earliest announcements of the Secretary of State for Energy and Climate Change in the new UK Government in 2010 emphatically ruled out public subsidies for new nuclear power stations.[11]

The issue is important, because if nuclear power is an important contributor to UK energy security and emissions reduction, and if private companies decide that it is not in fact financially viable without public subsidy (as has been the case in the past, and may still be the case), then without public subsidy new nuclear stations will not be built and the contribution of nuclear power will not be delivered.

Policies for reducing carbon emissions from energy demand

In addition to decarbonizing energy supply, policy may seek to reduce emissions by managing energy demand, using instruments such as carbon pricing (for example, carbon taxation or environmental tax reform), subsidies or tax reductions for low carbon equipment (for example, boilers, insulation, cars, combined heat and power [CHP]) or behaviours, carbon rationing (Personal Carbon Allowances, emission trading), or a wide range of regulations, voluntary agreements, or information instruments.

The three major sources of emissions on the demand side are industry and commerce, transport and homes. The most significant government document to have been produced recently in this area is its strategy for household energy management (DECC, 2010b), which seeks to go well beyond earlier Government approaches in this area to include both financing mechanisms and supply chain development, as well as an area-based approach in collaboration with local authorities, and support for home insulation on a 'whole house' basis, rather than just the cavity wall and loft insulation that have been the main components of schemes to date.

Economic instruments which have been used in respect of energy demand include the following.

- The *climate change levy*, an energy tax on business introduced in 2001, which in 2005 was forecast to reduce carbon emissions by 3.5MtCO$_2$ by 2010 (HMT, 2005, p171).
- *Fuel taxes.* Sterner (2007, p3201) estimates that the difference in fuel taxes between Europe and the USA, which results in European consumer prices of road fuels being about three times higher than those in the US, has resulted in European CO$_2$ emissions from road fuels being about half what they would be at the US price. The average new car fuel efficiency in Europe is also about 25–50 per cent higher than in the US (EEA, 2005).
- Other taxes such as the CO$_2$-graded Vehicle Excise Duty and the Air Passenger Duty.
- The *CRC Energy Efficiency Scheme* (formerly called the Carbon Reduction Commitment),[12] which began in 2010. This is a mandatory energy efficiency scheme, involving emissions trading, aimed at improving energy efficiency and cutting emissions in large public and private sector organizations.

Regulatory climate policy instruments on the demand-side have included the following.

- The *Carbon Emissions Reduction Target*, CERT, formerly called the Energy Efficiency Commitment (EEC), currently due to run to 2012, whereby energy suppliers are required to install energy efficiency measures in their customers' homes, up to a certain level of imputed carbon saving. This is a regulatory approach that gives considerable freedom of implementation to the suppliers on whom it falls, in contrast to more traditionally prescriptive regulations such as Building Regulations (see below). Most recently, in some areas this obligation has been folded into the Community Energy Saving Programme (CESP), which seeks to achieve these energy efficiency improvements through partnerships between energy suppliers, local authorities and other local bodies. This approach will be extended as part of the new household energy management strategy.
- *Warm Front and Warm Zones*, two schemes for installing subsidized energy efficiency measures, especially in the homes of relatively poor people.
- *Building Regulations* for new buildings, which are intended to reduce carbon emissions, such that by 2016 all new homes, and by 2019 all new buildings, will need to be 'zero carbon'. The effectiveness of regulations

depends on the adequacy of their implementation and enforcement. There is evidence that, in respect of Building Regulations, this leaves something to be desired (EST, 2004).

It is interesting that two of the Government's most significant policies for carbon reduction, CERT and the RO, place obligations on energy suppliers, which, as noted in respect of CERT above, they can meet in a variety of ways. A characteristic of both these obligations is that they do not involve public expenditure (they are funded by energy consumers), and neither of them is particularly visible, so that they do not raise public awareness of the objectives they are intended to achieve. It remains to be seen whether the much greater carbon reductions sought in the future can be achieved without generating greater public awareness and support.

Voluntary agreements in respect of energy demand have included the following.

- *Climate Change Agreements* (CCAs), which give an 80 per cent rebate from the Climate Change Levy to energy-intensive sectors that sign up to energy efficiency targets. CCAs were estimated to have reduced carbon emissions by 4.5 million tonnes of carbon in their first target period of 2001–2003 (HMT, 2005, p171).
- EU fuel efficiency agreements for new vehicles. Because the voluntary targets for fuel efficiency improvements by 2008, agreed between the European Commission and the major vehicle manufacturers, were not met, the new targets negotiated out to 2015 are mandatory.

The principal demand-side information policies in the UK are related to labelling, which is now required for a wide variety of white goods and, most recently, vehicles and buildings. Figure 3.2 shows how this has worked for fridge freezers, with the most efficient A-rated fridge freezers increasing to around 80 per cent of the market over a period of about five years.

Labels (Energy Performance Certificates, EPCs) have also recently been introduced for homes,[13] and there are ongoing trials of so-called 'smart meters' which give consumers real-time information about their energy consumption (and can potentially allow for much wider innovations in the energy system, such as real-time variable pricing, and various forms of demand management), in the context of a Government commitment to introduce smart meters into all homes by 2020 (at an estimated cost of some £7bn).

Finally, many climate policies, especially on the demand side, are implemented in 'packages' of policy measures affecting different actors, with such names as Market Transformation,[14] which includes:

- EU energy labelling;
- marketing campaigns (e.g. Energy Efficiency Recommended branding and advertising) by the Government and its agencies (e.g. Energy Saving Trust [EST]);

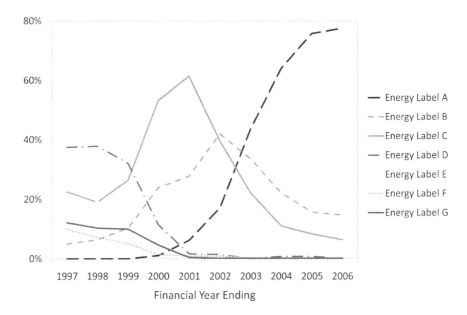

Source: Lees, 2006

Figure 3.2 *Development of the UK fridge freezer market by energy rating*

- consumer advice from Energy Efficiency Advice Centres;
- media coverage on climate change;
- retail staff training and point of sale material from the EST;
- EU Minimum Performance Standards;
- CERT funding for incentives for consumers to purchase energy efficient models of appliances.

Similarly, EU Integrated Product Policy includes Sustainable Consumption and Production (itself a package of different policy approaches), state aid, voluntary agreements, standardization, environmental management systems, eco-design, labelling and product declarations, greening public procurement, encouragement of green technology, and legislation in areas including waste and chemicals.[15]

Policies for energy security

A secure energy system gives access to energy services in the quality and quantity that energy users need and want, when they want them, and at a price they can afford. No energy system can be 100 per cent secure. In general, increasing the security of an energy system will require investment; that is to say, energy security comes at a cost.

As a concept, energy security is a more complex issue than is perhaps at first apparent. The Centre for Strategic and International Studies (CSIS) at the World Resources Institute identifies eleven factors which contribute to energy security: diversity of energy sources; diversity of suppliers; import levels; security of trade flows; geopolitics and economics; reliability; risk of nuclear proliferation; market and/or price volatility; affordability; energy intensity (energy used per unit of gross domestic product); and feasibility (CSIS, 2009, p1). Chapters 4 and 6 will reframe the concept of energy security more in terms of the *resilience* of the energy system. This section discusses briefly the policy approach to energy security that has been taken by the UK Government, mainly over the last ten years.

Following energy privatization and the liberalization of UK energy markets, with the UK through the 1990s still being largely or wholly self-sufficient in oil and gas, the dominant perception in UK policy circles was that liberalized energy markets would provide energy security. In the European Union, UK policy-makers projected the same message, and the UK model of privatization and independent regulation has not been uninfluential with the European Commission and other EU member states; but, in a world of volatile geopolitics, it was never wholly convincing on the wider European stage, which lacked the UK's self-sufficiency in oil and gas, and the UK itself has also gradually moved away from it as its energy self-sufficiency has evaporated, as will be seen.

The first sign of a more active policy approach came in 2001, when the DTI and Ofgem set up the Joint Energy Security of Supply working group (JESS) 'to assess risks to Britain's future gas and electricity supplies' (JESS, 2003, p2). JESS produced annual reports, the sixth of which was published in April 2006. The Energy Review (DTI, 2006) followed two months later, and prompted a further JESS report in December 2006, which stated that the Review had 'identified the need for new arrangements for the provision of forward-looking energy market information and analysis relating to security of supply'. Although it had 'reinforced the Government's commitment to a market-based approach to deliver security of supply ... it also identified two key security of supply challenges for the UK: managing increased dependence on oil and gas imports, and ensuring that the market delivers timely investment' (JESS, 2006, p4). A series of consultations was undertaken on a number of areas where it was perceived that further work was needed.

The results of the consultations fed into the Energy White Paper (EWP) of 2007 (DTI, 2007a). This reiterated the Government's commitment to an open European energy market, but also identified the need to improve the framework for investment in energy infrastructure in the UK, for both electricity and gas, through the generation of more information, planning reform, a more supportive policy framework, and the provision for further investment in electricity transmission and distribution networks.

The commitment in the 2007 EWP to provide more market information led in October 2007 to the publication of the first Energy Markets Outlook (BERR, 2007a), a far more substantial document than the JESS reports of earlier years. This reiterated the EWP identification of 'continuing commitment to competitive energy markets and effective, independent regulation as the most effective way of delivering secure and reliable energy supplies' (BERR, 2007a, p8), but public confidence in this approach was undermined by tightness in the gas market, and consequent price volatility, in the winters following 2004–2005 (see Figure 3.3), the closure of the Rough gas storage field due to a fire in February 2006, and the projected closure of a large number of nuclear and coal power stations in the years 2010–2020, the latter due to the provisions of the EU Large Combustion Plants Directive.

The next Energy Markets Outlook (DECC, 2008) report identified as further issues relevant to energy security the long-term availability and price of oil (following the oil price of nearly $150 per barrel earlier that year) and the implications for the electricity system of the large amount of renewable electricity generation (with its associated intermittency) to which the UK was committed as a result of the EU Renewables Directive.

Evidence of continuing unease with the market approach to security came in 2009, both from academic analysts (see, for example, Baker et al, 2009) and from the UK Government itself, when the former Energy Minister Malcolm Wicks was commissioned to conduct a review of international energy security. The resulting report recognizes explicitly that 'the energy security challenges presented by a dramatically changing global economic, geopolitical and energy landscape, combined with the urgent need to tackle climate change, are new and require us to re-assess our approach' (Wicks, 2009, p3).

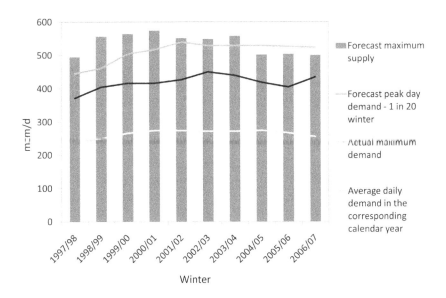

Source: BERR, 2007b, Chart 1, p6

Figure 3.3 *Gas capacity versus demand and supply*

In his personal introduction to his report, Wicks is more explicit still about the need for a new approach:

> *My conclusion is that the era of heavy reliance on companies, competition and liberalization must be re-assessed. The time for market innocence is over. We must still rely on companies for exploration, delivery and supply, but the state must become a more active interventionist where necessary. This is critical in Europe, when, despite progress, full liberalization of energy has not been achieved, and when key states are strong players in energy decision-making. Moreover, internationally, independent private-sector oil companies control smaller proportions of global oil and gas reserves, as nation states use national companies to develop these national resources. Many countries use political influence to gain access to energy supplies* (Wicks, 2009, p1).

From this it is clear that, while markets are perceived still to have an important role in delivering energy security, the UK Government by 2009 had little faith that what might be called the market approach to UK energy security, that has held sway for two decades, could respond effectively to the new energy challenges ahead. The UK Coalition Government elected in 2010 endorsed this view with a commitment 'to reform energy markets to deliver security of supply and investments in low carbon energy' (HMG, 2010, p16).

How this will work out in practice and in detail is still unclear. One significant development was Ofgem's Project Discovery (Ofgem, 2009b; 2010a), which, not dissimilarly to some aspects of UKERC's Energy 2050 project reported in this book, projected out to 2050 different energy futures for the UK energy system, in a 2×2 matrix structure, the axes of which were low to high economic growth and low to high environmental commitment. These scenarios were then subjected to a number of 'stress tests', including re-direction of LNG supplies, temporary loss of the gas import facility at Bacton, no wind at peak periods, investment delay and oil price shocks. From these scenarios Ofgem concluded that:

> *there are reasonable doubts as to whether the current arrangements will deliver security of supply and environmental objectives at least not without consumers paying substantially more than they would otherwise need to. For this reason, Ofgem does not consider that leaving the current arrangements unaltered is in the best interests of consumers* (Ofgem, 2010b, p14).

The same publication then proposed five policy packages with different governance arrangements with increasing levels of policy intervention.

Another significant development, perhaps prompted by the Ofgem work, was the entry of the Treasury into this territory with the publication in March 2010 of its Energy Market Assessment (HMT, 2010). This too presents a number of options for the reform of electricity markets. While it rules out a return to the era of a single, centralized purchaser of electricity, it also rules out exclusive reliance on energy and carbon prices to deliver secure low carbon energy. The future of the electricity market in the UK is therefore a mixed market, with the balance between market and government forces still the subject of analysis. Because the electricity and gas markets are inextricably linked (not least through the extent of gas-fired power generation), it seems unlikely that reform of the electricity market will leave the arrangements for gas governance untouched. As of late 2010, energy market arrangements as a whole therefore seem scheduled for the biggest change in more than two decades.

Conclusion

The achievement of low carbon and secure energy supplies for the UK has emerged as a major policy priority for the UK over the last decade. There is a new recognition that the energy price increases that this will entail will need to be managed as far as possible to mitigate their impacts on UK competitiveness and vulnerable households, but they cannot be avoided.

Climate policy has been the major driver of energy policy over the past ten years, and has seen enormous innovation, as is clear from the selection of policies for emissions reduction that have been discussed above, many of which would not have been dreamed of in earlier years. These are the kinds of policies which will have to be more widely applied to achieve the targets underlying the various scenarios described in subsequent chapters. However, these policies have so far yielded limited results. There is little doubt that the Government would have missed its 2010 target of a 20 per cent reduction in carbon emissions from 1990's level by a large margin, had it not been for the approximately 10 per cent reduction in carbon emissions caused by the recession in 2008 and 2009, which was neither foreseen nor desired. It therefore seems likely that while the policies have been innovative, they have not been applied stringently enough (CCC, 2009), and, no doubt, some barriers to policy effectiveness have still not been identified and tackled. This remains a challenge if emissions of GHGs are to be reduced to the required extent in the decades that follow.

A further issue has been the transparency of climate policy. It was noted above that some of the flagship policies (for example, CERT and the RO) have been implemented such that their implications are almost invisible to the general public. Doubtless this was because it was feared that some of those implications (most obviously the energy price rises) would be unpalatable to voters. But this has also meant that the public is largely unaware of what is being done to reduce emissions and is therefore less aware of the seriousness with which the Government takes the issue of climate change as a whole. It is doubtful whether this disengagement of the public from climate policy can generate the sort of commitment to and understanding of the need for emissions reduction that will be required for the much larger proportional cuts in emissions that need to be delivered in the future.

Finally, although climate policy has largely been in the energy policy driving seat since 2000, this has changed, and energy security is now at least an equal co-pilot, a situation recognized institutionally by the creation in 2008 of the Department of Energy and Climate Change to address jointly the issues of climate change and energy security. As will become apparent in the chapters which follow, these issues do not always make easy bedfellows, and tensions between them, and between other policy objectives such as energy affordability, may be expected in the future. It was a core objective of the Energy 2050 project to illuminate and seek to address some of these issues and tensions. The following chapters will show the extent to which it has done so.

Notes

1 See www.scotland.gov.uk/News/Releases/2010/01/06141510, accessed 7 June 2010
2 DTI – Department of Trade and Industry; BERR – Department of Business, Enterprise and Regulatory Reform; DECC – Department of Energy and Climate Change
3 See www.ofgem.gov.uk/SUSTAINABILITY/Pages/Sustain.aspx, accessed 7 June 2010
4 See Ekins and Lees (2008) for a discussion of these Directives
5 See http://ec.europa.eu/environment/air/transport/co2/co2_home.htm, accessed 7 June 2010
6 Based on a provisional estimate of 2009 emissions, see www.decc.gov.uk/en/content/cms/statistics/climate_change/gg_emissions/uk_emissions/2009_prov/2009_prov.aspx, accessed 7 June 2010
7 See www.decc.gov.uk/en/content/cms/what_we_do/change_energy/tackling_clima/ccas/uk_ets/uk_ets.aspx, accessed 5 May 2010, though link since discontinued
8 See www.fitariffs.co.uk/?gclid=CIqm74-mu6ECFRg-lAod9A08_A, accessed 5 May 2010
9 The consultation closed on 22 February 2010, and was subject to parliamentary scrutiny at the time of the General Election. See DECC, 2009b, and www.energynpsconsultation.decc.gov.uk/, accessed 7 June 2010. At the time of writing it was not clear how the new UK Government would proceed with the NPSs
10 See http://infrastructure.independent.gov.uk/, accessed 5 May 2010. One of the early actions of the new UK Government was to abolish the IPC, intending to move such planning decisions within Government
11 See www.timesonline.co.uk/tol/news/politics/article7127202.ece, accessed 7 June 2010
12 See www.decc.gov.uk/en/content/cms/what_we_do/lc_uk/crc/crc.aspx, accessed 5 May 2010. From the Coalition Government's decision in the October 2010 Comprehensive Spending Review to retain the revenues from the Scheme (as opposed to recycle them to the participants), it is not clear whether the Scheme remains a trading scheme or has been converted into a tax.

13 EPCs are being retained despite the abolition by the new UK Government of the Home Information Packs (HIPs) of which they were a part, see www. communities.gov.uk/housing/buyingselling/homeinformation/, accessed 7 June 2010, though link since discontinued

14 See www.mtprog.com/, accessed 5 May 2010

15 See http://ec.europa.eu/environment/ipp/, accessed 5 May 2010

References

AEA Technology (2010) *UK Greenhouse Gas Inventory, 1990 to 2008: Annual Report for Submission under the Framework Convention on Climate Change*, Report to Department of Energy and Climate Change, April, AEA Technology, Harwell, www.airquality.co.uk/reports/cat07/1004301344_ukghgi-90-08_main_chapters_Issue3_Final.pdf, accessed 5 May 2010, though link since discontinued

Anandarajah, G., Strachan, N., Ekins, P., Kannan, R. and Hughes, N. (2008) *Pathways to a Low-Carbon Economy: Energy systems modelling*, UKERC Energy 2050 Research Report 1, Ref. UKERC/RR/ESM/2008/03, November, UKERC, London

Baker, P., Mitchell, C. and Woodman, B. (2009) The Extent to which Economic Regulation Enables the Transition to a Sustainable Electricity System, UKERC, Report 2009: REF UKERC/WP/ESM/2009/013, UKERC, London, http://ukerc.rl.ac.uk/UCAT/cgi-bin/ucat_query.pl?GoButton=Year&YWant=2009, accessed October 31 2010

BERR (Department for Business, Enterprise and Regulatory Reform) (2007a) *Energy Markets Outlook*, October, BERR, London, www.berr.gov.uk/files/file41995.pdf, accessed 8 June 2010, though link since discontinued

BERR (Department for Business, Enterprise and Regulatory Reform) (2007b) *Introduction to Gas Market Liquidity*, October, BERR, London, www.bis.gov.uk/files/file41843.pdf, accessed 8 June 2010

BERR (Department for Business, Enterprise and Regulatory Reform) (2008) *Meeting the Energy Challenge: A White Paper on Nuclear Power*, CM7296, January, The Stationery Office, London, www.berr.gov.uk/files/file43006.pdf, accessed 5 May 2010

CCC (Committee on Climate Change) (2008) *Building a low-carbon economy – the UK's contribution to tackling climate change*, Climate Change Committee inaugural report, CCC, London, www.theccc.org.uk

CCC (Committee on Climate Change) (2009) *Meeting carbon budgets – the need for a step change*, Progress Report to Parliament, CCC, London, www.theccc.org.uk

CSIS (The Centre for Strategic and International Studies) (2009) 'Evaluating the Energy Security Implications of a Carbon-Constrained U.S. Economy', January, CSIS, Washington DC, http://csis.org/files/media/csis/pubs/090130_evaluating_energy_security_implications.pdf, accessed 8 June 2010

DECC (Department of Energy and Climate Change) (2008) *Energy Markets Outlook*, December, DECC, London, www.bis.gov.uk/files/file49406.pdf, accessed 8 June 2010

DECC (Department of Energy and Climate Change) (2009a) *The UK Renewable Energy Strategy*, Cm7686, July, The Stationery Office, London, www.decc.gov.uk/en/content/cms/what_we_do/uk_supply/energy_mix/renewable/res/res.aspx, accessed 5 May 2010

DECC (Department of Energy and Climate Change) (2009b) *Consultation on Draft National Policy Statements for Energy Infrastructure*, November, DECC, London, www.energynpsconsultation.decc.gov.uk/home/, accessed 7 June 2010

DECC (Department of Energy and Climate Change) (2009c) *Annual Report on Fuel Poverty Statistics*, DECC, London

DECC (Department of Energy and Climate Change) (2010a) *Renewable Heat Incentive: Consultation on the proposed RHI financial support scheme*, February, DECC, London, www.decc.gov.uk/en/content/cms/consultations/rhi/rhi.aspx, accessed 5 May 2010

DECC (Department of Energy and Climate Change) (2010b) *Warm homes, greener homes: a strategy for household energy management*, March, DECC, London, www.decc.gov.uk/en/content/cms/what_we_do/consumers/saving_energy/hem/hem.aspx, accessed 5 May 2010

DETR (Department of the Environment, Transport and the Regions) (2000) *Climate Change: the UK Programme*, November, Cm 4913, The Stationery Office, London

DTI (Department for Trade and Industry) (2003) *Our Energy Future – Creating a Low Carbon Economy*, Energy White Paper, Cm 5761, The Stationery Office, London

DTI (Department for Trade and Industry) (2006) *The Energy Challenge: Energy Review*, July, DTI, London, www.decc.gov.uk/en/content/cms/publications/energy_rev_06/energy_rev_06.aspx, accessed 5 May 2010

DTI (Department of Trade and Industry) (2007a) *Energy White Paper: Meeting the Energy Challenge*, Cm7124, DTI, London, www.berr.gov.uk/files/file39387.pdf, accessed 5 May 2010

DTI (Department of Trade and Industry) (2007b) *The Future of Nuclear Power: the Role of Nuclear Power in a Low-Carbon Economy. Consultation Document*, May, DTI, London, http://webarchive.nationalarchives.gov.uk/20071204135946/http://www.berr.gov.uk/consultations/page39704.html, accessed 5 May 2010

EC (European Commission) (2008) 'The EU Climate and Energy Package', EC, Brussels, http://ec.europa.eu/environment/climat/climate_action.htm, accessed 5 May 2010

EEA (European Environment Agency) (2005) 'Market based instruments for environmental policy in Europe', EEA Technical Report No 8/2005, Copenhagen, Denmark, http://reports.eea.europa.eu/technical_report_2005_8/en/index_html

EEA (2006) *Using the Market for Cost-Effective Environmental Policy: Market-based Instruments in Europe*, EEA Report No.1/2006, European Environment Agency, Copenhagen

Ekins, P. and Etheridge, B. (2006) 'The Environmental and Economic Impacts of the UK Climate Change Agreements', *Energy Policy*, vol 34 (15), pp. 2071–2086

Ekins, P. and Lees, E. (2008) 'The Impact of EU Policies on Energy Use in and the Evolution of the UK Built Environment', *Energy Policy*, vol 36, pp. 4580–4583, http://dx.doi.org/10.1016/j.enpol.2008.09.006

EST (Energy Saving Trust) (2004) *Assessment of energy efficiency impact of Building Regulations compliance,* BRE client report 219683, November 2004, Energy Saving Trust, London, www.est.org.uk/partnership/uploads/documents/Houses_airtightness_report_Oct_04.pdf

EU (European Union) (2009) 'Directive 2009/29/EC of the European Parliament and of the Council of 23 April 2009 amending Directive 2003/87/EC so as to improve and extend the greenhouse gas emission allowance trading scheme of the Community', *Official Journal of the European Union*, June 5, EU, Brussels, http://eur-lex.europa.eu/LexUriServ/LexUriServ.do?uri=OJ:L:2009:140:0063:0087:EN:PDF, accessed 5 May 2010

Helm, D. (2003) *Energy, the Market and the State*, revised edition, Oxford University Press, Oxford

HMG (Her Majesty's Government) (2006) *Climate Change: The UK Programme 2006*, Cm 6764, The Stationery Office, London, www.decc.gov.uk/en/content/cms/what_we_do/change_energy/tackling_clima/programme/programme.aspx, accessed 5 May 2010

HMG (Her Majesty's Government) (2008) *The Climate Change Act*, Office of Public Section Sector Information, London, www.opsi.gov.uk/acts/acts2008/ukpga_20080027_en_1, accessed 18 March 2010

HMG (Her Majesty's Government) (2009a) *The UK Renewable Energy Strategy*, Cm7686, July, The Stationery Office, Norwich

HMG (Her Majesty's Government) (2009b) *The UK Low Carbon Transition Plan*, July, The Stationery Office, Norwich

HMG (Her Majesty's Government) (2010) *Warm Homes, Greener Homes: A Strategy for Household Energy Management*, March, Department of Energy and Climate Change, London

HMT (Her Majesty's Treasury) (2005) *Budget 2005: Investing for our Future*, HM Treasury, London

HMT (Her Majesty's Treasury) (2010) *Energy Market Assessment*, HM Treasury, London, www.direct.gov.uk/prod_consum_dg/groups/dg_digitalassets/@dg/@en/documents/digitalasset/dg_186447.pdf, accessed 8 June 2010

IEA (International Energy Agency) (2008a), *Energy Technology Perspectives 2008 – Scenarios and strategies to 2050*, International Energy Agency, Paris

IEA (International Energy Agency) (2008b) *Deploying Renewables: Principles for Effective Policies*, International Energy Agency, Paris

IPCC (Intergovernmental Panel on Climate Change) (2007) *Climate Change 2007: Mitigation*, Contribution of Working Group III to the Fourth Assessment Report of the Intergovernmental Panel on Climate Change, Cambridge University Press, Cambridge

JESS (Joint Energy Security of Supply working group) (2003) *Second Report*, February, DTI, London, www.berr.gov.uk/files/file10729.pdf, accessed 8 June 2010

JESS (Joint Energy Security of Supply working group) (2006) *Long-term Security of Energy Supply*, December, DTI, London, www.bis.gov.uk/files/file35989.pdf, accessed 8 June 2010

Jordan, A., Wurzel, R. and Zito, A. (eds) (2003) *'New' Instruments of Environmental Governance? National Experiences and Prospects*, Frank Cass, London

Kemp, R. (1997) *Environmental Policy and Technical Change: a Comparison of the Technological Impact of Policy Instruments*, Edward Elgar, Cheltenham

Lees, E. (2006) *Evaluation of the Energy Efficiency Commitment 2002–05*, report to DEFRA from Eoin Lees Energy, www.defra.gov.uk/Environment/energy/eec/pdf/eec-evaluation.pdf

Ofgem (Office of Gas and Electricity Markets) (2009) *Energy Supply Probe*, Ofgem, London, www.ofgem.gov.uk/Markets/RetMkts/ensuppro/Pages/Energysupplyprobe.aspx, accessed 5 May 2010

Ofgem (Office of Gas and Electricity Markets) (2010) *Project Discovery: Options for delivering secure and sustainable energy supplies*, Ofgem, London, www.ofgem.gov.uk/MARKETS/WHLMKTS/DISCOVERY/Documents1/Project_Discovery_FebConDoc_FINAL.pdf, accessed 8 June 2010

PIU (Performance and Innovation Unit) (2002) *The Energy Review*, February, Cabinet Office, London

SDC (Sustainable Development Commission) (2007) *Lost in Transmission? The Role of Ofgem in a Changing Climate*, September, SDC, London, www.sd-commission.org.uk/pages/ofgem-and-the-energy-system.html, accessed 11 May 2010

Stern, N. (2007) *The Economics of Climate Change: The Stern Review*, Cambridge University Press, Cambridge

Sterner, T. (2007) Fuel taxes: an important instrument for climate policy, *Energy Policy*, 35: pp3194–3202

Strachan, N., Pye, S. and Kannan, R. (2009) *The Iterative Contribution and Relevance of Modelling to UK Energy Policy*, Energy Policy, vol 37 (3), pp850–860

ten Brink, P. (ed.) (2002) *Voluntary Environmental Agreements: Process, Practice and Future Use*, Greenleaf Publishing, Sheffield

UNFCCC (UN Framework Convention on Climate Change) (2009) *Copenhagen Accord*, Draft Decision of the President, December 18, FCCC/CP/2009/L.7, UNFCC, Bonn, http://unfccc.int/resource/docs/2009/cop15/eng/l07.pdf, accessed 5 May 2010

Wicks, M. (2009) *Energy Security: a National Challenge in a Changing World*, August, Department of Energy and Climate Change, London, www.decc.gov.uk/en/content/cms/what_we_do/change_energy/int_energy/security/security.aspx, accessed 8 June 2010

4
Energy Futures: The Challenges of Decarbonization and Security of Supply

*Jim Skea, Gabrial Anandarajah, Modassar Chaudry,
Anser Shakoor, Neil Strachan, Xinxin Wang
and Jeanette Whitaker*

Introduction

Chapter 2 made it abundantly clear that current and past trends show weak progress towards the UK's decarbonization goal, and increasing energy insecurity associated with import dependence and reliance on a limited range of supplies. Chapter 3 identified a stirring in the policy world that is leading to new approaches better fitted for meeting future challenges. However, the gap between policy aspiration and real change on the ground remains striking.

This chapter sets out a framework for thinking through the technical, economic and institutional feasibility of a low carbon resilient energy economy, the range of factors – social, environmental and technological – that need to be taken into account, and the means by which goals can be achieved. Reflecting a 'whole systems' approach to energy analysis, a wide definition of a national energy system, is adopted:

> *the set of technologies, physical infrastructure, institutions, policies and practices located in and associated with a country which enable energy services to be delivered to its consumers.*

For the UK, this definition covers all of the equipment along the energy supply chain that is located domestically – extraction of non-renewable resources, electricity generation, energy conversion, transportation, transmission, storage, distribution and end-use equipment. It excludes physical infrastructure located outside the national boundaries. It covers institutions (government at the national, regional and local levels, other statutory and non-statutory public bodies and private companies), policies, regulatory frameworks (economic and environmental) and operating practices. Although the analysis in this book focuses on the UK, the same approach could be applied to other countries, or even to sub-national or multinational regions.

The definition builds on the concept of delivering 'energy services' to consumers rather than energy *per se*. This allows for the fact that technologies and practices on the energy demand side can help to ensure a reliable supply of energy services. For example a well-insulated house will make the occupants more resilient against supply interruptions during cold weather. At a more sophisticated level, advanced technologies such as micro-generation and private grids could also promote resilience. This goes beyond the simple 'security of supply' aspects of energy security and resilience.

A country's energy system is bounded by what is within the sphere of influence of national institutions. It excludes energy infrastructure outside the national jurisdiction as well as supra-national (EU) or international institutions (e.g. International Energy Agency). There is necessarily some blurring at the edges. For example, any EU member state has an influence over EU energy policy, but it must share responsibility with the European Commission, the European Parliament and 26 other member states. Its influence cannot therefore be said to be decisive. However, the energy system does include infrastructure in which the country has a major share, for example, gas and electricity inter-connectors.

It is also helpful to draw some boundaries in terms of internal policies. For example, homeland security and defence policy can be seen as having an impact on the resilience of a national energy system. However, such policies relate to a range of sectors and other types of critical infrastructure such as water, transport and telecommunications. These are therefore excluded from the definition, although it may be useful for the energy research community to interact with those operating in other policy domains.

The chapter begins by setting out the framework for thinking about the twin goals of decarbonization and energy security. As described in Chapter 1, the concept of 'resilience' is used to take forward the energy security analysis. The approach to issues such as lifestyle change, innovation policy ('technological acceleration') and environmental impacts is also discussed.

This book has been informed by rigorous quantitative analysis. After the framework is set out, the key models and other analytical tools that have underpinned the work are described. These include the whole system MARKAL model and more detailed models covering individual sectors (e.g. households) and different network industries (e.g. gas and electricity). Their strengths, weaknesses and the questions that they are fit to answer are discussed.

The next stage is to describe a 'reference' scenario for the UK energy system out to 2050. This *Reference* scenario is the starting point for all of the subsequent analysis. It is not a forecast. It establishes a benchmark from which costs can be measured and a counterfactual from which new policies and approaches can be inferred. In a world where policy is developing rapidly, there is a degree of arbitrariness as to how a *Reference* scenario should be constructed. The policies that were in place at the time of the UK's 2007 Energy White Paper were chosen as the starting point. This provides a coherent and consistent starting point and can be used to assess the appropriateness of subsequent policy initiatives. The basic premise underlying the *Reference* scenario is that markets and institutions operate so as to maximize welfare for UK citizens, *without taking into account decarbonization goals or security constraints*. If this were a forecast it would clearly be unrealistic. However, this scenario has a useful place within an analytical schema that compares and contrasts different energy futures.

Finally, the chapter draws out the gap between the *Reference* scenario and the policy objectives and approaches discussed in Chapter 3. This points the way towards the range of scenarios for the UK's energy future covered in Chapters 5–11.

Energy systems, decarbonization and resilience

Progress in reducing the energy sector's impact on the climate can be measured through a single metric, tonnes of CO_2 equivalent emitted. The energy security agenda responds to anxieties and insecurities about a range of contingencies which are often not well thought through, or are ill-defined. These include adequacy of investment in electricity generation capacity, loss of critical infrastructure whether through deliberate action or by accident, or politically motivated interruptions to supply in global markets. For security, no single metric is adequate by itself.

It has become a commonplace notion, prevalent in both policy-making and academic circles (e.g. Grubb et al, 2006), that technologies and measures that reduce CO_2 emissions contribute to energy security and vice versa. This is clearly the case for energy efficiency which reduces both CO_2 emissions and dependence on imported energy. But fossil fuels will play an important role in the energy mix for some time to come. Energy sources such as coal may contribute to diversity of supply and thereby enhance security, but deploying them (without carbon capture and storage technology) works against climate change policy goals. Increasingly ambitious plans to deploy intermittent renewable energy pose challenges for the reliability of electricity supplies without corresponding investment in 'back-up' capacity.

Energy security and resilience

This book does not attempt to consider geo-political aspects of the global energy system. Instead, in addressing energy security issues for the UK, the focus is on the ability of the energy system to respond to shocks and disturbances. This capacity to respond to disturbances is termed the *resilience* of the energy system. Although the concepts of 'security' and 'security of supply' are frequently (and often rather loosely) used in the energy domain, the concept of 'resilience' has had relatively little usage. The links between resilience and security are explored below.

The advantage of focusing on resilience as the key concept is that it can be seen as an intrinsic characteristic of the energy system itself. It does not require any analysis of the underlying causes of a particular shock, for example a prolonged interruption of gas supply. It is only necessary to know that a particular kind of shock is possible.

The concept of 'resilience' has acquired a long pedigree in other fields, notably ecology, since the seminal work of Holling (1973).[1] The substantial amount of systematic thinking that has been invested in this field over the last 35 years is well worth considering. Resilience is seen to be a key concept in ecology because:

> *A resilient system, in a desirable state, has a greater capacity to continue providing us with the services that support our quality of life while being subjected to a variety of shocks* (Walker and Salt, 2006).

The analogy with the delivery of energy services and the maintenance of quality of life is a helpful one. A classic definition of 'ecological resilience' focuses on maintaining existence of function and is described as:

> *the capacity of a system to absorb disturbance; to undergo change and still retain essentially the same function, structure and feedbacks* (Walker et al, 2004).

The ability to bounce back is key to this definition. A second type of definition refers to *engineering resilience*. It considers ecological systems to exist close to a stable steady state, where resilience is defined as:

> *the ability to return to the steady state following a disturbance, and focuses on maintaining efficiency of function. Here the major measure is return time where the speed of the bounce-back is the most important factor* (Holling, 1996).

In terms of an energy system, both the ability to bounce back and the speed of bounce-back will be important for industry and consumers. Building on the definition of the energy system suggested above, the following definition of energy system resilience has been adopted:

> *Resilience is the capacity of an energy system to tolerate disturbance and to continue to deliver affordable energy services to consumers. A resilient energy system can speedily recover from shocks and can provide alternative means of satisfying energy service needs in the event of changed external circumstances.*

In the field of climate change impacts and adaptation, resilience has been characterized as the 'flip side of vulnerability' (IPCC, 2001). The concept of vulnerability is also helpful in the energy field. IPCC described climate vulnerability as being 'a function of the sensitivity of a system to changes in climate (the degree to which a system will respond to a given change in climate)'. The following working definition of the vulnerability of an energy system has been adopted:

Vulnerability is the sensitivity of an energy system to external distur-bance or internal malfunction. A vulnerable energy system lacks the capacity to recover speedily from shocks and may not be able to satisfy energy service needs affordably in the event of changed external circumstances.

Stirling (2009a) has made an interesting attempt to relate the concepts of resil-ience and energy security. In his framework, resilience is one of four aspects of energy security (Figure 4.1). Security is characterized as the ability to mitigate threats which may take the form of transient disruptions ('shocks') or more permanent shifts ('stress').

In thinking about security and resilience, this book focuses purely on tran-sient threats ('shocks'). Under Stirling's framework, enduring shifts ('stress') could include, for example, pressure to decarbonize the energy economy. The attempt to use 'energy security' as an organizing framework for all aspects of energy system development is interesting, but is not compatible with the approach taken here.

key aspects of energy security

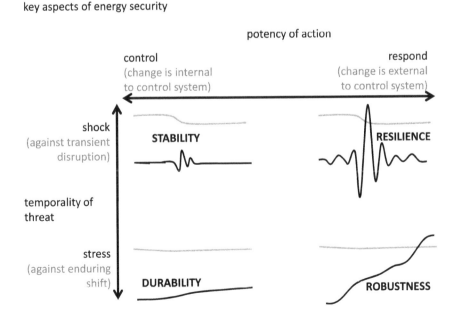

Source: Stirling, 2009a

Figure 4.1 *Conceptualizing energy security*

According to Stirling, resilience is the ability to respond to shocks that are external to the system. However, given the broad definition of the energy system adopted above, the definition of resilience adopted here also encompasses what Stirling describes as stability, the ability to respond to shocks that are internal to the energy system. This would apply, for example, with reference to the adequacy of electricity generation capacity to meet levels of demand caused by low probability outages of plant.

Shocks, resilience and vulnerability

The shocks against which an energy system needs to be resilient are varied in nature. There may be knowledge about the nature of some of these shocks ('known unknowns') but others have yet to be imagined ('unknown unknowns'). Three broad types of incertitude can be identified (Stirling, 2009b).

- *Risk:* the nature of the shock can be defined and a quantitative probability can be attached to it. Unplanned outages of electricity generating equipment fall into this category.
- *Uncertainty:* the nature of the shock can be described but no probability can be attached to it. An example would be the interruption of gas supplies from Russia.
- *Ignorance:* the nature of the shocks cannot be characterized and no probabilities can be attached. Some types of terrorist attack could fall into this category.

Broadly speaking, energy policy has dealt far more rigorously with those types of incertitude to which probabilities can be attached and that are more amenable to a formal risk-based assessment.

Table 4.1 describes a range of possible external events that would threaten the operation of the UK energy system, locating them within the risk-uncertainty-ignorance framework.

Some of these events are already routinely factored into energy decision-making, notably those associated with technical failures and weather. Energy price volatility refers to measurable price variability over periods of months or years. A small but growing literature is using portfolio approaches from finance theory to address energy policy (Bazilian and Roques, 2008).

Table 4.1 *Events impacting on the energy system*

Type of event	Type of incertitude
Technical equipment failure, unplanned outages	Risk
Weather-related risks	Risk
Volatility in global energy prices	Risk
Energy price 'shocks'	Uncertainty
Interruption of a major supply source	Uncertainty
Attack on energy infrastructure	Ignorance

The *vulnerability* of the energy system can be seen as falling into two main areas, each implying different management approaches. The coverage of the two areas and the range of the management approaches reflect our definition of the energy system which focuses on the delivery of energy services to consumers.

1   *The availability and cost of primary energy supplies.* Vulnerability in this area raises the prospect of disruptions to supplies and fluctuating prices. Vulnerability can be managed partly through domestic energy policy and the choice of primary energy supplies (as measured by supply diversity, import dependence, etc.). Measures to reduce vulnerability include reducing energy demand to minimize the economic impacts of supply interruption or price fluctuations, stand-by electricity generation capacity, installing multiple-fuel capabilities or maintaining energy stocks at industrial or commercial premises. Vulnerability can also be managed through foreign or defence policies, or through energy-sharing agreements struck via the EU or International Energy Agency (IEA), but these are beyond the scope of this work.

2   *Transformation, conversion, storage and distribution systems* which allow primary energy to be converted and made available to final consumers when and where they want it. Domestic energy policy has a major role to play here by ensuring the reliability of transmission and distribution systems and the provision of sufficient margins between potential supply and demand. Current UK energy policy centres on this area of vulnerability, but homeland security policies also have an important role to play in protecting critical infrastructure from attack. Note that a wide conception of energy policy is required as it can be argued, for example, that markets are sufficient to produce adequate margins between potential supply and demand, that is, that the optimum policy in this respect is non-intervention.

Reducing the vulnerability of the energy system from the perspective of final consumers is key. Vulnerabilities for final consumers run right through the energy system. The important policy questions concern which types of measure, applied at different points in the energy system, can most easily and cost effectively be deployed to protect consumers. In addressing these questions, consideration must be given as to which types of incertitude (risk, uncertainty, ignorance) specific measures address. A measure that reduces quantifiable risk should be assessed differently from a measure about which there is a state of 'ignorance'.

The scenario framework

Chapter 6 systematically considers the synergies and trade-offs between energy security and decarbonization. This section sets out the scenario framework and explores the linkages between decarbonization, resilience, lifestyle change, technology acceleration, scale and global uncertainties. The two primary areas of policy concern – decarbonization and resilience – are used to define a set of four 'core' scenarios, as shown in Figure 4.2.

The *Reference* scenario assumes no policies other than those in place at the time of the 2007 Energy White Paper. The *Low Carbon* scenario assumes that the UK is on a pathway to an 80 per cent reduction in CO_2 by 2050. The *Resilient* scenario ignores CO_2, but incorporates a set of measures mitigating against different types of energy shock. The final *Low Carbon Resilient* scenario combines the two attributes.

Five sets of factors are held constant across the four core scenarios.

1   The international context, specifically the cost and availability of fossil fuels.
2   The cost and performance of energy demand- and supply-side technologies.
3   The way energy investment decisions are made, i.e. the financial rates of return ('hurdle rates') that must be achieved before investments take place.
4   The evolution of people's lifestyles. Input assumptions about economic growth, lifestyle and energy service demands are based on historical trends.
5   Energy consumers' preferences. The assumption of unchanged consumer preferences means that people respond to changing prices and the availability of technologies in the same way as they have done in the past.

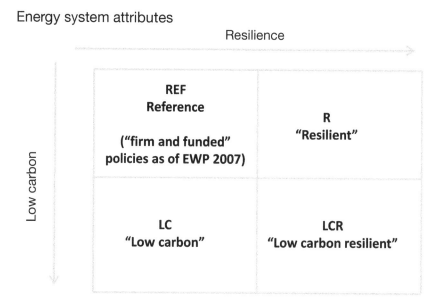

Figure 4.2 *The UKERC core scenarios*

The factors that change across the core scenarios are:

- final energy use resulting from changes in energy service demands and the take-up of conservation and efficiency technologies in response to price signals and other policy incentives;
- actual investment in energy supply, infrastructure and technologies;
- the policy framework for post-2007 initiatives.

The *Reference* (REF) scenario assumes that concrete policies and measures in place at the time of the 2007 Energy White Paper continue into the future and that no additional measures are introduced. For example, the CO_2 allocations for the EU Emissions Trading Scheme in Phase II (2008–2012) are assumed to continue from 2013 onwards. The scenario does not include the UK's 80 per cent carbon reduction goal, intermediate targets for 2020 or the effects of the 2008 EU Renewable Energy Directive. The *Reference* scenario represents neither a prediction nor a preferred future. It provides a baseline from which to assess the actions and costs associated with achieving policy goals.

The *Low Carbon* (LC) scenario assumes the introduction of a range of policies leading to an 80 per cent reduction in UK carbon dioxide emissions by 2050 relative to 1990, with intermediate milestones at 2020 and 2030. The 2020 milestone is broadly compatible with the third carbon budget (2018–2022) set under the 2008 Climate Change Act. The policies assumed are those which will result in the goal being achieved with the lowest possible impact on welfare, subject to the five factors above remaining unchanged.

The *Resilience* (R) scenario takes no account of the carbon reduction goal but assumes additional investment in infrastructure, demand reduction and supply diversity with a view to making the energy system more resilient to external shocks. The detailed indicators defining a 'resilient' energy system are set out in Chapter 6.

The *Low Carbon Resilient* (LCR) scenario combines the carbon and resilience goals.

The core scenarios act as the starting point for a set of scenario variants (Table 4.2) which explore a wider range of issues facing decision-makers. Even by themselves, however, the core scenarios address important policy questions.

- What additional steps must be taken to secure deep cuts in carbon emissions by 2050?
- How will emission reductions be distributed across sectors, fuels and technologies if costs are minimized?
- What is the additional cost of investing in resilience compared to carbon emission reductions?
- Will investing in an energy system that is resilient to shocks make it easier or more difficult to reduce carbon emissions? Will reducing carbon emissions make the energy system more resilient?

Table 4.2 *List of scenarios and scenario variants*

Scenario identifier	Scenario name	Notes
CORE SCENARIOS (Chapters 4, 6)		
REF	Reference	Includes 'firm and funded' policies at the time of the 2007 Energy White Paper
LC	Low Carbon	80% CO_2 reduction by 2050 and 26% by 2020 from a 1990 baseline
R	Resilient	Reference scenario with constraints on final energy demand and supply diversity
LCR	Low Carbon Resilient	Combines the constraints in LC and R
CARBON PATHWAYS (Chapter 5)		
LC-40	Faint-heart	40% CO_2 reduction by 2050 and 15% by 2020
LC-60	Medium carbon	60% CO_2 reduction by 2050 and 26% by 2020
LC-90	Super ambition	90% CO_2 reduction by 2050 and 32% by 2020
LC-EA	Early action	80% CO_2 reduction by 2050 and 32% by 2020
LC-LC	Least-cost path	Optimized carbon pathway using the 2010–2050 budget from *LC-EA*
LC-SO	Socially optimal least-cost path	Optimized carbon pathway using the 2010–2050 budget from *LC-EA* and a social discount rate
ACCELERATED TECHNOLOGY DEVELOPMENT (Chapter 7)		
LC-Acctech	LC Acctech	LC core scenario with seven key technologies accelerated
LC-Renew	LC Renew	LC core scenario with four renewable technologies accelerated
LC-Acctech-no FC	LC Acctech, with no FC acceleration	All technologies accelerated except fuel cells
LC-Acctech-no CCS	LC Acctech, with no CCS	Seven key technologies accelerated with carbon capture and storage not allowed
LC-Acctech-later CCS	LC Acctech, with later CCS	Seven key technologies accelerated with carbon capture and storage commercialization delayed until 2030
LC-60-Acctech	LC-60-Acctech	Seven key technologies accelerated based on the *LC-60* scenario
LC-60-wind	LC-60-wind	Acceleration of wind technology based on the *LC-60* scenario. Similar scenarios for other individual technologies

Scenario identifier	Scenario name	Notes
ENERGY LIFESTYLES (Chapter 9)		
REF-LS	Reference lifestyle	REF core scenario with lifestyle change
LC-LS	Low carbon lifestyle	LC core scenario with lifestyle change
SOCIO-ENVIRONMENTAL SENSITIVITIES (Chapter 10)		
DREAD	DREAD	LC with unfamiliar technologies constrained
ECO	ECO	LC with technologies that impinge on ecosystem services constrained
NIMBY	NIMBY	LC with technologies with high local impact constrained
GLOBAL UNCERTAINTIES (Chapter 11)		
LC-HI	high fossil prices	LC core scenario with high imported fossil prices
LCR-HI	high fossil prices	LCR core scenario with high imported fossil prices
LC-CC	central cost credits	LC core scenario with international credit purchases allowed
LCR-CC	central cost credits	LCR core scenario with international credit purchases allowed
LC-HI-LC	high prices, cheap credits	LC core scenario with high imported fossil prices and low cost international credit purchases ('best case')
LCR-NB	no imported biomass	LCR core scenario with no imported biomass ('worst case')

In reality, there is considerable uncertainty about the five factors held constant across the core scenarios. Some could be changed through conscious social or policy choice. Chapters 5–11 go beyond the restrictive assumptions embedded in the core scenarios to explore a wider range of possibilities. These possibilities cover both social and policy choices and global uncertainties and contingencies over which UK decision-makers have no control.

Chapter 5, by developing variants on the *Low Carbon* scenario, assesses the impacts of setting more and less ambitious carbon emission reduction targets for 2050 and the impacts of different pathways to an 80 per cent reduction by 2050, that is, delaying or accelerating action.

In Chapter 6, the energy system is 'stress-tested' by looking at the impact of various high-impact, low-probability events which test its resilience. These events are superimposed on each of the four core scenarios. Mitigating measures that reduce the impact of the events are also considered.

Chapter 7 assesses the acceleration of the development of particular technological options through research and innovation policies and how this affects the ease with which decarbonization goals can be met. Variants on the *Low Carbon* scenario are used to conduct the analysis.

Chapter 8 addresses the development of a more distributed energy system, specifically one that places more reliance on micro-generation at the household level.

Chapter 9 considers futures in which people voluntarily adopt lower carbon lifestyles that are reflected in changes in the way in which they travel and use energy in the home. Variants of the *Reference* and *Low Carbon* scenarios are considered.

Chapter 10 addresses the possibility that certain technological options will be closed off because they prove not to be socially acceptable on environmental grounds. Three different 'flavours' of environmentalism that are variants on the *Low Carbon* scenario are considered.

Chapter 11 considers a wider range of possibilities regarding the availability and prices of fossil fuels in international markets and access to international CO_2 credits. Table 4.2 summarizes the full set of scenario variants described in the book. Finally, Chapter 12 considers the overall implications of the analysis for policy, research and action.

Scenario analysis and modelling tools

Modelling overview

Quantitative modelling of the UK energy system has been carried out using three separate models with complementary characteristics. These models have been 'soft-linked'; that is, the output of one model has been used as input to another rather than attempting to solve the models simultaneously.

The three models used have been: the MARKAL model, developed and supported by the International Energy Agency (IEA) via the Energy Technology and Systems Analysis Program (ETSAP) (Loulou et al, 2004); the Wien Automatic System Planning (WASP) model developed by the International

Atomic Energy Agency (IAEA) which determines electricity generation expansion plans at the national level (IAEA, 2001); and the Combined Gas and Electricity (CGEN) model (Chaudry et al, 2008). There are several variants on the MARKAL model. The variant used here is MARKAL-MED ('MARKAL elastic demand') which models consumer responses to energy price changes. The CGEN model determines geographically specific investments in new gas and electricity infrastructure (gas pipelines, terminals and storage facilities; transmission lines; and power stations) given assumptions about final demand for gas and electricity, and patterns of investment in electricity generation capacity at the national level.

The common feature of the models is that they are all optimization models programmed to minimize the total discounted cost over time of the particular part of the energy system that they cover (although the version of MARKAL used here, MARKAL-MED, maximizes welfare, rather than minimizing cost, as is described below). In general, a real discount rate of 10 per cent is used, reflecting the rate of return required by an investor exposed to market risk. However, the use of a social discount rate (3.5 per cent) is also explored where this is considered relevant.

Table 4.3 summarizes key characteristics of the three models. Note that the scope of the MARKAL model includes parts of the energy system also covered by the WASP electricity model. However, the WASP model addresses the electricity system in considerably more detail and can cast light on issues, especially those related to reliability, that MARKAL cannot. Specifically: WASP has a far more detailed characterization of the load duration curve (which summarizes electricity demand at different levels throughout the year); WASP's use of mixed integer programming means it takes account of the 'lumpiness' of investment in power plant; and WASP can use a more sophisticated set of indicators to ensure reliability of electricity. Whereas MARKAL assumes a fixed margin of electricity generation plant in excess of peak demand (strictly speaking a capacity margin over the highest load step), WASP uses three reliability indicators: the number of years in a century in which demand is expected not to be met fully; a 'value of lost load' (VOLL) measure in £/kWh; and a 'loss-of-load expectation' (LOLE), which is the number of hours in which load is not met in any year. In general, this results in WASP building more capacity than MARKAL, especially when there are large quantities of intermittent renewable energy on the system.

Each of the core scenarios is characterized using the three models operating in tandem. Figure 4.3 shows the way that information flows between the three models. MARKAL is run first. Depending on technology assumptions, price assumptions and other constraints such as those on carbon, it determines the extent to which energy service demands are met through electricity, gas or other fuels. Electricity demand is then passed to WASP which generates a more refined picture of the need for and operation of generation capacity at the national level. The CGEN model then takes electricity and gas demands directly from MARKAL and national electricity generation capacity from WASP in order to determine the location of generation plant and other infrastructure.

Table 4.3 *Key characteristics of the modelling tools*

	MARKAL-MED	WASP	CGEN
Scope	The entire UK energy system	Electricity generation at the national level	Gas and electricity infrastructure including geographical distribution
Objective	Maximizing discounted welfare by deploying available technologies in order to meet energy service demands which adjust in response to energy price changes	Minimizing discounted cost by investing in and dispatching plant in order to meet specified levels of electricity demand and levels of reliability	Minimizing discounted cost by locating new plant and infrastructure to meet specified final gas and electricity demands
Method	Linear programming	Mixed integer programming	Non-linear optimization
Key inputs	Baseline energy service demands The availability and cost of energy resources (energy supply curves) Detailed characterization of technology performance and costs Elasticity of energy service demands with respect to price Key constraints such as carbon emissions	Annual and peak electricity demand Profiles of existing and committed plant Performance and cost of new plant, reliability measures	Final gas and electricity demands Geographical characterization of the gas and electricity systems Costs and performance of plant and infrastructure
Key outputs	Energy demand and supply by sector, energy source and technology	Levels of investment in and operation of plant by type	Selection, location and operation of plant and infrastructure

Energy system: the MARKAL model

MARKAL (acronym for MARKet ALlocation) is a widely applied bottom-up, dynamic, linear programming (LP) optimization model. This energy model framework has long been used in the UK for exploring longer-term costs and technological impacts of climate policy through a scenario-based approach (Strachan et al, 2009a). A comprehensive description of the UK model, its applications and core insights can be found in Strachan et al (2008), and model documentation in Kannan et al (2007).

MARKAL portrays the entire energy system from imports and domestic production of fuel resources, through fuel processing and supply, explicit representation of infrastructures, conversion of fuels to secondary energy carriers (including electricity, heat and hydrogen), end-use technologies and energy service demands of the entire economy. As a perfect foresight partial equilibrium optimization model, MARKAL minimizes discounted total system cost by considering the investment and operation levels of all the interconnected

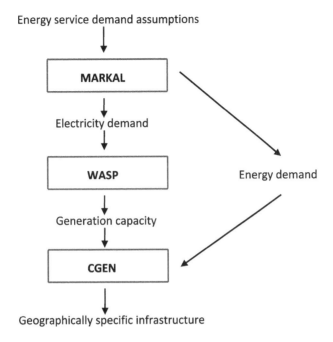

Figure 4.3 *Operation of the three models in tandem*

system elements. The inclusion of a range of policies and physical constraints, the implementation of all taxes and subsidies, and calibration of the model to base-year capital stocks and flows of energy, enables the evolution of the energy system under different scenarios to be plausibly represented.

The UK MARKAL model hence provides a systematic exploration of least-cost configurations to meet exogenous demands for energy services. These may be derived from standard UK forecasts for residential buildings (Shorrock and Uttley, 2003), transport (DfT, 2005), the service sector (Pout and MacKenzie, 2006), and industrial sub-sectors (Fletcher and Marshall, 1995). Generally these sources entail a projection of low energy growth, with saturation effects in key sectors.

One key set of input parameters is resource supply curves (BERR, 2008), which give baseline costs of energy resources. Multipliers are used to translate these into higher cost supply steps for primary fuels and costs for imports of refined fuels. A second key input is dynamically evolving technology costs. Future costs are based on expert assessment of technology vintages or, for less mature electricity and hydrogen technologies, via exogenous learning curves derived from an assessment of learning rates (e.g. McDonald and Schrattenholzer, 2002) combined with global forecasts of technology uptake (e.g. European Commission, 2006). Cost reductions from learning for less mature technologies are, generally, not related to their deployment in the UK, but imposed exogenously, as the relatively small UK market is assumed to be a 'price taker' for globally developed technologies.

UK MARKAL is calibrated in its base year (2000) to data within 1 per cent of actual UK resource supplies, energy consumption, electricity output, installed technology capacity and CO_2 emissions (all from BERR, 2007). In addition, considerable attention is given to the near-term (2005–2020) convergence of sectoral energy demands and carbon emissions with the econometric outputs of the government energy model (BERR, 2008). The model then solves from years 2000 to 2070 in 5-year increments. All prices are in £(2000). Substantial efforts have been made in respect of the transparency and completeness of the model structure and assumptions, including through a range of stakeholder events (for example Strachan et al, 2007), expert peer review, and publication of the model documentation (Kannan et al, 2007).

The standard version of MARKAL optimizes (minimizes) the total energy system cost by choosing the investment and operation levels of all the inter-connected system elements. The participants of this system are assumed to have perfect inter-temporal knowledge of future policy and economic develop-ments. In order to approximate the real-world operation of the energy system, MARKAL relies on a wide range of assumptions, which are key to the model results. For example, for the generation of electricity, coal, nuclear and gas power plants are selected on the basis of capital costs, operation and main-tenance costs, conversion efficiency and fuel input costs. On the basis of such assumptions, MARKAL delivers an economy-wide solution of cost-optimal energy market development.

An important point to stress is that MARKAL is not a forecasting model. It is not used to predict the future energy system of the UK in 50 years' time. Instead it offers a systematic tool to explore the trade-offs and tipping points between alternative energy system pathways, and the cost, energy supply and emissions implications of these alternative pathways. The results detailed and discussed in Chapters 5–11 illustrate the complexity of insights that is gener-ated from a large energy system model. They should be viewed and interpreted as different plausible outcomes from a range of input parameters and model-ling assumptions. There is no attempt to assign probabilities to the most likely outcome or 'best' model run. Equally there is no attempt to assign probabilities to individual model parameters.

The strengths of the UK MARKAL energy system model include:

- a well understood least-cost modelling paradigm (efficient markets);
- a framework to evaluate technologies on the basis of different cost assump-tions, to check the consistency of results and explore sensitivities to key data and assumptions;
- transparency, with open assumptions on data, technology pathways, constraints, etc.;
- depiction of interactions within the entire energy system (e.g. resource supply curves, competing use for infrastructures and fuels, sectoral tech-nology diffusion);

- incorporation of possibilities for technical energy conservation and efficiency improvements;
- the ability to track emissions and energy consumption across the energy system, and model the impact of constraints on both;
- the ability to investigate long timeframes (in this case to 2050) and novel system configurations, without being constrained by past experiences or currently available technologies, thus providing information on the phasing of technology deployment;
- and through the elastic demand version of MARKAL, MARKAL-MED, demand-side responses to price changes.

The principal disadvantages or limitations of the MARKAL energy system model include:

- the model is highly data intensive (characterization of technologies and the reference energy system);
- by cost optimizing, it effectively represents a perfect energy market and neglects barriers and other non-economic criteria that affect decisions; one consequence of this is that, without additional constraints, it tends to over-estimate the deployment of nominally cost-effective energy efficiency technologies;
- being deterministic the model cannot directly asses data uncertainties, which have to be investigated through separate sensitivity analyses;
- limited ability to model behaviour (partially addressed by MARKAL-MED in respect of price changes);
- there is no spatial disaggregation and hence no representation of the siting of infrastructures and capital equipment;
- there is limited temporal disaggregation, so that the model cannot be used to explore such issues as the daily supply–demand balancing of electricity, heat and other energy carriers.

The MARKAL-MED model

For the work described in this book, an *elastic demand* version of the MARKAL model (MARKAL-MED) was used to account for the response of energy service demands to prices. This is implemented at the level of individual energy service demands using linear programming. The UK model does not represent trade and competitiveness effects, and as a partial equilibrium energy-economic model does not include government revenue impacts, and hence does not provide an assessment of macro-economic implications (e.g. GDP).

A simplified representation of energy supply and elastic demands is given in Figure 4.4. In MARKAL-MED, demand functions are defined which determine how each energy service demand varies as a function of the market price of that energy service. Hence, each demand has a constant own-price elasticity (E) in a given period. The demand function is assumed to have the following functional form:

$ES/ES_0 = (p/p_0)^E$

Where: ES is a demand for some energy service;

ES_0 is the demand in the reference case;

p is the marginal price of each energy service demand;

p_0 is the marginal price of each energy service demand in the reference case;

E is the (negative) own-price elasticity of the demand.

In this characterization, ES_0 and p_0 are obtained by running standard MARKAL. ES_0 is the energy service demand projection as defined by the user exogenously (as a function of social, economic and technological drivers); p_0 is the marginal price of that energy service demand determined endogenously by running the reference case. A simple calibration process ensures that the MARKAL-MED reference case is consistent with the reference case run in the standard model.

The standard MARKAL model optimization, when energy service demands are unchanging, i.e. are a straight vertical line on the horizontal axis, is on (discounted) energy system cost, i.e. the minimum cost of meeting all energy services. With non-changing demands, this is equivalent to the area between the supply curve and the horizontal line from the equilibrium price. In MARKAL-MED, these exogenously defined energy service demands have been replaced with demand curves (actually implemented in a series of small steps). Following calibration to a reference case that exactly matches the standard MARKAL reference case, MARKAL-MED now has the option of increasing or decreasing demands as final energy costs fall and rise respectively. Thus demand responses combine with supply responses in different scenarios (e.g. one with a CO_2 constraint).

The objective function which MARKAL-MED maximizes is the sum of producer surplus and consumer surplus, the combined area between the demand function and the supply cost curve in Figure 4.4. The sum of consumer and producer surplus (economic surplus) is considered a valid metric of social welfare in microeconomic literature, giving a strong theoretical basis to the computed equilibrium.

The economic surplus is affected by annualized investment costs; resource imports, export and domestic production costs; taxes, subsidies, emissions costs; and fuel and infrastructure costs as before in the standard model. In addition, the MARKAL-MED model accounts for welfare losses from reduced demands. If consumers give up some energy services that they would otherwise have used if prices were lower, they experience a loss in utility.

Under different scenarios, transfers between producer surplus and consumer surplus are possible. In general, if a scenario has higher prices than in the *Reference* case (e.g. as the result of a CO_2 constraint) it is likely that producer surplus will rise while consumer surplus falls. The opposite is likely to occur if prices fall: then consumer surplus rises while producer surplus falls. However in a higher price scenario, total economic surplus (producer plus consumer) will always be lower, while under lower prices the economic surplus will always be higher.

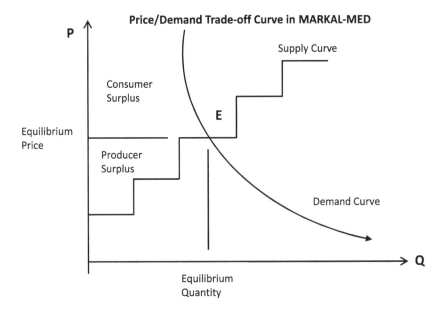

Figure 4.4 *Representation of MARKAL-MED supply–demand equilibrium*

Electricity generation: the WASP model

The Wien Automatic System Planning (WASP) model (version IV) developed by the International Atomic Energy Agency (IAEA) is designed to determine medium- to long-term economically optimal expansion policy for a power generation system within user-specified constraints. WASP has been distributed to more than 75 countries and has become the standard approach to investment planning in many of the IAEA and World Bank's member countries (ESMAP, 2007). Many IAEA member states apply the model in their national and regional studies to analyse the issues of electric power system expansion planning. One recent study has been commissioned by the European Union to identify and prioritize investment in power generation and related electricity infrastructure for the Balkan region (PWC, 2004). The WASP model is used to explore each of the core scenarios described in Chapter 6.

Timely and optimal development of adequate electricity generating capacity is vital for maintaining security of supply while meeting desired policy objectives regarding energy supplies, environment and affordable energy prices. Over-investment in generating capacity increases the reliability of the system but the average cost of electricity will also increase because the costs of that excess capacity will ultimately be borne by the customers. On the other hand, under-investment will result in some portion of demand not being served. If the economic costs of this unserved energy are significant and are added to the generation cost, this summed cost of generation also increases as the degree of under-investment becomes more severe. Thus, an appropriate level of reliability for the generating system is required which depends on a large number of system characteristics.

The optimal expansion plan for a power generating system in the WASP model is evaluated in terms of minimum discounted total costs within the system reliability constraints given by the planner. System reliability is evaluated on the basis of three indices: reserve margin, loss-of-load expectation, and unserved energy. Each possible sequence of power generating units added to the system (expansion plan or expansion policy) meeting the constraints is evaluated by means of a cost function (the objective function) which is composed of capital investment costs, salvage value of investment costs, fuel costs, fuel inventory costs, non-fuel operation and maintenance costs and cost of the energy not served.

The model applies probabilistic simulation to evaluate the electricity generation system's production costs and costs associated with unserved energy and reliability. It uses a linear programming technique for determining optimal dispatch policy satisfying exogenous constraints on environmental emissions, fuel availability and electricity generation by specified plants. The dynamic programming method is then applied for comparing and optimizing the costs of alternative system expansion policies that would serve future electricity demand with the desired level of system reliability (IAEA, 2001).

The model provides options for introducing constraints on environmental emissions, fuel usage and energy generation. These constraints are handled by a multiple group-limitation technique wherein a group of plants is constrained and plants can be included in more than one type of constraint. Environmental emissions for each year and for each period within a year are based on the electricity generated by each plant and the user-specified characteristics of fuels used. These options have been extremely useful for real-life planning in view of the increasing importance of environmental concerns as well as due to the fact that in many cases the availability of some fuels for power generation may be limited or energy generation from some plants may need to be restricted.

In the WASP model the changing nature of the load from one year to another is taken into account by specifying the peak demand forecast for each year. In order to consider the seasonal changes of the load characteristics, the year is sub-divided into a number of equal periods. For long-range planning studies, such as the ones presented here, the chronological hourly load curves are transformed into load duration curves for each period and scaled to represent future electricity demand projections. The load duration curve, with the area under the curve representing the electricity demand, characterizes the load in each period of every year. The demand load factor (the ratio of average to peak demand) is assumed to stay constant.

The WASP model can be applied in two modes.

- Evaluation of a *Fixed Plan* – This mode is applied to investigate the reliability, cost and environmental performance of a predefined expansion plan.
- Search for the *Optimal Plan* – In this mode the model is allowed to determine the economically optimal expansion plan or plans within user-defined

constraints. However, if required, the appropriate year and a range of additions to capacity of candidate technologies can be defined.

The model has been applied in the latter mode, using the future UK electricity demand projections obtained from the output of the MARKAL-MED model. As noted above, MARKAL-MED determines the future electricity sector demand based on an optimization (in respect of welfare) of the overall energy system for the designed energy scenarios.

Thermal plants are described by maximum and minimum capacities, heat rate at minimum capacity and incremental heat rate between minimum and maximum capacity, maintenance requirements (scheduled outages), failure probability, emission rates and specific energy use, capital investment cost (for expansion candidates), variable fuel cost, fuel inventory cost (for expansion candidates), fixed component and variable component of (non-fuel) operating and maintenance costs and plant life. The schedule of annual maintenance of the plants in the system can also be specified.

Some recent studies have introduced new approaches to model wind generation in WASP (Koritarov, 2005). For this book, wind generation has been represented in a simplified way, similar to the treatment of run-of-river hydro plant in previous studies. The average behaviour of wind power output (available wind energy and its seasonal variation) from future on- and offshore wind farms is based on historical wind speed data.

The key outcomes of the application of the model under the given constraints include:

- build schedule; i.e. which capacities are installed to ensure an appropriate level of reliability, the best combination among the different technologies at hand now and in future, the appropriate time to incorporate new plant in the system;
- costs: investment costs and cost of system operation that includes fuel, operation and maintenance costs, and cost of energy not served;
- expected generation of plants;
- fuel requirements;
- emissions.

Gas and electricity infrastructure: the CGEN model

The Combined Gas and Electricity (CGEN) model describes the UK's gas and electricity infrastructure. The objective of the CGEN model is to minimize total discounted costs related to the combined operation and expansion of the gas and electricity networks whilst meeting demand requirements over the entire planning horizon. As for the WASP model, the CGEN model is used to explore each of the core scenarios described in Chapter 6.

The model consists of a DC load flow analysis for the electricity network and detailed modelling of the gas network, including facilities such as gas storage and compressor stations. The interaction between the two networks is through gas turbine generators connected to both networks (Chaudry et al, 2008).

CGEN is a geographical model, thus, the connections of gas pipes and electricity transmission wires in a network are explicitly modelled. This geographical element allows a realistic picture of network flows and the physical constraints that are present in both networks. The components modelled within CGEN are illustrated in Figure 4.5.

The components are arranged into distinct categories, describing energy supply, energy transportation (networks), generation technologies, and energy end use.

Resource supply

This includes bounds on the availability of primary energy supplies (gas, coal, oil, etc.) and electricity imports. Gas import inter-connectors are modelled as gas pipes with maximum transport capacities.

Networks

The gas network includes the detailed modelling of pipelines, compressors and storage facilities. The gas flow in a pipe is determined by employing the Panhandle 'A' gas flow equation (which calculates the gas flow rate given the pressure difference between upstream and downstream nodes). A DC power flow model is used to represent the electricity network. The DC power flow formulation enables the calculation of MW power flows in each individual transmission circuit. Gas turbine generators provide the linkage between gas and electricity networks. They are considered as energy converters between these two networks. For the gas network, the gas turbine is looked upon as a gas load. Its value depends on the power flow in the electricity network. In the electricity network, the gas turbine generator is a source.

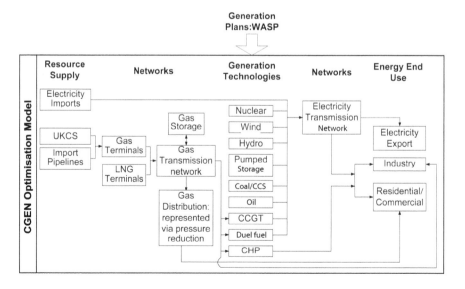

Figure 4.5 *CGEN optimization model*

Generation technologies

CGEN includes representations of all the conventional generation technologies such as combined cycle gas turbine (CCGT), coal and nuclear. Generation technologies are described by a number of characteristics, such as ramp up/down rates, maximum generation output and thermal efficiency. CGEN introduces new power generation plants according to the generation capacity plan schedule developed in WASP. To capture the spatial nature of the wind resource, wind power is modelled using average wind load factors for different locations around the UK. The geographic wind power variation influences energy balancing in the electricity network, and the location of new wind power generation capacity.

Energy end use

Gas and electricity energy demand assumptions are fed into CGEN from MARKAL-MED. The demand is split into residential and industrial and commercial components for gas and electricity. Gas used for electricity generation is determined endogenously within CGEN.

Location of generation plants

CGEN does not endogenously build generation capacity to satisfy future demand. Generation plans from the WASP model are used as an input into the CGEN model. Since WASP is a non-geographical generation planning model, it does not provide location-specific information on new generation capacity. Using the concept of minimizing costs (operational plus infrastructure), CGEN optimally places these generation plants around the electricity network. Figure 4.6 describes an example of how CGEN would deal with the location of a CCGT plant. CGEN calculates the cost associated with placing the CCGT plant at every location. For instance, if the CCGT plant were to be located in Scotland this would incur a reinforcement cost of £A for new gas pipes and £B for electricity transmission lines. However, if the CCGT plant were to be located in the southeast the total investment cost would be lower at £C plus £D. CGEN would choose the southeast option. However, once operational costs are included, the southeast option may well turn out to be more expensive. The CCGT plant is placed at a location that minimizes both gas and electricity operational and infrastructure expansion costs.

Transmission capacity

For both gas and electricity networks transmission capacity is added to satisfy peak demand requirements. Figure 4.7 illustrates how the optimization routine within CGEN will explore all possible solutions to satisfy peak demand. This ranges from building additional network capacity to the re-dispatching of energy (e.g. substituting cheap gas from Scotland with expensive gas from LNG terminals in the south in order to bypass transmission bottlenecks). The model will select the cheapest solution over the entire time horizon.

Electricity network expansion

Electricity transmission capacity expansion is assumed to be carried out on a radial network.

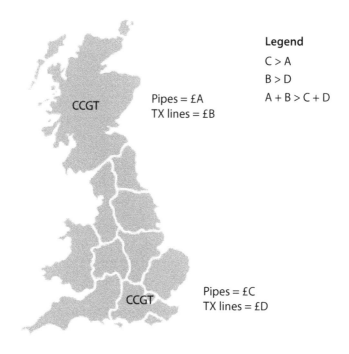

Legend
C > A
B > D
A + B > C + D

Pipes = £A
TX lines = £B

CCGT

Pipes = £C
TX lines = £D

CCGT

Figure 4.6 *Location of a CCGT plant*

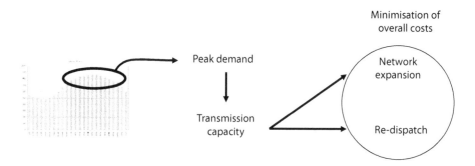

Minimisation of
overall costs

Peak demand

Transmission
capacity

Network
expansion

Re-dispatch

Figure 4.7 *Network infrastructure expansion*

Gas network expansion

Gas pipe capacity expansion is based on building additional pipes parallel to existing pipes. The Panhandle 'A' equation, which describes the physical properties of gas in a pipeline, is used to determine flow rates. In addition to increasing gas pipeline capacity, CGEN allows capacity expansion of import pipelines, LNG terminals, storage facilities and compressor stations.

The key outputs of the model are:

- location of electricity generation capacity;
- the volume and location of investment in electricity transmission capacity and gas infrastructure (inter-connectors, pipelines, LNG terminals and storage facilities);
- utilization of electricity generation capacity and infrastructure;
- system costs;
- the amount of energy unserved and its monetary value.

CGEN can operate in two modes. In the *planning mode*, infrastructure additions are made to an initial network over a time horizon in order to satisfy demand and network related constraints (pressure, electricity flow constraints, etc.). CGEN was used in this mode to model the 'core' scenarios.

In the *operational mode*, a user-specified network is used to test various scenarios (price increases, shocks, etc.). CGEN was used in this mode to assess the benefits of investment in infrastructure aimed at increasing resilience.

CGEN has been used in both the planning and operational modes to conduct the analysis of energy system resilience presented in Chapter 6.

Key assumptions in the core scenarios

Table 4.4 provides an overview of the assumptions which are held constant across the core scenarios. This is very much a business-as-usual world in terms of economic growth and consumer behaviour. It is assumed that global energy prices (Table 4.5) ramp upwards in the future but that the very high levels seen in 2008 (over $100/barrel of oil) are not sustained.

Table 4.4 *Key assumptions in the core scenarios*

GDP growth	2% per annum
World energy prices	The assumed prices are translated into global supply curves. Some domestic energy may be available at a lower cost. For detail see Table 4.5
Discount rates	10% real
'Hurdle' rates	Higher discount rates are applied in sectors with high transaction costs and consumer inertia
Technology performance and cost	Exogenous 'global' data, based on detailed stakeholder consultations

Table 4.5 *World energy price assumptions*

		2000	2005	2010	2015	2020	2030	2040	2050
Crude oil	2005$/bbl	31.38	50.62	57.50	55.00	55.00	60.00	70.00	70.00
Gas	2005$/MMBTU	4.85	7.46	6.75	6.75	7.00	7.64	8.91	8.91
Coal	2005$/tonne	36.54	61.14	55.00	55.00	55.00	60.00	65.00	65.00
Crude oil	2005£/GJ	3.39	4.56	4.49	4.40	4.51	4.92	5.74	5.74
Gas	2005£/GJ	3.20	4.10	3.21	3.29	3.50	3.82	4.45	4.45
Coal	2005£/GJ	0.90	1.25	0.98	1.00	1.03	1.12	1.21	1.21

Assumptions about energy service demands, which reflect a range of demographic, economic and social aspects, are a critical driver of energy system costs. Industrial energy service demands reflect recent international trends (McKenna, 2008). Transport energy service demands reflect growth rates assumed by government (DfT, 2005) coupled with assumptions about saturation effects, notably in the domestic aviation sector (IPPR/WWF, 2007). Specific assumptions are made about the seasonality of energy service demands (Stokes et al, 2004; Abu-Sharkh et al, 2006).

Energy service demands for electricity and gas appliances in the residential sector have been specified exogenously using results from the Environmental Change Institute's UK Domestic Carbon Model (UKDCM) (Layberry, 2007). Efficiency and fuel switching options have been ruled out in MARKAL-MED as they are assumed to have been taken into account already in the UKDCM model. However, MARKAL-MED can still reduce demands through behavioural responses to price changes.

Table 4.6 shows the assumed elasticities of energy service demands with respect to price. These are long-run elasticities due to the MARKAL-MED model's five-year time periods and the assumption of perfect foresight. They are derived from three key sources:

1   other MARKAL modelling teams outside the UK (Loulou and van Regemorter, 2008);
2   the Cambridge Econometrics MDM-E3 macro-econometric model (Dagoumas, 2008);
3   the BERR energy model (Oxford Economics, 2008).

While the model is driven by the assumption of a 10 per cent global discount rate, additional and more restrictive hurdle rates for investment are applied to transport technologies and to conservation technologies in buildings to reflect market (non-cost) barriers, consumer preferences and risk factors that limit the purchase of new energy technologies (Train, 1985). The hurdle rate is applied only to annualized capital investment, effectively increasing the capital cost of the affected technologies.

Table 4.6 *Price elasticities of energy service demands*

Sector and Description		Price Elasticity
Industry and agriculture	Chemicals	−0.49
	Metals	−0.44
	Other industry	−0.32
	Pulp and paper	−0.37
	Combined agriculture	−0.32
Residential	Electrical appliances	−0.31
	Gas appliances	−0.33
	Space and water heating	−0.34
Services	Cooking	−0.23
	Cooling	−0.32
	Electrical appliances	−0.32
	Space and water heating	−0.26
	Lighting	−0.32
	Refrigeration	−0.25
Transport	Air	−0.38
	Bus	−0.38
	Car	−0.54
	Rail (freight)	−0.24
	Heavy/light goods vehicles	−0.61
	Rail (passenger)	−0.24
	Shipping (domestic)	−0.18
	Two-wheelers	−0.41

Hurdle rates of 25 per cent, 20 per cent and 15 per cent are applied, graded on dates of commercial availability, the severity of perceived market barriers and the uncertain requirements of new infrastructures. All building conservation technologies, and all personal electric and hydrogen transport vehicles have a 25 per cent hurdle rate. Public transport modes using hydrogen see a 20 per cent discount rate. Other advanced personal road transport options have a hurdle rate of 25 per cent, except for hybrid technologies which are closer to market and are implemented with a 15 per cent hurdle rate.

A very large number of specific detailed assumptions about technology performance, costs and resource availability are required to drive the MARKAL model. Only key assumptions and broad approaches are noted here. More detailed information about the cost and efficiency assumptions for nuclear, carbon capture and storage, hydrogen fuel cells, wind, solar PV, marine and biomass technologies can be found in Winskel et al (2009). With respect to carbon capture and storage (CCS), a theoretical storage potential of $21GtCO_2$ is assumed based on cost curves derived from more detailed reservoir descriptions covering aquifers, enhanced oil recovery and storage in depleted oil and/or gas fields (DEFRA, 2007).

On- and offshore wind resources are disaggregated to take account of diurnal and seasonal availability (Sinden, 2007). Capital costs associated with offshore wind transmission are assumed to be £4000/MW/km, with offshore wind located 0km, 60km, 120km and 180km from the shore in successive tranches. No additional onshore transmission costs are assumed for offshore wind or wave.

The impact of intermittent renewable technologies on the electricity system has been treated by assuming a set of *capacity credits* which measure the extent to which intermittent sources can replace continuous sources without reducing the system's ability to meet peak demands. Table 4.7 shows the capacity credits associated with different levels and types of intermittent generation. These are derived from Gross et al (2006).

Based on Jablonski et al (2008), the main biomass chains have been broken out into wood, ligno-cellulosic crops, bio-pellets (high and low quality), first and second generation bio-oils, biodiesel, ethanol, methanol, biogases, biomethane and wastes. The range of bio-delivery options (oils, pellets, etc.) are disaggregated to the industrial, residential and service sectors. Biomass boilers, utilizing both solid and liquid fuels, are included for all buildings sectors. The model covers enhanced co-firing using different biomass fuels. In the transport sector, the extensive bio-energy chains include the option of bio-kerosene fuel chains and technologies for domestic aviation.

For space and water heating applications in the residential and service sectors, micro-generation costs are assumed to fall through technology learning. Capital costs fall at 2.5 per cent per year until 2020, adding up to a 45 per cent cost reduction between 2000 and 2020. The MARKAL-MED model includes the option of installing heat pumps and night storage electric heating in the residential sector as an alternative to conventional boilers. The market share of night storage heating is limited to 30 per cent.

Table 4.8 shows the main car technologies included in MARKAL-MED. The model includes plug-in hybrid vehicles with both night and daytime charging options, reflecting the potentially important aggregate and temporal interactions between the transport and electricity sectors. Fuel-flexible hybrid cars that can

Table 4.7 Intermittent generation capacity credit

Intermittent capacity	Incremental capacity credit	Applied to
0–5 GW	28%	Onshore wind (tranches T1–T7), Offshore wind tranche 1
5–15 GW	18%	Onshore wind (tranches T8–T9), Tidal, Wave (tranches T1–T2), Offshore wind tranches T2 (adjusted for 15GW limit)
15–50 GW and above	8.6%	Offshore wind (tranches T3–T4), Severn Barrage

Table 4.8 *Main car technologies included in MARKAL-MED*

Technology	Description
Car – Diesel/biodiesel ICE	Internal combustion engine (ICE)
Car – Diesel/biodiesel Hybrid	Hybrid technology (ICE and electric motor)
Car – Diesel/biodiesel Plug-in	Plug-in hybrid technology (ICE and electric motor); battery can be charged by plugging-in
Car – Petrol ICE	Internal combustion engine (ICE)
Car – Petrol Hybrid	Hybrid technology (ICE and electric motor)
Car – Petrol Plug-in	Plug-in hybrid technology (ICE and electric motor); battery can be charged by plugging-in
Car – E85	Fuel-flexible hybrid (petrol and ethanol) car – up to 85% ethanol
Car – Battery	Battery-operated electric car
Car – Hydrogen ICE	Hydrogen with Internal combustion engine technology
Car – Hydrogen FC	Hydrogen fuel cell technology
Car – Methanol	Methanol with Internal combustion engine technology

Note: Only car technologies are included in the table for simplicity, but many of the same categories apply also for Buses, Light Goods Vehicles (LGV) and Heavy Goods Vehicles (HGV)

take up to an 85 per cent ethanol mix ('E85 cars') are included, and battery costs for electric vehicles have been reviewed to make later year vintages more directly comparable to conventional and other technologies.

Assumptions about hydrogen distribution infrastructures are based on Strachan et al (2009b). This has focused on a scale and distance approach to the costs and efficiencies of gaseous and liquid hydrogen options for the full range of transport demands. In addition hydrogen chains for storage and transportation are included for stationary electricity generation and CHP applications. Liquid hydrogen distribution for both liquid internal combustion engine and gaseous fuel cell technologies is also covered.

Reference scenario results

This section briefly sets out the results of the *Reference* scenario, describing a world which is in many ways an extrapolation of the trends reviewed in Chapter 2. However, patterns of energy service demands change, and different technologies are deployed as a result of changing relative energy prices and access to technologies and fuels that were not available in previous decades. It

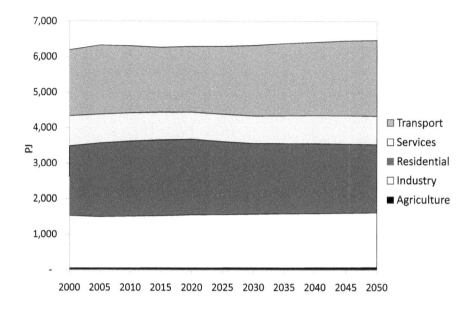

Figure 4.8 *Final energy demand by sector*

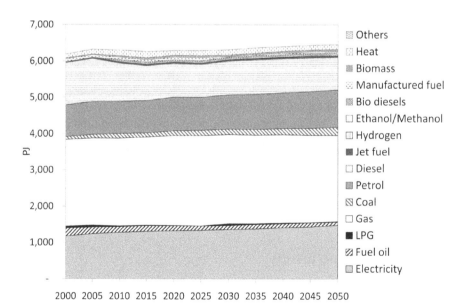

Figure 4.9 *Final energy demand by fuel*

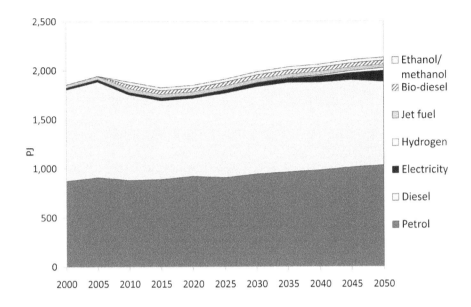

Figure 4.10 *Transport fuel demand*

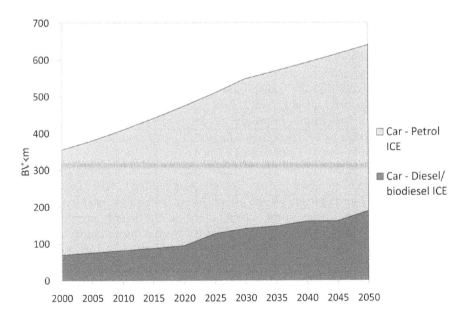

Figure 4.11 *Passenger cars distance travelled*

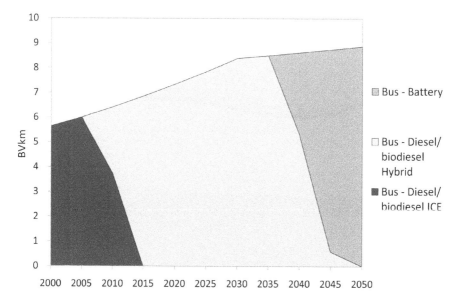

Figure 4.12 *Buses distance travelled*

is critical to remember that this is not a forecast but a reference scenario used for benchmarking purposes. It assumes no additional policy effort to deal with the decarbonization and security agendas. Already, policies are being put in place which will lead to a break with past trends.

Figure 4.8 shows that final energy demand is projected to be virtually flat throughout the projection period. However, this is the result of two offsetting trends, namely growth in the transport sector and a corresponding decline in residential energy demand after 2020. Figure 4.9 shows final energy demand broken down by fuel. Again there are no dramatic changes, but electricity demand is projected to be 25 per cent higher in 2050 than in 2000, with its market share rising from 19 per cent to 23 per cent.

Digging a little deeper into demand patterns, Figure 4.10 shows transport energy demand broken down by fuel. This shows a modest early switch to bio-fuels followed by an expansion in the use of electricity for transport from 2030 onwards. This latter switch is at the expense of diesel. The underlying reasons are demonstrated in Figures 4.11 and 4.12 which show vehicle kilometres travelled by passenger cars and buses respectively. The distance travelled by passenger cars is projected to grow by 80 per cent between 2000 and 2050 though, as a result of improved vehicle efficiency, fuel demand grows by only 30 per cent. For passenger cars, diesel use doubles between 2000 and 2050 whereas petrol use grows only slowly. Diesel use includes a considerable amount of biodiesel.

However, for buses, two transitions are projected between 2000 and 2050. In the period up to 2020, conventional engines are substituted by diesel hybrids leading to a substantial drop in fuel demand due to efficiency gains. Diesel use

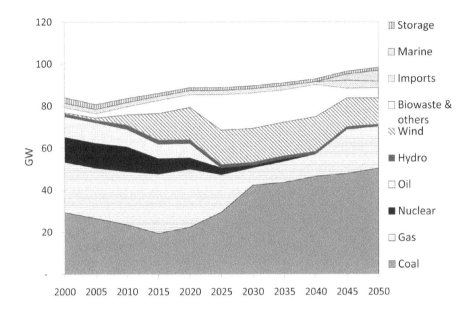

Figure 4.13 *Installed electricity generation capacity mix*

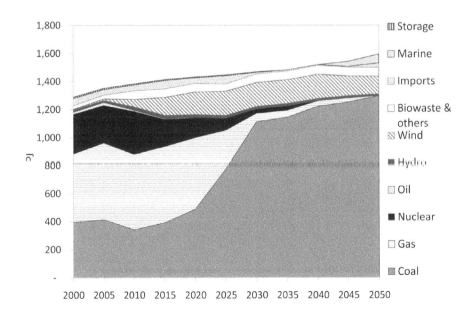

Figure 4.14 *Electricity generation*

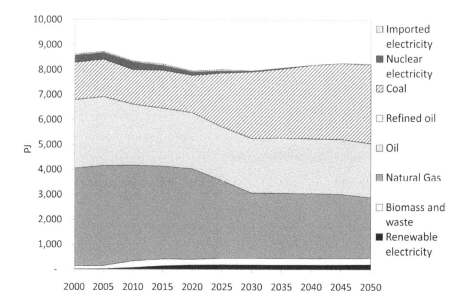

Figure 4.15 *Primary energy demand*

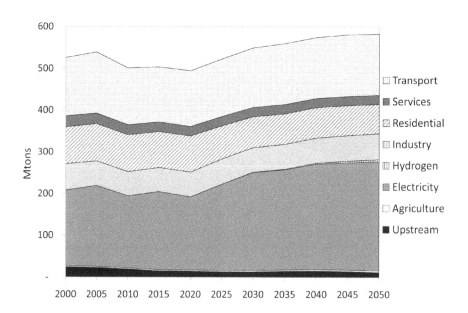

Figure 4.16 *CO₂ emissions by sector*

by buses is 40 per cent lower in 2020 than in 2000. From 2035 onwards there is then a further transition to battery electric buses. This explains the overall drop in demand for diesel in transport shown in Figure 4.10. Heavy goods vehicles make the same early transition to hybrid engines as buses, but there is no subsequent take-up of battery electric vehicles in this sector.

There are no radical changes in demand patterns in the residential sector, although electricity use grows by about 50 per cent between 2000 and 2050 as a result of increased use in appliances. Gas remains the dominant fuel for heating and condensing boilers dominate from the early 2020s. The use of solid fuels for home heating peters out and there is no penetration of either solar energy or heat pumps.

Moving to the electricity sector, Figure 4.13 shows that there is a significant change in installed generation capacity. Nuclear and oil are phased out by around 2030 while gas capacity is considerably reduced. The striking features are investment in coal-firing, wind capacity and biomass capacity. In the 2040s, biomass starts to be phased out, allowing gas capacity to grow again and marine renewables to enter the market. These trends are largely driven by exogenous assumptions about coal and other fossil fuel prices. Because coal is by far the cheapest fuel, its domination of the generation mix is even larger than that of its installed capacity (Figure 4.14). By 2050, 80 per cent of electricity generated is coming from coal. Gas-fired stations move from being base-load plant in 2000 through to a zero load factor in 2050; that is, their only role is to provide capacity support.

Final energy demand and electricity generation feed through into the pattern of primary energy demand shown in Figure 4.15. This shows a decline in natural gas use in the 2020s as it retreats into heating markets, compensated by a rise in coal use for electricity generation. Renewables – mainly wind and biomass – gain a small market share while nuclear is phased out during the 2030s.

The gap between the *Reference* scenario and policy aspirations

The expansion of coal use has an inevitable impact on CO_2 emissions which, far from declining, rise to $583 MtCO_2$ by 2050 (Figure 4.16). The increase is entirely due to electricity sector emissions, with emissions from other sectors remaining stable. Overall, primary energy is slightly more diverse in 2050 than it is now, with coal, oil and gas all having broadly similar market shares. However, diversity declines significantly in the electricity generation sector because of the dominance of coal. With the UK's domestic fossil fuel reserves in rapid decline, this is also a world in which the UK would become almost entirely dependent on imports to meet its energy needs.

The clear message is that in no respect does this *Reference* scenario meet the UK's needs in terms of decarbonization and security of supply. The following chapters assess what is needed to get the UK energy system on track to meet policy aspirations.

Note

1   This section has been informed by helpful discussions concerning resilience and ecosystem services with UKERC's Energy and Environment team based at the Centre for Ecology and Hydrology, Lancaster, 2004–2009. UKERC's 'whole systems' approach to research has allowed us to get 'read over' from one area of science to another when considering concepts such as resilience.

References

Abu-Sharkh, S., Arnold, R.J., Kohler, J., Li, R., Markvart, T., Ross, J., Steemers, K., Wilson, P. and Yao, R. (2006) 'Can micro-grids make a major contribution to UK energy supply?' *Renewable and Sustainable Energy Reviews*, vol 10, pp78–127

Bazilian, M. and Roques, F. (2008) *Analytical Methods for Energy Diversity and Security*, Elsevier, Oxford

BERR (Department of Business Enterprise and Regulatory Reform) (2007) *Digest of UK Energy Statistics*, Department of Business Enterprise and Regulatory Reform, London, www.decc.gov.uk/media/viewfile. ashx?filepath=statistics/publications/dukes/dukes06.pdf&filetype=4

BERR (2008) *Updated energy projections*, Department of Business Enterprise and Regulatory Reform, London

Chaudry M, Jenkins, N. and Strbac, G. (2008) 'Multi-time period combined gas and electricity network optimisation', *Electric Power Systems Research*, vol 7 (78), pp1265–79

Dagoumas, A. (2008) MDM-E3 price elasticities, personal communication

DEFRA (Department for Environment, Food and Rural Affairs) (2007) *MARKAL Macro analysis of long run costs of climate change mitigation targets*, Analysis by AEA Energy and Environment, Didcot, Oxfordshire, www.defra.gov.uk/environment/climatechange/research/index.htm

DfT (Department for Transport) (2005) *National Transport Model (NTM)*, Department for Transport, London

ESMAP (Energy Sector Management Assistance Program) (2007) *Risk Assessment Methods for Power Utility Planning*, Special Report 001/07, The World Bank, March

European Commission (2006) *World Energy Technology Outlook – WETO H2*, Report No. EU22038, Directorate General for Research, Brussels

Fletcher, K. and Marshall, M. (1995) 'Forecasting regional industrial energy demand: The ENUSIM end-use model', *Regional Studies*, vol 29 (8), 801–811

Gross, R., Heptonstall, P., Anderson, D., Green, T., Leach, M. and Skea, J. (2006) *The Costs and Impacts of Intermittency: An Assessment of the Evidence on the Costs and Impacts of Intermittent Generation on the British Electricity Network*, UKERC Report ISBN 1903144043, UK Energy Research Centre, London

Grubb, M., Butler, L. and Twomey, P. (2006) 'Diversity and security in UK electricity generation: the influence of low carbon objectives', *Energy Policy* vol 34, pp4050–4062

Holling, C.S. (1973) 'Resilience and stability of ecological systems', *Annual Review of Ecology and Systematics*, vol 4, pp1–23

Holling C.S. (1996) 'Engineering resilience versus ecological resilience' in Schulze, P. (ed.) *Engineering within ecological constraints*, National Academy Press, Washington DC

IAEA (International Atomic Energy Agency) (2001) *Wien Automatic System Planning (WASP) Package Version-IV, User's Manual*, International Atomic Energy Agency, Vienna, Austria

IPCC (Intergovernmental Panel on Climate Change) (2001) *Climate Change 2001: Working Group II: Impacts, Adaptation and Vulnerability*, Section 1.4.1., Intergovernmental Panel on Climate Change, Geneva

IPPR/WWF (Institute for Public Policy Research/World Wildlife Foundation UK) (2007) *2050 Vision: How can the UK play its part in avoiding dangerous climate change?*, Institute for Public Policy Research/World Wildlife Foundation UK, London, www.ippr.org.uk/publicationsandreports/publication.asp?id=572

Jablonski, S., Pantaleo, A., Bauen, A., Pearson, P., Panoutsou, C. and Slade, R. (2008) 'The potential demand for bioenergy in residential heating applications (bio-heat) in the UK based on a market segment analysis', *Biomass and Bioenergy*, vol 32 (7), pp635–653, doi:10.1016/j.biombioe.2007.12.013

Kannan, R., Strachan, N., Pye, S. and Balta-Ozkan, N. (2007) UK MARKAL model documentation, UK Energy Research Centre, London, www.ukerc.ac.uk/

Koritarov, V. (2005) *A Methodology for the Evaluation of Wind Power Using the WASP-IV Computer Model*, World Bank, Washington DC

Layberry, R. (2007) *UKDCM model documentation*, UKERC Energy 2050 technical paper, ECI, University of Oxford, Oxford

Loulou, R., Goldstein, G. and Noble, K. (2004) *Documentation for the MARKAL family of models*, Energy Technology Systems Analysis Program, Paris, www.etsap.org/

McDonald, A. and Schrattenholzer, L. (2002) 'Learning curves and technology assessment', *International Journal of Technology Management*, vol 23 (7/8), pp718–745

McKenna, N. (2008) *Review of UK industrial energy service demands*, UKERC Energy 2050 technical paper, University of Bath, Bath

Oxford Economics (2008) *Review of the BERR Energy Demand Model*, Oxford Economics, Oxford, www.theccc.org.uk/pdfs/Final_Report_Dec_2008.pdf

Pout, C. and MacKenzie, F. (2006) *Reducing Carbon Emissions from Commercial and Public Sector Buildings in the UK*, BRE Client Report for Global Atmosphere Division, Watford

PWC (Pricewaterhouse Coopers, MWH and Atkins) (2004) *REBIS: GIS Generation and Transmission, Volume 3: main report, The European Union's*

CARDS programme for the BALKAN region, Pricewaterhouse Coopers, London and European Commission, Brussels, http://siteresources.world-bank.org/INTECAREGTOPPOWER/Home/20551044/Volume%201%20 -%20Exec%20sum_final.pdf

Shorrock, L. and Utley, J. (2003) *Domestic energy fact file 2003*, Buildings Research Establishment, Watford

Sinden, G. (2007) 'Characteristics of the UK wind resource: Long-term patterns and relationship to electricity demand', *Energy Policy*, vol 35 (1), pp112–127

Stirling, A (2009a) *What is Energy Security? Uncertainties, dynamics, strategies*, Presentation to Sussex Energy Group Seminar 'UK energy security: What do we know, and what should be done?', Coin Street Neighbourhood Centre, South Bank, London, 27 January 2009, www.external.stir.ac.uk/postgrad/course_info/manage/economics/energy-management.php

Stirling, A. (2009b) 'Risk, Uncertainty and Power', *Seminar*, vol 597, pp33–39

Stokes, M., Rylatt, M. and Lomas, K. (2004) 'A simple model of domestic lighting demand', *Energy and Buildings*, vol 36 (2), pp103–116

Strachan, N., Kannan, R. and Hughes, N. (2007) *Workshop on the UK MARKAL-Macro Model and the 2007 Energy White Paper*, 21 June 2007, BERR Conference Centre, London

Strachan, N., Kannan, R. and Pye, S. (2008) *Scenarios and Sensitivities on Long-term UK Carbon Reductions using the UK MARKAL and MARKAL-Macro Energy System Models*, UKERC Research Report 2, UK Energy Research Centre, London, www.ukerc.ac.uk/

Strachan, N., Pye, S. and Kannan, R. (2009a) 'The Iterative Contribution and Relevance of Modelling to UK Energy Policy', *Energy Policy*, vol 37 (3), pp850–860

Strachan, N., Balta-Ozkan, N., Joffe, D., McGeevor, K. and Hughes, N. (2009b) 'Soft-linking energy systems and GIS models to investigate spatial hydrogen infrastructure development in a low carbon UK energy system', *International Journal of Hydrogen Energy*, vol 34 (2), pp642–657

Train, K. (1985) 'Discount Rates in Consumers' Energy-Related Decisions', *Energy: The International Journal*, vol 10 (12), pp1243–1253

Walker, B. and Salt, D. (2006) *Resilience Thinking: Sustaining ecosystems and people in a changing world*, Island Press, Washington DC

Walker, B., Holling, C.S., Carpenter, C. and Kinzig, A. (2004) 'Resilience, adaptability and transformability in social-ecological systems' *Ecology and Society*, vol 9 (2), pp5–14

Winskel, M., Markusson, N., Moran, B., Jeffrey, H., Anandarajah, G., Hughes, N., Candelise, C., Clarke, D., Taylor, G., Chalmers, H., Dutton, G., Howarth, P., Jablonski, S., Kalyvas, C., and Ward, D. (2009) *Decarbonizing the UK Energy System: Accelerated Development of Low Carbon Energy Supply Technologies*, UKERC Energy 2050 Research Report No. 2, UKERC, London, www.ukerc.ac.uk/Downloads/PDF/U/UKERCEnergy2050/TAcceleration_Draft.pdf

5
Pathways to a
Low Carbon Economy

Gabrial Anandarajah, Paul Ekins and Neil Strachan

Introduction

This chapter presents the results and analysis of different carbon pathways, entailing different levels of reduction of carbon emissions, for the UK through to 2050. The results come from the application of the MARKAL-MED model, the structure and working of which are briefly described in Chapter 4. This chapter focuses on the decarbonization pathways themselves, their energy-economic system implications, choices between technologies, and interactions between technological changes and behavioural responses to price changes. Following the presentation of the results, there is a discussion of the insights and conclusions to which they lead, including the policy implications of seeking to attain these pathways to a low carbon economy. The full set of modelling results is given in the Appendix to Anandarajah et al (2009).

Scenario design

The MARKAL-MED model has been run for a Reference (REF) scenario and a total of seven low carbon pathways. These are listed in Table 5.1 and their associated emission pathways shown in Figure 5.1. These scenarios are designed for relevance to the UK policy process for the near- and long-term targets of the Committee on Climate Change, the latter of which were included in the Climate Change Act (2008), and the former of which were adopted by the UK Government along with the financial budget in April 2009.

A first set of 'carbon ambition' scenarios (LC-40, LC-60, LC, LC-90) focus on ever more stringent 2050 CO_2 reduction targets ranging from 40 to 90 per cent reductions. These scenarios also have intermediate (2020) targets of 15 to 32 per cent reductions by 2020 (from a 1990 base year). A second set of 80 per cent reduction scenarios (LC-EA, LC-LC, LC-SO) have the same cumulative emissions but test the sensitivity of the results to slightly different specifications. LC-EA involves 'early action' and is constrained to achieve a 32 per cent reduction by 2020. LC-LC is a least-cost scenario without this 'early action' constraint. LC-SO has a 'social' discount rate of 3.5 per cent (HMT, 2006).

As described in Chapter 4, the majority of the scenarios – REF, LC-40, LC-60, LC, LC-90, LC-EA, LC-LC – employ a 'market' discount rate of 10 per cent to trade off action in different time periods as well as annualize technology capital costs. This 10 per cent market discount rate is higher than a risk-free portfolio investment return (which could be around 5 per cent) and accounts for the higher return that investors require to account for risk. In addition the model uses technology-specific 'hurdle' rates on future transport technology and on building conservation and efficiency options. These hurdle rates apply only to, and effectively increase, the capital costs of these efficiency technologies, in order to simulate the barriers to investment in them. Set at 15, 20 and 25 per cent, these hurdle rates represent information unavailability, non-price determinants for purchases and market imperfections (e.g. principal and/or agent issues between landlords and tenants). In the LC-SO scenario, which uses a 3.5 per cent discount rate, technology hurdle rates are reduced proportionally – that is, a previously doubled hurdle rate of 20 per cent is now still doubled but only to 7 per cent. These lower technology hurdle rates imply changes in consumer preferences or government policies that reduce barriers to the take-up of the technologies concerned. LC-EA, LC-LC and LC-SO are hereafter referred to as the cumulative emissions scenarios.

Table 5.1 *Carbon pathway scenarios*

Scenario	Emission reduction targets	Cumulative carbon budgets/targets	Cumulative emissions GTCO$_2$	Notes/Comments
REF	–	–	30.03	The Reference scenario
LC-40	15% by 2020 40% by 2050	–	25.67	A low carbon reduction target scenario
LC-60	26% by 2020 60% by 2050	–	22.46	A medium carbon reduction target scenario
LC	26% by 2020 80% by 2050	–	20.39	A 2050 target scenario in line with current UK legislation
LC-90	32% by 2020 90% by 2050	–	17.98	A high carbon reduction target scenario
LC-EA	32% by 2020 80% by 2050	–	19.24	Same 2050 target as LC, but with early action
LC-LC	None	Budget (2010–2050) similar to LC-EA	19.24	Same cumulative emissions as LC-EA, but a least-cost path (no forced early action)
LC-SO	None	Budget (2010–2050) similar to LC-EA	19.24	Same cumulative emissions as LC-EA, but with a social discount rate

Scenario results

The scenario pathways in this chapter are derived by imposing an annual carbon emission constraint from 2015 to reach the 2020 (if any) and 2050 emissions targets, as shown in Table 5.1. The results consist of time trends through to 2050, with occasional comparisons for the years 2020, 2035 and 2050. The 2020–2050 period trajectories for the LC-40, LC-60 and LC scenarios follow a straight line trajectory. In the LC-EA and LC-90 scenarios carbon emissions decline exponentially to ensure that the annual percentage reduction in late periods is not excessive. As explained above, LC-LC and LC-SO have the same cumulative emissions as the LC-EA 80 per cent reduction case. The full results for all the scenarios are given in an Appendix to Anandarajah et al (2009). This chapter presents the results for the main variables of interest.

CO₂ emissions

If no new policies or measures are enacted, energy related CO_2 emissions (in the REF scenario) in 2050 would be $583MtCO_2$, which is 6 per cent higher than the 2000 emission level of $551MtCO_2$ and only 1.5 per cent lower than the 1990 emission level of $592MtCO_2$. Existing policies and technologies would bring down the emissions in 2020 to about $500MtCO_2$, achieving about a 15 per cent reduction from 1990's level, which falls well short of the target reduction implied by the legislated carbon budget relevant for 2020 (34 per cent reduction in GHGs, 29 per cent in CO_2, from 1990's level). From 2020–2050, economic and energy service demand growth overwhelms near-term efficiency and fuel-switching measures (which are partially driven by the effects of the EU-ETS price, and the electricity and transport renewables obligations), and CO_2 emissions rise.

Figure 5.1 provides annual CO_2 emission levels under different scenarios over the projection period.

For the CO_2 mitigation scenarios where annual emissions constraints are not imposed (LC-LC and LC-SO), these two scenarios choose the optimal emissions path with the same cumulative emissions level as LC-EA (Figure 5.2). As expected, UK MARKAL-MED results show *later* action for the LC-LC scenario as the model tries to delay reductions (which require investment) as far as possible (owing to the 10 per cent discount rate and hence lower costs assigned to reductions in later periods). For the LC-SO scenario (at 3.5 per cent discounting), the model undertakes *earlier* decarbonization as the overall objective function now gives more weight to costs imposed later in the time horizon. As LC-SO focuses on earlier emission reductions it requires a reduction of only 70 per cent in 2050. On the other hand, the LC-LC scenario suggests that the UK can go beyond an 80 per cent target in 2050 as its later action cuts UK emissions by 89 per cent in 2050. The flexibility offered by the overall cumulative emissions constraint, rather than imposed annual reductions, is reflected by a slightly lower discounted system cost in LC-LC, about £700 million lower than in LC-EA (with the same cumulative emissions).

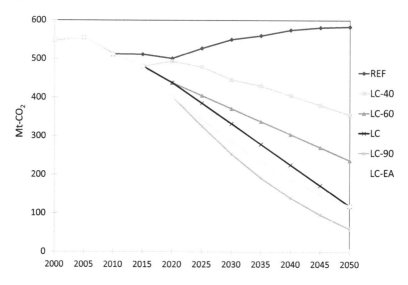

Figure 5.1 *CO₂ emissions under scenarios with different annual carbon constraints*

For nearer-term emissions reductions (2020), the CO_2 emissions constraint in 2020 is imposed in LC-60 and LC (26 per cent) and LC-90 and LC-EA (32 per cent). Among the cumulative emissions scenarios, the LC-SO scenario has the lowest emission level in 2020, emitting 39 per cent lower CO_2 emissions than in 1990 while the LC-LC cuts only by 32 per cent.

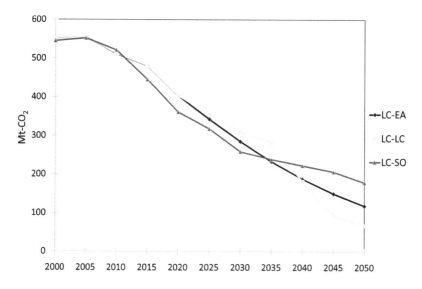

Figure 5.2 *CO₂ emissions under the cumulative emissions scenarios*

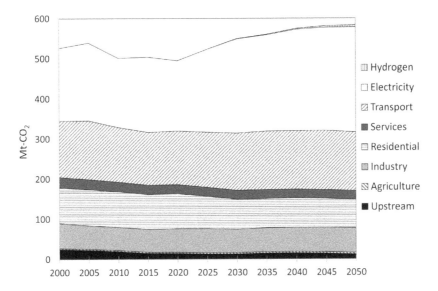

Figure 5.3 *Sectoral CO₂ emissions during 2000–2050 in REF*

Sectoral CO₂ emissions

The power sector has a relatively high share of total CO_2 emissions in REF, followed by the transport, residential and industrial sectors (Figure 5.3). The contribution of the power sector to total CO_2 emissions increases from 35 per cent in 2020 to 45 per cent in 2050 while the transport and residential sectors show slight reductions. The increased level of renewable electricity (especially wind) due to the Renewables Obligation brings down the power sector emissions during 2005–2020 while replacement of retiring existing nuclear and gas power plants by high CO_2 emitting coal plants during 2025–2030 radically increases the power sector CO_2 emissions between 2020 and 2030.

Figure 5.4 presents the sectoral CO_2 emissions in REF, LC-40, LC-60, LC and LC-90 for the selected years 2035 and 2050. Decarbonization is foremost in the power sector until the middle or end of the projection period. Then major efforts switch to the residential and/or transport sector. Service sector and upstream emissions are also heavily decarbonized in the LC and LC-90 cases in 2050 as the residual emissions budget shrinks. Residential and transport sectors work harder to meet the relatively higher early mitigation target in LC-90, reducing their emissions respectively by 67 and 47 per cent in 2035 as compared to REF.

To meet the 80 per cent target in LC, the power sector CO_2 emissions are reduced by 93 per cent compared to REF in 2050. The figures for the residential, transport, services and industrial sector are 92, 78, 47 and 26 per cent respectively. Since the industrial sector is only moderately decarbonized,[1] in 2050 it is the prime contributor to the remaining CO_2 emissions in LC and LC-90, followed by the transport sector.

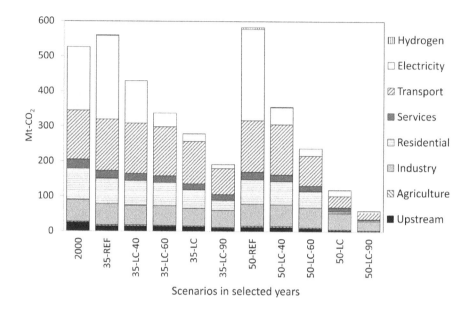

Figure 5.4 *Sectoral CO$_2$ emissions in years 2000, 2035, 2050: carbon ambition scenarios*

Note: *As with electricity, it is the production of hydrogen that is associated with carbon emissions, rather than its use*

End-use sectors have their lowest CO$_2$ emissions in LC-90, which has the highest mitigation target of 90 per cent in 2050. Conversely, the model meets the modest 40 per cent CO$_2$ reduction target in LC-40 by decarbonizing the power sector (with limited reductions in industry and service sectors) in 2035 and then further decarbonizing the power sector in 2050.

Sectoral CO$_2$ emissions under the LC and cumulative emissions scenarios (LC, LC-EA, LC-LC and LC-SO) are presented in Figure 5.5. In these cases, the general pattern of early decarbonization focused on the power sector is augmented in the LC-EA and especially the LC-SO scenarios where a focus on earlier action means the transport sector works harder. Although all the end-use sectors contribute to meet the CO$_2$ targets beside the power sector, in 2035 the residential sector plays a major role in LC-EA and LC-LC and the transport sector plays a major role in LC-EA and LC-SO.

In 2050, power sector CO$_2$ emissions are almost at the same low level in all the 80% CO$_2$ mitigation scenarios, and decarbonization effort is shifted from the power sector to end-use sectors, especially the residential and transport sectors. In the LC-LC scenario, the industry and services sectors are also heavily decarbonized in 2050 as the total CO$_2$ emission reduction is 89 per cent. A point here to be noted is that decarbonization of end-use sectors results in shifting to greater levels of low carbon electricity.

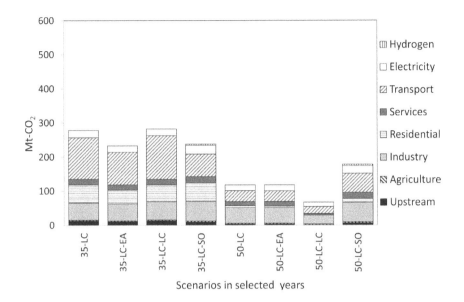

Figure 5.5 *Sectoral CO₂ emissions in years 2035, 2050: LC and cumulative emissions scenarios*

All the end-use sectors have their lowest CO$_2$ emission level in 2050 under the tightly decarbonized LC-LC scenario. Residential, upstream and services sectors combined emit only 5MtCO$_2$ while power, transport and industry sectors emit 13MtCO$_2$, 26MtCO$_2$ and 20MtCO$_2$ respectively in 2050 under the LC-LC scenario.

Energy-economic system implications

Primary energy demand

Total primary energy demand during 2000–2050 under all the carbon miti-gation scenarios is presented in Figure 5.6. Despite the fact that final energy demands increase slightly during 2000–2050 to meet the UK's growing energy service demands (see Figure 5.10), the primary energy demands are well below the 2000 level during 2000–2050 in REF. This is due to the improvement in efficiency of energy process and conversion technologies (power plants) and the increased share of renewables (notably wind). Primary energy demand decreases until 2020. The increased level of renewable electricity replacing oil and its products, due to the Renewables Obligation, reduces primary energy demand until 2020. Thereafter, the use of coal for power generation (replacing nuclear and gas) slightly increases primary energy demand in REF.

Primary energy demand in 2035 and 2050 by fuel/resource types under REF, LC-40, LC-60, LC and LC-90 is presented in Figure 5.7. In REF, fossil fuels (coal, oil and gas) dominate the primary energy supply in the early years. The

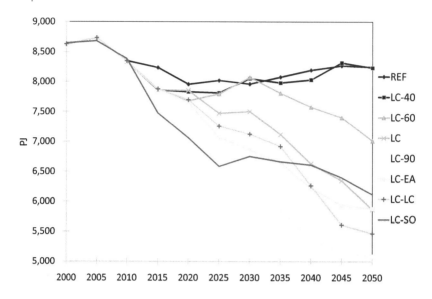

Figure 5.6 *Primary energy demand under different scenarios*

UK's primary energy demand in 2000 was met by fossil fuels with 81 per cent from natural gas and oil. But the trend changes as the share of coal increases in the medium and (especially) long term. This is largely due to the replacement of retiring gas power plants by coal plants. Further, there is no nuclear electricity in primary energy demand after 2035 as plants are retired and replaced by coal plants in REF. Demand reduction and efficiency improvements reduce carbon emissions and primary energy demand in the early years in REF.

Decarbonization essentially defines the future of energy supply. When the carbon reduction target is increased, fossil fuels are replaced by nuclear and renewable electricity[2] and by biomass. Biomass is mainly imported (Figure 5.9) and heavily used in the transport sector, and to a lesser extent in the residential and service sectors. A very large amount of biomass is selected at higher mitigation targets, especially in LC and LC-90, where biomass is the dominant resource, supplying one-third of total primary energy demand in 2050 (Figure 5.7).

When the carbon target becomes more stringent (40, 60, 80 and 90 per cent), very large reductions in primary energy demand are possible by 2050 as nuclear and renewables (especially wind) play a major role in the power sector, alongside efficiency improvement and demand reduction in the end-use sectors.[3]

In 2020, there is no significant change in total primary energy demand in LC-40, LC-60, LC and LC-90 as compared to that in REF. This is mainly due to the decarbonization of the power sector by coal-CCS plants to meet the mitigation target in 2020. Further, as coal-CCS plays a major role in meeting the 40 per cent mitigation target in LC-40, there is no big change in the primary energy demand mix as compared to REF except in the early years when CCS technology is not available. In 2050 in REF and LC-40 over 93 per cent of the

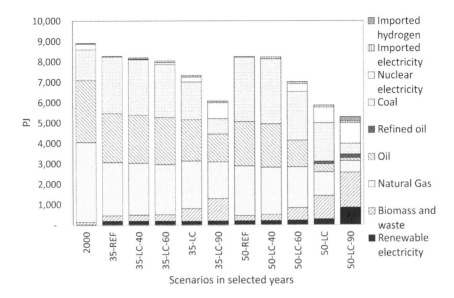

Figure 5.7 *Primary energy demand in selected years under different scenarios*

primary energy demand is supplied by fossil fuels, with coal accounting for over 38 per cent. In 2050, natural gas is mainly used in the residential sector and is the sole contributor to the residential sector's CO_2 emissions.

Figure 5.8 shows primary energy demand under the LC and cumulative emissions (LC-EA, LC-LC and LC-SO) scenarios. Primary energy demand in LC-LC and LC-SO is similar to the CO_2 emissions pattern. The early action under LC-EA and LC-SO involves lower primary energy than in LC in 2035, while LC-LC, in which carbon reductions occur later, demands lower primary energy than in LC in 2050. In LC-SO, the share of nuclear and biomass is relatively low in 2035 and large amounts of oil are replaced by electricity (imported and nuclear) and imported hydrogen, mainly produced by electrolysis. Among the cumulative emissions scenarios, primary energy demand has its lowest value in 2050 under LC-LC, where nuclear plays a major role in meeting the CO_2 target.

Coal is largely used in the power sector and with CCS in all CO_2 mitigation scenarios. When the carbon target is increased, fossil fuels are replaced by biomass and nuclear. By 2050 LC-LC requires a large amount of biomass, accounting for 30 per cent of total primary energy, to meet its stringent CO_2 target. In LC-LC, half of primary energy is supplied by biomass and nuclear in 2050. When demand for biomass is increased, supply is shifted from imports to domestic sources, owing to conservative assumptions on the imports of sustainable biomass to which the UK has access.[4] Hence a large proportion of biomass comes from expensive domestic resources in LC, LC-EA and LC-LC (Figure 5.9). Interestingly, a considerable amount of hydrogen (139PJ) is supplied in LC-SO for transport.

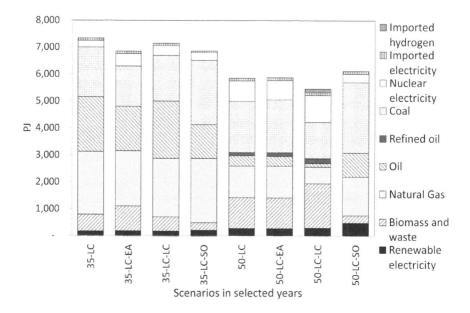

Figure 5.8 *Primary energy demand in selected years in LC and cumulative emissions scenarios*

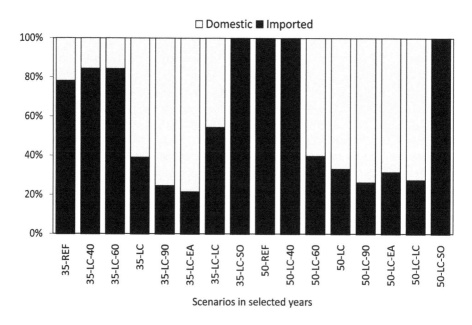

Figure 5.9 *Share of biomass provided by imports*

Final energy demand

Though primary energy demand is lower in the REF scenario during 2000–2050 than in 2000, final energy demand slightly increases during the period (Figure 5.10). However, its growth rate is much lower when compared to the growth in energy service demands. This is due to the increased efficiency of end-use devices, and energy conservation measures, especially in buildings.

The UK economy is decarbonized mainly by end-use efficiency improvements and fuel switching, energy conservation measures and demand reduction. This leads to reductions in final energy demand in each successive CO_2 reduction target in LC-40, LC-60, LC and LC-90 throughout the period except meeting the target of 90 per cent (LC-90) in 2050 (Figure 5.10). Even though LC-90 meets a 90 per cent target in 2050, it demands slightly more final energy than in LC, which meets only an 80 per cent CO_2 reduction target in 2050. This reflects the fact that mitigating CO_2 emissions by fuel switching does not always mean reducing final energy demand. Mitigation is also possible and cost effective with a less carbon-intensive fuel with lower energy efficiency.

Figure 5.11 shows final energy demand by fuel types for selected years in REF, LC-40, LC-60, LC and LC-90. Gas is the dominant fuel in the base year as well as in 2035, accounting for more than one-third of final energy demand in all scenarios. Overall, although the share of gas is decreasing over time in the low carbon scenarios, gas and electricity still dominate final energy demand in all scenarios except LC-90 in 2050. The share of electricity in total final energy demand in REF is only 19 per cent in 2000, but its share increases continuously throughout the period, reaching 23 per cent in 2050. Petrol and diesel together meet about one-third of final energy demand, with diesel having

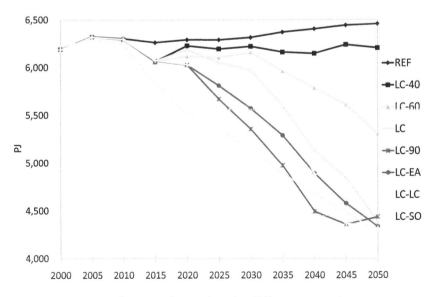

Figure 5.10 *Final energy demand under different scenarios*

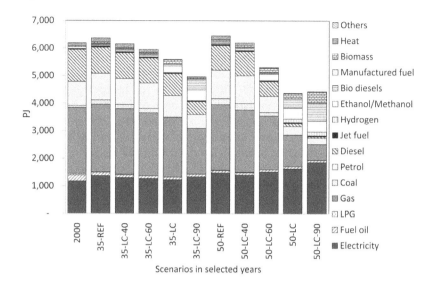

Figure 5.11 *Final energy demand by fuel under different scenarios*

a slightly higher share in the early and middle periods. Bio-energy (biodiesel and ethanol) plays a considerable role in LC-90 in 2050. The transport sector consumes a large amount of bio-energy (ethanol and biodiesel) leading to greater final energy demand in LC-90 as compared to LC as the efficiency of biodiesel-based vehicles is relatively low compared to hybrid plug-in vehicles. Further, a large amount of biomass is used in the service sector for heating. Note that the remaining (high efficiency) gas will be a major contributor to residential and service sector CO_2 emissions, along with transport (including aviation) and industrial liquid fuels.

Final energy demand by fuel type under LC, LC-EA, LC-LC and LC-SO is shown in Figure 5.12. The two mitigation scenarios (LC-LC and LC-SO) which are run with cumulative CO_2 constraints show completely different final energy demand patterns. In 2035 LC-LC involves more final energy use than LC-EA, because emission reduction in LC-LC occurs later in the projection. In contrast, LC-SO demands less final energy than LC-EA in 2050 as well as 2035 despite the fact that its annual CO_2 mitigation level in 2035 is similar to LC-EA. The reason for the low final energy demand in the medium term in LC-SO is the relatively low energy demand in the transport sector as the sector is decarbonized by shifting to electricity (hybrid plug-in) and hydrogen vehicles. High-capital cost hydrogen vehicles become relatively cheaper in LC-SO as the annualized cost is lower due to the lower social discount and hurdle rates. This early decarbonization means that by 2050 bio-fuels are not directly used for transport modes in LC-SO, in marked contrast to the other scenarios.

Natural gas is mainly used in the industrial sector, followed by the residential and service sectors. By 2050, the residential and service sectors use a very low amount of natural gas in LC-LC. Natural gas is replaced by biomass

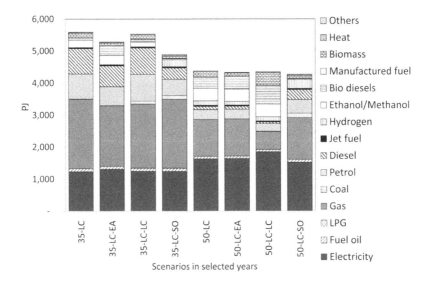

Figure 5.12 *Final energy demand by fuel in LC, LC-EA, LC-LC and LC-SO*

in the service sector and by electricity in the residential sector. Some available inexpensive gas is used for gas-fired power generation with CCS in 2050, under LC-LC. In LC-SO, a large amount of gas goes to boilers for heating.

Calculations from the detailed results show that none of the scenarios would meet the UK target from the EU's Renewables Directive of at least 15 per cent of UK final energy deriving from renewables by 2020 – it reaches just 5 per cent in 2020 under all carbon abatement scenarios. By 2050, the share of renewables in final energy demand rises to 49 per cent in LC-90. The respective figures in 2050 for the other scenarios are 6 per cent in REF, 7 per cent in LC-40, 13 per cent in LC-60, 27 per cent in LC, 27 per cent in LC-EA, 39 per cent in LC-LC and 12 per cent in LC-SO.

Electricity generation

In REF electricity generation increases by 24 per cent during 2000–2050 to meet continuously increasing electricity demand in end-use sectors. Over two-thirds of total electricity generation comes from fossil fuels (coal and gas) in REF in 2020 (Figure 5.13). In the absence of significant CO_2 pricing, high carbon content coal becomes the dominant fuel for electricity generation, gradually replacing gas and nuclear over the years, and generating more than 80 per cent of the total electricity supplied in 2050 (Figure 5.14). Since coal is responsible for almost all CO_2 emissions from the power sector in 2050, decarbonization of the power sector in the other scenarios involves decarbonizing coal generation by coal-CCS and/or replacing coal generation with nuclear and renewable generation such as wind, biomass, marine and solar. In 2020, the early decarbonization requirements of the electricity sector are achieved by replacing coal plants with coal-CCS plants in all mitigation scenarios.

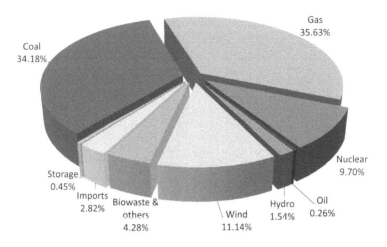

Figure 5.13 *Electricity generation fuel mix in 2020 in REF*

To meet CO_2 reduction levels in 2020, end-use sectors also contribute to meeting the carbon target by means of efficiency improvements and demand reductions, alongside the decarbonization of the power sector. As the power sector is decarbonized by capturing the carbon from coal plants, there is no big change in the fuel mix of power generation in 2020 in all carbon mitigation scenarios.

Electricity generation mixes under REF, LC-40, LC-60, LC and LC-90 are shown in Figure 5.14 for 2035 and 2050. Total electricity generation increases or decreases in the mitigation scenarios compared to REF depending on the electricity demand. In 2035, electricity generation decreases in line with the increasingly ambitious targets of LC-40, LC-60 and LC (but not in LC-90), as decarbonization is achieved through increased end-use efficiency and demand reduction in end-use sectors. However, as decarbonization targets tighten from 2035 to 2050, the shift to low carbon electricity in end-use sectors outweighs the efficiency improvements and demand reduction and leads to electricity demand increasing as the required carbon reductions rise. Thus the overall demand for electricity is the outcome of drivers in different directions: the shift to electricity increasing it, but efficiency improvements and demand reductions reducing it.

Coal-CCS is an early choice (i.e. in 2020) for the decarbonization of power generation in all the carbon reduction scenarios. However, coal-CCS only reduces CO_2 emissions by 90 per cent, and as increasing carbon reduction requirements drive carbon emissions down to very low levels in the power sector (almost complete decarbonization in 2050 in LC-90) coal-CCS is gradually replaced by nuclear and wind as the 10 per cent residual CCS emissions limit the use of coal-CCS plant. A large amount of electricity (more than one-third) is generated from wind in LC-90 in 2050.[5]

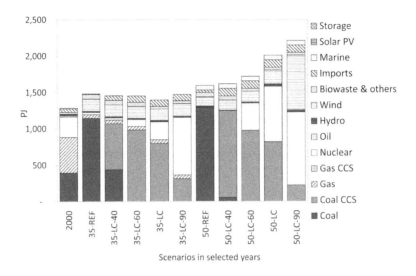

Figure 5.14 *Electricity generation mix under different scenarios*

Figure 5.15 presents the electricity generation mix in LC-EA, LC-LC and LC-SO. There is no big difference in overall levels of electricity generation in 2050 among these scenarios. But in 2035, early action LC-SO requires a larger amount of electricity as it reduces CO_2 emissions by 60 per cent, including through the use of plug-in electric vehicles and electric heat boilers. Electricity

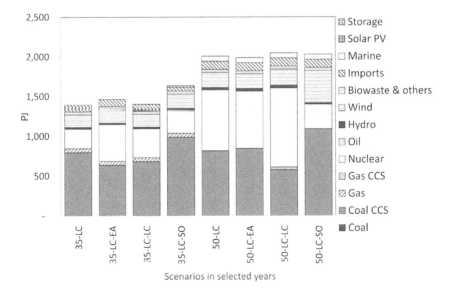

Figure 5.15 *Electricity generation mix in LC, LC-EA, LC-LC and LC-SO*

demand under LC-SO in 2050 is met by a large expansion of wind (a zero-carbon technology that is commercialized relatively early). As explained in endnote 5, and shown in Figure 5.28, this necessitates a very large expansion in overall electricity capacity for peak constraints. Wind expansion is mainly from offshore wind as all cost-effective onshore wind is already selected in REF itself. The contribution of intermittent renewables such as wind, marine and solar to peak load is limited. Therefore, the selection of renewables (wind power plants) to meet the carbon target needs a large amount of reserve capacity from gas plants (see Figure 5.28).

Marginal cost of CO_2

As described in Chapter 4, MARKAL-MED is a least-cost optimization model that computes the marginal cost of CO_2 abatement and welfare costs to meet the CO_2 constraints based on a range of input assumptions. In these scenarios all costs are given in the prices of the year 2000, shown as £(2000).

The scenarios described in this book cover only mitigation options available in the UK. If the price of carbon credits from other countries were below the marginal cost of abatement it would be cheaper to purchase credits and forgo some domestic mitigation actions. This possibility is explored in Chapter 11.

The marginal prices shown in Figure 5.16, 5.17 and 5.18 indicate that marginal emission prices (abatement costs) rise as the annual CO_2 constraint tightens across scenarios and through time. In 2035 marginal CO_2 prices rise from £13/tCO_2 in LC-40 to £133/tCO_2 in LC-90, and by 2050 this range is £20/tCO_2 to £300/tCO_2. This illustrates the difficulty of achieving very deep CO_2 reductions.

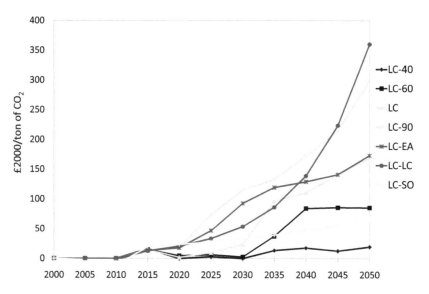

Figure 5.16 *Marginal price of CO_2 emissions under different scenarios*

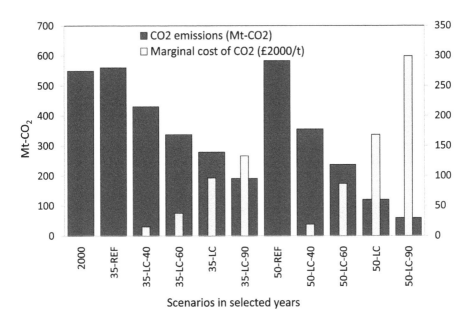

Figure 5.17 *CO₂ emissions and marginal cost of CO₂ under different scenarios*

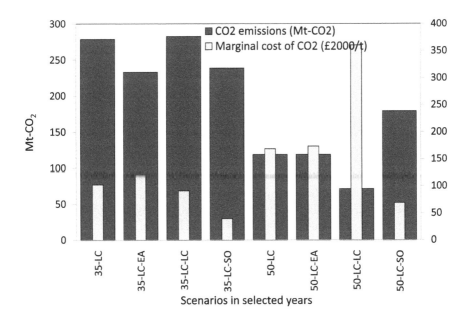

Figure 5.18 *CO₂ emissions and marginal cost of CO₂ under LC and cumulative emissions scenarios*

The cumulative emissions scenarios which choose the least-cost path from 2010 through 2050 with no intermediate targets (LC-LC and LC-SO), again follow the logic of later and earlier action depending on the weight implied by the discounting process. Costs in LC-SO (which involves early action) are £24/tCO$_2$ and £66/tCO$_2$ in 2020 and in 2050 respectively, while in (late action) LC-LC the costs are respectively £21/tCO$_2$ and £360/tCO$_2$. The low costs in LC-SO reflect the fact that the lower technology-specific hurdle rates imply a lower cost of capital. An interpretation of these lower hurdle rates is that consumer preferences change and/or government works to remove uncertainty, information gaps and other non-price barriers.

Demand reduction

It is the demand for energy services that drives the demand for energy. As carbon constraints are applied and more expensive low carbon technologies have to be used, energy service demands are reduced, with an associated societal loss in utility. The optimal solution of MARKAL-MED's objective function involves the maximization of combined producer and consumer surplus. This is reduced when carbon constraints reduce energy service demands. Demand reduction levels for selected sectors and transport energy service demands under different scenarios in 2050 are shown in Figure 5.19. Demand reduction levels are higher in 2050 than in 2035 as the CO$_2$ reduction constraint is tighter. Agriculture, industry, residential and shipping have higher demand reductions than the air, car and HGV (heavy good vehicles) transport sectors.

A demand function is constructed based on the price elasticity and reference prices of REF. The level of demand reduction then depends on both the price elasticity of demand and the prices of alternative technologies and fuels available to meet the particular energy service demand. For a particular energy service demand, if the alternatives are only available with a relatively high incremental cost, then the demand reduction level would be high, other things being equal. For example, the price elasticity of demand is relatively low for transport shipping (−0.17) and relatively high for transport HGV (−0.61). However, demand reduction is relatively higher for transport shipping than transport HGV as transport shipping has no alternative technologies in the UK MARKAL model other than diesel, which is a high carbon content fuel. HGV on the other hand has many alternative technologies such as diesel ICE, diesel hybrid, hydrogen ICE and hydrogen fuel cells. Similarly, car demand also has a relatively high price elasticity (−0.45), but because of the availability of relatively low cost alternative technologies, the demand reduction level is low.

Demand reductions in the agriculture, industry, services and residential sectors are combinations of reduced individual energy service demands of the component parts of those sectors. In particular, relatively high elasticities and restricted technology options for residential demand (notably direct electricity and gas use) and industrial sectors (notably chemicals) result in substantial reductions in energy service demands. Reaching 20–25 per cent reductions in service demands implies both significant behavioural change and an industrial reorientation process concerning energy usage.

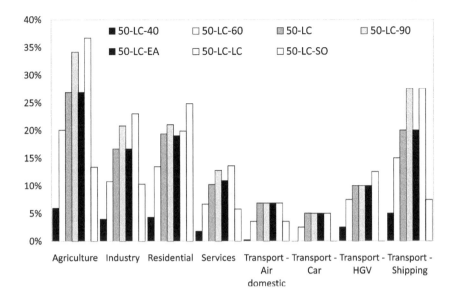

Figure 5.19 *Selected energy service demand reductions in 2050 under different scenarios*
Note: *The bars for each sector are in the order given by the key to the scenarios.*

As expected, demand reduction levels are lowest in LC-40 for all sectors. The level of demand reduction increases with the increasingly stringent mitigation targets in LC-40, LC-60, LC and LC-90 in 2035 and 2050. But demand reductions under LC-LC and LC-SO in 2050 are dissimilar as the mitigation pattern is different for these scenarios. The relatively lower weight for near-term costs in LC-LC results in the model not taking up the immediately available options for demand reductions, although this is reversed by 2050 when the LC-LC scenario is decarbonizing to a very great extent. Demand reductions in 2050 under LC-SO are generally lower (as the model places more weight on late-period demand welfare losses), except in the residential sector. In terms of early demand reductions for LC-SO, this is seen in residential electricity and gas energy service demands where demands are sharply reduced, as an alternative to (relatively expensive) power sector decarbonization. Interestingly, no demand reductions are envisaged in personal transport, where the LC-SO scenario undertakes very significant technological substitution.

Welfare
Although the reduction of energy service demands is an immediately available option to reduce demand for energy and consequently CO_2 emissions, it has a negative impact in lost utility from not having the benefit of this additional

energy use. When combined with reductions in producer surplus, the resulting metric (the combined loss of consumer and producer surplus) is the social welfare loss.[6] This is a far superior metric of a change in social welfare than changes in energy system costs, because it takes account of the welfare cost of demand reduction. Energy system cost, however, may be reduced by demand reduction if this serves to reduce the size of the overall energy system.

As shown in Figure 5.20, by 2050, overall welfare losses (in £(2000)) in the carbon ambition scenarios range from £5 billion for 40 per cent reductions to £52 billion for 90 per cent CO_2 reductions, showing significant increases in welfare loss – including a near doubling of costs for a 60 per cent versus an 80 per cent reduction – represent a key decision variable when deciding on more stringent UK emission reduction targets. The low welfare losses in the LC-SO scenario are a reflection of the lower capital costs implied by the social discount rate, and lower technology hurdle rates. The precise split between producer and consumer surplus is dependent on the ability of producers to pass through additional CO_2 emission costs.

Key sectoral and energy technology trade-offs

Sectoral energy demand and technologies
Final energy demand by end-use sector is presented in Figure 5.21 for selected years under the REF, LC-40, LC-60, LC and LC-90 scenarios. The transport, residential and industry sectors have relatively high energy demands while the agriculture sector has the lowest energy demand (50–70PJ/annum during 2000–2050). Overall in REF, sectoral energy demands in transport, industry and agriculture are increasing during the projection period, while the residential and services sectors' energy demand is lower in 2050 than in 2000.

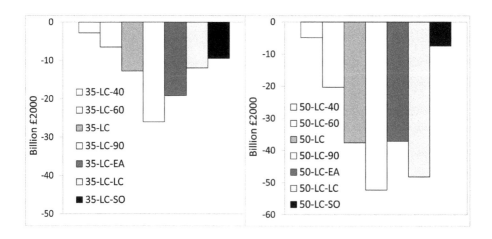

Figure 5.20 *Change in social welfare under different scenarios*

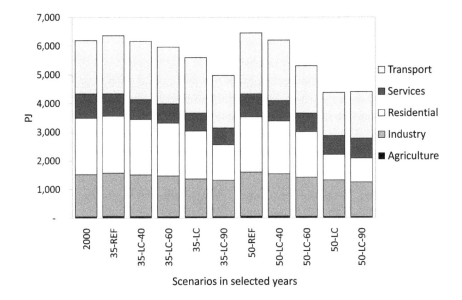

Figure 5.21 *Sectoral energy demand under different scenarios*

The sectoral energy demand and technology mix depend on the level of decarbonization required. When looking at the decarbonization of end-use technologies, the residential sector is decarbonized by shifting to electricity (from gas) as well as technology switching from boilers to heat pumps for space heating and hot water heating. The transport sector is decarbonized by shifting to hybrid plug-in, ethanol, hydrogen and battery operated vehicles. The service sector is decarbonized by shifting to biomass (in LC-LC) and electricity. In addition to efficiency, fuel switching and technology shifting, the elasticity of demand also plays a major role in reducing CO_2 emissions by reducing energy service demands, and therefore final energy demand.

Despite the fact that the residential, services and transport sectors have been heavily decarbonized to meet the carbon targets, the residential sector, compared to the transport and other end-use sectors, shows increasingly large reductions in final energy demand (Figure 5.21) as the targets become more stringent. The reason for the low energy demand in the residential sector is that here decarbonization is mainly by shifting from gas to electricity. The associated electrical devices have relatively high efficiency, especially heat pumps (Figure 5.24) for space and water heating, and therefore relatively low energy demand. In the case of the transport sector, bio-fuels also play a role in decarbonization in addition to the switch to electricity (Figure 5.23). Compared with the residential sector, demand reductions in the transport sector are relatively low, especially for cars, which consume two-thirds of the transport sector energy demand in REF.

Sectoral final energy demand in LC, LC-EA, LC-LC and LC-SO is presented in Figure 5.22. In LC-SO and LC-EA the residential and transport sectors decarbonize relatively early, as shown in Figure 5.5. Demand reduction in the residential sector plays a considerable role in this process, mainly in space heating, water heating and electricity use (Figure 5.22 shows considerably less demand in 2035 in LC-EA than in LC). Decarbonization technologies in the residential sector are electric boiler night storage, and heat pumps (Figure 5.24) for water heating. In the transport sector, early decarbonization involves shifting to hybrids and hybrid plug-ins from 2020 in LC-SO and shifting to E85 cars, which use a fuel that is 85 per cent bio-ethanol, and battery buses from 2030, and a very low amount of shifting to rail-electric in the LC-EA. LC-LC, which decarbonizes later, meets the constraint by a large amount of nuclear replacing coal-CCS, which has residual emissions of 10 per cent as explained earlier, and also by means of transport demand reduction (HGV and shipping) and by shifting to bio-energy in the service and transport sectors (Figure 5.23).

Though the service sector is heavily decarbonized (by 94 per cent) in LC-90 in 2050 (Figure 5.4), there is little change in the sector's final energy demand as decarbonization is mainly through the replacement of gas boilers with biomass boilers. The service sector consumes 373PJ of biomass mainly in boilers in 2050 (Figure 5.23). Enhanced biomass use in the transport sector in the most stringent scenarios (especially LC-EA, LC-LC and LC-90) provides a further example of how the model can decarbonize while final energy use increases. This is also the case in LC-SO although here the low carbon technologies are hydrogen and electric vehicles rather than biomass.

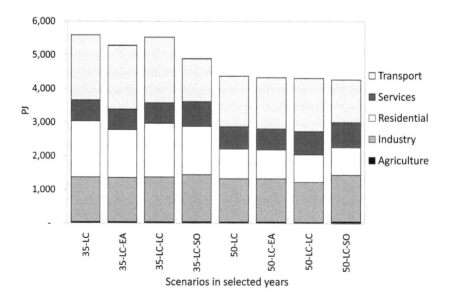

Figure 5.22 *Sectoral energy demand under LC, LC-EA, LC-LC and LC-SO scenarios*

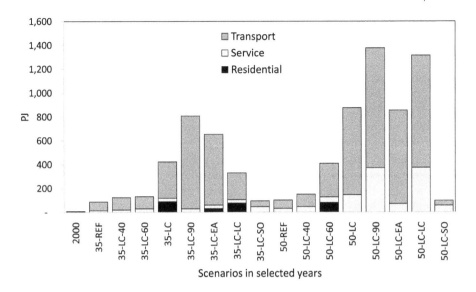

Figure 5.23 *Sectoral bio-fuel energy demand under different scenarios*

A heat pump acts as a refrigerator in reverse, and uses relatively little electricity to pump large amounts of heat to where it is needed. Though heat pumps are capital intensive, large numbers of them are selected by the model for space heating, but also for water heating, replacing gas boilers in the residential sector. They become cost effective from 2030 in LC-EA, from 2035 in LC, LC-LC, and LC-SO and from 2045 in LC-60 (Figure 5.24). Due to their

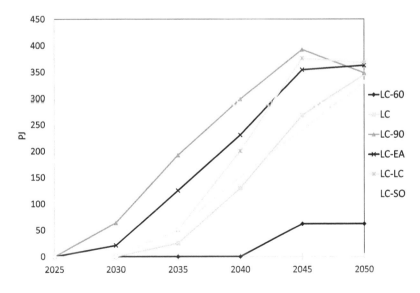

Figure 5.24 *Electricity demand for heat pumps under different scenarios*

high penetration, heat pumps consume large amounts of electricity, equivalent to about 350PJ in 2050 in LC-60, LC, LC-90, LC-EA and LC-LC.

Transport fuel demand

In the transport sector the model scenarios give a detailed breakdown of the uptake of different vehicle technologies, including those with greater energy efficiency, although MARKAL-MED only distinguishes between differently fuelled vehicles, rather than vehicles of the same type (e.g. petrol ICEs) with different energy efficiency – improved vehicle efficiency within types has to be imposed exogenously as part of the technology characterization.

Cars are the biggest energy consumers in the UK transport sector, accounting for over half the transport sector energy demand in REF (Figure 5.25). This is mainly due to the high demand for transport services in terms of passenger-km in the base years as well as the expected high growth rate during the period. Further, cars tend to have a low occupancy, leading to high energy consumption per passenger-km. Goods transport vehicles (HGV and LGV) are responsible for at least 27 per cent of transport energy demand. In REF petrol and diesel IC engines cars are selected while for two-wheelers only petrol engines are selected. For buses, there are complete transitions from diesel to diesel hybrid during 2010–2015 and then from hybrid to battery-operated electric buses during 2040–2045 in REF. Hybrid (diesel) vehicles replace diesel-based HGV and LGV during 2010–2015 and thereafter there is no technological change or fuel switch for goods vehicles in REF.

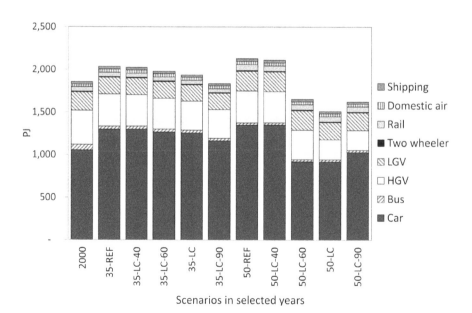

Figure 5.25 *Transport sector energy demand by modes under different scenarios*

The transport sector in 2035 in LC-40, LC-60, LC and LC-90 shows relatively little reduction in energy demand (Figure 5.25) and relatively little decarbonization (Figure 5.4). This is least true in LC-90, where the transport sector has to work harder, but still decarbonization is mainly by technology shifting rather than demand reduction, in this case to E85 cars (55 per cent) and, to a smaller extent, to petrol plug-in cars (11 per cent). In 2050, a significant difference in energy demand can be observed in the more ambitious scenarios (i.e. not LC-40) as the transport sector is decarbonized in the latter part of the period. Though transport sector CO_2 emissions are lowest in LC-90, energy demand is higher than in LC. This is due to the larger consumption of biodiesel and ethanol in LC-90 and the greater penetration of more efficient plug-in hybrid cars in LC and LC-60.

With regard to the cumulative emissions scenarios, Figure 5.26 shows that early CO_2 reductions in LC-SO mean relatively low transport energy use in 2035 when compared to other scenarios (LC, LC-EA, LC-LC). As in LC, biodiesel and/or ethanol decarbonize the transport sector in LC-EA and LC-LC, in addition to electric (hybrid) cars (petrol and diesel) and goods vehicles (HGV and LGV). Demand for biodiesel and/or ethanol fuels in the different scenarios is more or less proportional to the transport sector decarbonization level (see Figure 5.23 and Figure 5.4), while the demand for electricity stays more or less the same in LC-EA and LC-LC, in the range of 200–250PJ in 2050. The transport sector also consumes a small amount of hydrogen in LC, LC-EA and LC-LC, mainly for HGV.

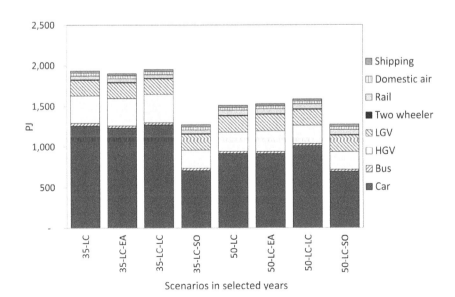

Figure 5.26 *Transport sector energy demand by modes in LC,*
LC-EA, LC-LC and LC-SO

A large reduction in transport final energy demand occurs in LC-SO, especially in 2035, with more efficient hydrogen fuel cell vehicles replacing petrol and diesel vehicles. In the LC-SO scenario 218PJ and 279PJ of hydrogen are used in 2035 and 2050 respectively for goods vehicles, especially HGV. In addition to hydrogen, in LC-SO considerable electricity is used for the decarbonization of the transport sector, amounting to 140PJ in 2035 and 220PJ in 2050. For this scenario, comparison of Figures 5.19, 5.23 and 5.26 illustrates the interactions and trade-offs in responding to carbon constraints between energy service demand reduction, energy demand reduction from higher efficiency vehicles, and the use of zero-carbon fuels.

In both the LC-EA and the LC-LC scenarios battery buses are used from 2030, plug-in hybrids from 2040, ethanol (E85) cars from 2035 and hydrogen (HGV) from 2050. In LC-SO, hydrogen and battery cars have been selected in 2050 and no ethanol cars have been selected. Battery buses and hydrogen fuel cell HGVs have been used from 2030. Battery and hydrogen LGVs are selected in 2050 under LC-SO. The diversity of different technologies in different scenarios indicates both the range of broadly competitive options in the transport sector, and the effect of the change in the discount rate, which also has a significant impact on the welfare and CO_2 marginal costs.

Electricity generation technologies

Electricity generation in REF is mainly from coal, gas, nuclear and wind. Small amounts of oil, hydro and bio-waste generations are also selected. Marine renewables become cost effective from 2045. Figure 5.27 shows installed capacity in REF by fuel type.

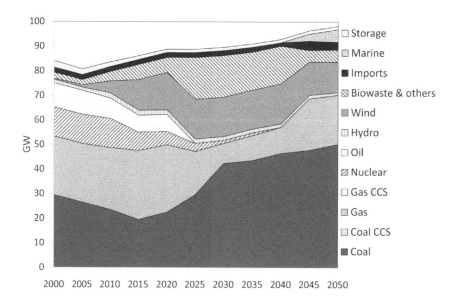

Figure 5.27 *Installed capacity in REF during 2000–2050*

Existing coal plants dominate in the early part of the projection period, accounting for 67 per cent of installed base load capacity in 2020. Existing nuclear technologies (advanced gas-cooled reactor, magnox reactor and PWR) operate in the early years until they are retired. The share of nuclear plants in base load capacity decreases substantially between 2010 and 2035. Coal plants (pulverized fuel technology) gradually replace existing coal and nuclear power plants from 2020. Their capacity gradually increases from 17GW in 2020 to 50GW in 2050. Investment in new capacity of combined cycle gas turbine (CCGT) plant rises from about 1GW in 2010 to 13GW in 2050. Existing CCGT (20.5GW installed capacity in 2000), coal plants, and gas and oil fired steam turbines are utilized until they are retired.

Gas turbines and gas engines are selected from 2010 and 2015 respectively for the non-base load gas plants. Wind, particularly onshore wind, plays a major role for non-base load, with over 12GW during 2015–2050. In the middle part of the period, a large quantity of sewage and landfill gas internal combustion engines (ICE) are also selected, their capacity increasing from 2.5GW in 2015 to 13GW in 2025. As the share of base load plants, in total installed capacity, is relatively high at the end of the projection period, the capacity of the sewage gas plants declines to 1GW in 2050. Further, 3GW and 5GW of tidal stream plants are selected in 2045 and 2050 respectively. There is also a slight decrease in wind capacity during the latter part of the projection period. A small amount of energy crops gasification, generation from steam turbines fired by municipal solid waste and agro-waste, and landfill gas IC engines is also selected in REF.

When CO_2 emissions are increasingly constrained (LC-40, LC-60, LC, LC-90), the MARKAL-MED model strongly decarbonizes the electricity sector, and there is a huge change in the capacity mix in the power sector (Figure 5.28). The decarbonization of end-use sectors by means of a shift to low carbon electricity, as well as the selection of intermittent plants, which need back-up capacity, increases the installed capacity level in the mitigation scenarios particularly during the latter part of the projection period.

There are several broadly competitive available low carbon options for power generation, including renewables, nuclear power, and carbon capture and storage (CCS) associated with coal- and gas-based fossil fuel power stations. However, decarbonization of the power sector begins with the deployment of CCS for coal plants in 2020 in all mitigation scenarios (Figure 5.28). Non-CCS coal in 2035 only remains in any quantity in LC-40, with its relatively low mitigation target. Coal-CCS is the main technology to meet the mitigation target in LC-40 and LC-60 in the later period. Coal-CCS decreases with the increased CO_2 reduction target level in LC and LC-90, as the carbon capture rate is only 90 per cent (i.e. there are 10 per cent residual emissions). Nuclear is selected at the expense of CCS to meet the carbon target in LC. A large amount of wind is selected with the LC-90 90 per cent target in 2050, together with a large capacity of back-up gas plants. The results are also affected by the assumption that learning effects cause the capital costs of technologies to fall over the period. For example, marine renewables become cheaper and are selected in 2045 because of relatively strong learning effects.

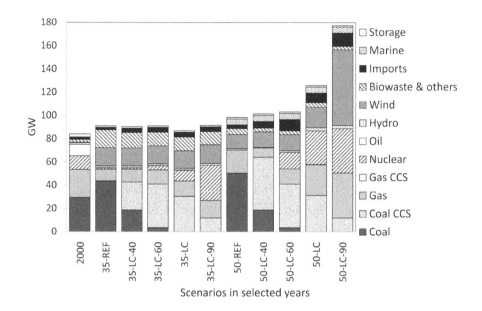

Figure 5.28 *Installed capacity under different scenarios*

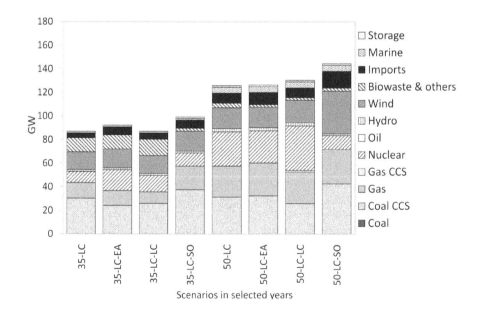

Figure 5.29 *Installed capacity under LC, LC-EA, LC-LC and LC-SO scenarios*

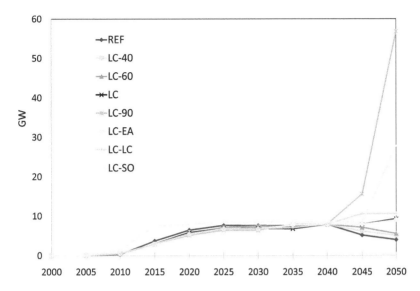

Figure 5.30 *Offshore wind installed capacity under different scenarios*

For two of the cumulative emissions scenarios (LC-LC and LC-SO), the required capacities in 2050 show a similar pattern (see Figure 5.29). The scenario LC-LC, with very high 2050 decarbonization (89 per cent), shows lower electricity generation and capacity than LC-90 owing to greater flexibility in the timing of its carbon reductions.

The maximum available capacity of onshore wind is selected in REF, while the deployment of offshore wind increases wind capacity in the mitigation scenarios. Figure 5.30 presents the deployment of offshore wind under different scenarios. As early action requires near-competitive technologies (and also as the social discount rate favours capital-intensive technology), a large amount of offshore wind is selected, reaching 28GW in LC-SO, and 57GW in the 90 per cent LC-90 scenario, when the construction rate over 2045–2050 is perhaps implausibly high. When MARKAL-MED selects more and more wind it has to increase the capacity of back-up plants for peak generation. The back-up plants are mainly gas-based CCGTs.

Insights and conclusions

This section considers what insights and conclusions may be derived from the scenarios, and what policies might be required to produce or approximate their outcomes.

There are two major sets of issues which mean that the scenarios are unlikely to represent the real evolution of the UK energy system to 2050. The first is to do with the inherent uncertainty around the costs and other parameters relating to the technologies in the model. We simply do not know and cannot know how these will develop over the next four decades. The numbers in the model are expert estimates, validated by peer review, but they are still very uncertain. Later chapters in the book change some of the key assumptions to see how this changes the results.

The second set of issues derives from the fact that the model's optimization procedure implies that, given the assumptions and constraints, decision-makers in the energy system have perfect foresight of events and developments through to 2050, they take decisions based only on market criteria, and markets work perfectly. Of course this is not the case in the real world.

It should therefore be clear that in no sense are any of the scenarios, even REF, predictions of what will happen if carbon constraints are applied with different levels of stringency. Instead they are quantitative aids to thought and analysis of different possible developments in the energy system given concerns to reduce carbon emissions from it. Generating such insights is the reason for undertaking energy systems modelling.

Tightening constraints on CO_2 emissions

The set of scenarios with increasingly ambitious CO_2 emission constraints (40, 60, 80 and 90 per cent reductions from 1990 levels by 2050) offer insights on decarbonization pathways, sectoral–technology–behavioural trade-offs, and resultant cost implications.

In REF, if new policies and/or measures are not taken, CO_2 emissions in 2050 would be 583MtCO$_2$: 6 per cent higher than 2000 levels and 1.5 per cent lower than 1990 levels. The policies and technologies assumed in REF (those in place in 2007) would bring down emissions in 2020 to about 500MtCO$_2$ – a 15 per cent reduction. However this is a considerably smaller reduction than the government target of a 29 per cent reduction by 2020. In the absence of a strong carbon price signal, the electricity sector is the largest contributor to CO_2 emissions driven by conventional coal-fired power plants, with substantial contributions also from the transport and residential sectors.

The power sector is a key sector for decarbonization, especially where this occurs early. This early electricity decarbonization, led in these scenarios by coal-CCS technologies, and combined with end-use conservation measures, is relatively low-cost. However, the scenarios exhibit considerable uncertainty over the dominant technology in any portfolio containing CCS, nuclear and wind, due to the close marginal costs and future uncertainties relating to these technologies. Specifically, when examining the investment marginal costs when CCS technologies are the preferred technology, across the scenarios from 2030–2050 further tranches of offshore wind would be competitive with a cost improvement of £56–£260/kWe installed: this represents only 5–25 per cent of capital costs. Nuclear's marginal investment costs are even closer to CCS, from £2–£218/kWe installed, depending on the scenario and time period. Note that electricity system operation and wider energy system trade-offs will also influence the optimal uptake of these technologies.

Decarbonization of the power sector begins with the deployment of CCS for coal plants in 2020–2025 in all mitigation scenarios. When the CO_2 target is increased, nuclear and wind are selected alongside CCS. Note that in the most ambitious scenarios (especially 90 per cent reductions), nuclear, in one sense a 'zero-carbon' source, gains at the expense of CCS (a 'low carbon' source). Since the contribution of increasing levels of (offshore) wind to peak load is limited, the balanced low carbon portfolio of plants requires large amounts (20GW) of gas plants (CCGT) as reserve capacity. Imported electricity is also selected for reserve margins, with waste generation (landfill and sewage gas plants) contributing to the generation portfolio. Under stringent CO_2 reduction scenarios, zero carbon electricity is also provided by marine renewables.

In the LC-40 scenario, electricity decarbonization via CCS provides the bulk of a 40 per cent reduction in CO_2 by 2050. To get deeper cuts in emissions requires three things:

1   deeper decarbonization of the electricity sector with progressively larger deployments of low carbon sources;
2   increased energy efficiency and demand reductions particularly in the industrial and residential sectors;
3   changing transport technologies to zero carbon fuel and more efficient vintages.

Note that as emissions targets tighten, final energy use falls in 2050 from around 6,500PJ in REF to around 4,500PJ. Upon reaching this level decarbonization measures that do not reduce energy use continue to be implemented.

Decarbonization remains foremost in the power sector until the middle or end of the planning horizon, depending on the stringency of the target, then major efforts switch to the residential and transport sectors. The exception to this is in LC-90 where transport and residential sectors must be heavily decarbonized by 2035. By 2050, to meet the 80 per cent target in LC, power sector emissions are reduced by 93 per cent compared to REF. The reduction figures for the residential, transport, services and industrial sectors are 92, 78, 47 and 26 per cent respectively. Hence residual CO_2 emissions are concentrated in selected industrial sectors, and in transport modes (especially aviation).

In 2035, overall electricity generation declines (while decarbonizing) with target stringency owing to the role of early end-use efficiency and demand changes. By 2050, electricity generation increases in line with the successively tougher targets. This is because the electricity sector has highly important interactions with transport (plug-in vehicles) and buildings (boilers and heat pumps), as these end-use sectors contribute significantly to later period decarbonization. As a result, electricity demand rises in all scenarios, and is roughly 50 per cent higher than the REF level in 2050 in most of the 80 per cent reduction scenarios.

The shift to electricity use in the residential sector (from gas) combines with technology switching from boilers to heat pumps for space heating and hot water heating. The service sector is similarly decarbonized by shifting to electricity (along with biomass penetration in the most stringent scenarios). Natural gas, used with increasing efficiency, is still used in residential and service sectors for space heating and is a major contributor to remaining emissions.

The transport sector is decarbonized via a range of technology options but principally first by electricity (plug-in hybrid), and later by bio-fuel vehicles in more stringent scenarios (LC, LC-90). There is a trade-off between options to reduce energy service demands, efficiency to further reduce final energy and use of zero-carbon transport fuels. For example, bio-fuels in stringent reduction scenarios do not reduce energy demand as their efficiency is similar to petrol and diesel vehicles. Different modes adopt alternative technology solutions depending on the characteristics of the model. Cars (the dominant mode, consuming two thirds of the transport energy) utilize plug-in vehicles and then ethanol (E85). Buses switch to electric battery options. Goods vehicles (HGV and LGV) switch to biodiesel then hydrogen (only for HGV).

These scenarios do not produce decarbonization trajectories that are compatible with the EU's renewables directive requiring at least 15 per cent of UK final energy from renewables by 2020. Major contributions of bio-fuels in transport and offshore wind increases in electricity production only occur in later periods following tightening CO_2 targets and assumed technology improvements.

Besides efficiency, fuel switching and technology shifting, price-induced demand reduction also plays a major role in reducing CO_2 emissions by reducing energy service demands (5–25 per cent by scenario). Agriculture, industry, residential and shipping have higher demand reductions than those of air, car and HGV (heavy good vehicles) in transport sectors. This is driven both by the elasticities of demand in these sectors, and by the existence of alternative (lower cost) technological options. The significant energy service demand reductions (up to 25 per cent) in key industrial and buildings sectors imply employment and social policy consequences that merit further consideration.

Higher decarbonization levels (LC-40 to LC-60 to LC to LC-90) produce a deeper array of mitigation options (probably with more associated uncertainty). Hence these scenarios produce a very wide range of economic impacts, with CO_2 marginal costs in 2035 in the range £13–£133/tCO_2 and in 2050 in the range £20–£300/tCO_2. This convexity in costs, as targets tighten, illustrates the difficulty in meeting more stringent carbon reduction targets.

Welfare costs (sum of producer and consumer surplus) in 2050 associated with the scenarios are in the range £5–£52 billion. Moving from a 60 per cent to an 80 per cent reduction scenario almost doubles welfare costs (£20 billion to £39 billion). As noted earlier, welfare cost is a marked improvement on energy systems cost as an economic impact measure as it captures the lost utility from the reduced consumption of energy.

Overall, however, the scenarios follow similar pathways, with additional technologies and measures being required as targets become more stringent and costs rapidly increase.

Changing key scenario parameters

The results of the cumulative emissions scenarios show how the decarbonization pathways are affected by changing some of the key parameters. Giving the model freedom to choose the timing of reductions, as in LC-LC, with the same cumulative emissions as the early-action scenario LC-EA, illustrates intertemporal trade-offs in decarbonization pathways. Thus in LC-LC the model chooses to delay mitigation options, resulting in later action than in LC-EA, but a CO_2 reduction of up to 89 per cent in 2050. This results in very high marginal CO_2 costs in 2050, at £360/tCO_2 higher even than in LC-90.

Conversely, a cumulative constraint with a lowered (social) discount rate (LC-SO) gives more weight to later costs and hence causes earlier decarbonization – with CO_2 reductions of 39 per cent in 2020 and only 70 per cent in 2050. Similar to the early-action scenario LC-EA, this LC-SO focus on early action gives radically different technology and behavioural solutions. In particular, effort is placed on different sectors (transport instead of power), different resources (wind as early nuclear technologies are less cost competitive), and increased near-term demand reductions.

Within the LC-SO transport sector the broadest changes are seen with bio-fuel options not being commercialized in mid-periods. Instead the model relies on much increased diffusion of electric hybrid plug-in and hydrogen vehicles (with hydrogen generated from electrolysis). As hydrogen and electric vehicles dominate the transport mix by 2050, this has resultant impacts on the power sector with vehicles being recharged during periods of low demand (night time). Note that the selection of these highly efficient but high capital cost vehicles is strongly dependent on the assumptions of lowered discount and technology specific hurdle rates.

The inter-temporal trade-off extends to demand reductions where the LC-LC scenario with an emphasis on later action sees its greatest demand reductions in later periods. In LC-SO in contrast, energy service demand reductions in 2050 are much lower (except in the residential sector) as the model places more weight on late-period welfare losses. Earlier in the projection period, however, residential electricity and gas energy service demands in LC-SO are sharply reduced in preference to (relatively expensive) power sector decarbonization.

In terms of welfare costs, the flexibility in LC-LC gives lower cumulative costs than the LC-EA scenario (which has the same cumulative CO_2 emissions). As already noted, the fact that the LC-SO scenario produces the lowest costs is a reflection of the optimal solution under social levels of discounting (and correspondingly reduced technology-specific hurdle rates). The implication of the reduced hurdle rates is that consumer preferences change and/or government works to remove uncertainty, information gaps and other non-price barriers.

Application of energy demand policies to MARKAL-MED scenarios

Reductions in energy service demands in MARKAL-MED result from the rising marginal cost of abatement as carbon emissions are reduced towards the targets in 2050. This cost in 2050 is in the ranges £20–£300/tCO$_2$ as the target reductions increase from 40 to 90 per cent in LC-40 through to LC-90. In the scenarios with the same cumulative emissions and discount rates (LC-EA, LC-LC) the carbon marginal abatement costs in 2050 are £173 and £360/tCO$_2$ respectively, with the latter illustrating the extra cost incurred by delaying decarbonization (and therefore having to cut 2050 emissions by 89 per cent). In terms of total discounted energy system cost, this is a lower-cost scenario than LC-EA, the early action scenario. Not surprisingly, final energy demands decrease with the reduced energy service demands associated with rising carbon target reductions (i.e. through LC-40, LC-60 and LC), but are very similar for the scenarios with an 80 per cent reduction target and the same discount rate (LC, LC-EA, LC-LC)

In policy terms the implication of these scenarios is that these energy service demand reductions have been incentivized through a carbon tax or carbon rationing (and trading) scheme, the tax being applied at a rate, or the trading scheme delivering a carbon price, at the level of the marginal cost of CO$_2$ abatement in the model. For comparison, it may be noted that at 2009–2010 rates of the Climate Change Levy (0.47p/kWh for electricity, 0.16p/kWh for gas and 1.28p/kWh for coal), this amounts to an implicit carbon tax of £8.86/tCO$_2$ for gas, £8.77/tCO$_2$ for electricity and £5.21/tCO$_2$ for coal (see Annex 5.1 for the data underlying this calculation). Duty on road fuels in 2008 was about 50p/l. If this is all considered as an implicit carbon tax (i.e. ignoring any other environmental consequences of road travel which the duty may be considered to seek to account for), this amounts to about £208/tCO$_2$. This means that in the optimal market of the MARKAL model, rates of fuel duty would need to be about doubled in real terms by 2050, while other fuels would need taxes to have been imposed at about the 2008 fuel duty rate by the same date, in order for the targets to be met. While these tax increases seem large, they are actually a fairly modest annual tax increase if they were imposed as an annual escalator over 40 years.

In addition to reduced energy service demands from the price effect, MARKAL delivers reduced final energy demand through the increased uptake of conservation and efficiency measures (the former result in energy savings without using energy themselves, e.g. building insulation, while the latter cause appliances, for example, to use energy more efficiently). Except in the service sector, the increased uptake of conservation measures in these scenarios is taken from the UKDCM model described in Chapter 4, rather than computed directly by MARKAL. In the service sector, conservation measures save 151 and 172PJ in the LC and LC-90 scenarios respectively, compared to 64 and 135PJ in the REF and LC-40 scenarios respectively. The relatively high uptake of the measures in LC-40 indicates their cost effectiveness compared to other measures. Such savings would require strong and effective policy measures. It may be that the CRC Energy Efficiency Scheme, an emission trading scheme for large business and public sector organizations which started in 2010, will provide the necessary incentives for installing the conservation measures.

The uptake of efficiency technologies in buildings is again taken from UKDCM, with the major exception of space and water heating applications. One MED model example here is heat pumps, which play a major role in all the 80 per cent and 90 per cent carbon reduction scenarios, as seen in Figure 5.24. The present level of installation, and of consumer awareness, of heat pumps is very low indeed, and their installation in buildings is by no means straightforward. To reach the levels of uptake projected in these scenarios, where there is significant deployment of heat pumps from 2025, policies for awareness-raising and training for their installation need to begin soon.

Energy service demands in the transport sector in 2050 are not greatly reduced as the carbon targets become more stringent (falling from about 890 billion vehicle kilometres in REF to about 842bv-km in LC and 840bv-km in LC-90), but the energy demand required to meet those energy service demands falls by considerably more, from 2130PJ in REF to 1511PJ in LC (but 1656PJ in LC-90, due to its larger consumption of biodiesel and ethanol, as explained above). This means that the efficiency of fuel use has improved from 0.42v-km/MJ in REF to 0.56v-km/MJ in LC.

Even more dramatic, however, is the improvement in REF over the year 2000 efficiency, which was only 0.26v-km/MJ. This was due to the large take-up in REF of HGV diesel-biodiesel hybrids. This switch from HGV diesel/biodiesel to HGV diesel-biodiesel hybrids results in an efficiency improvement in 2050 from 0.08 to 0.14v-km/MJ. REF also sees a substantial take-up of LGV battery-electric vehicles (BEVs) and petrol plug-ins, as well as improving energy efficiency across the vehicle fleet (for example, the efficiency of diesel/biodiesel ICE cars, which are taken up in all the scenarios, improves from 0.37v-km/MJ in the year 2000 to 0.51v-km/MJ in REF in 2050).

The development of these new vehicle types, and of more efficient existing vehicle types, will be partly incentivized by the carbon price, but is also likely to require an intensification of energy efficiency policies, such as the EU requirements to improve vehicle efficiency, and demonstration and technology support policies to facilitate the penetration of the new vehicle types. Such policies will be even more necessary to incentivize the development and take-up of the petrol plug-in and E85 cars, and the hydrogen HGVs, which have an efficiency of 0.25v-km/MJ, nearly twice as efficient as the HGV diesel-biodiesel hybrids they largely replace, that make an appearance in 2050 in the most stringent carbon reduction scenarios, LC and LC-90.

Application of energy supply policies to MARKAL-MED scenarios

These scenarios reveal that the single most important policy priority is to incentivize the effective decarbonization of the electricity system, because low carbon electricity can then assist with the decarbonization of other sectors, especially the transport and household sectors. In all the scenarios, major low carbon electricity technologies are coal-CCS, nuclear and wind. All the low carbon scenarios have substantial quantities of each of these technologies by

2050, indicating that their costs are broadly comparable and that each of them is required for a low carbon energy future for the UK. The policy implications are clear: all these technologies should be developed.

The development of each of these technologies to the required extent will be far from easy. Most ambitious in terms of the model projections is probably coal-CCS, which is taken up strongly from 2020 to reach an installed capacity of 12GW by 2035 in LC-90 and 37GW in 2035 in LC-60 (as explained above, the residual emissions from coal-CCS are a limiting factor in the most stringent scenarios). At present, even the feasibility of coal-CCS has not yet been demonstrated at a commercial scale. There would seem to be few greater low carbon policy priorities than to get such demonstrations on the ground as soon as possible. The European Commission intends to stimulate the construction and operation by 2015 of up to 12 CCS demonstration plants, so that commercial CCS can be deployed from 2020 (as the MARKAL model currently assumes). The UK has put in place plans to build up to four commercial-scale CCS demonstration plants. Even so, the timescale for CCS deployment by 2020 is extremely tight, even if no large problems are uncovered during the demonstration process, which is by no means assured. The availability and uptake of CCS as projected by the scenarios are therefore optimistic.

The UK Government is not proposing to build new nuclear power stations itself, but believes that energy companies should do so, with appropriate public safeguards (BERR, 2008a). The Government is therefore proposing a number of measures to 'reduce the regulatory and planning risk associated with investing in nuclear power stations' (BERR, 2008a, p124), without planning either to invest in new nuclear power stations or to give subsidies to those who do. The Government acknowledges that it is uncertain whether these measures will actually bring forward proposals for new nuclear power stations, because this would be a private investment decision dependent on such issues as 'the underlying costs of new investments, expectations of future electricity, fuel and carbon prices, expected closures of existing power stations and the development time for new power stations' (BERR, 2008a, p129). These are all matters of considerable uncertainty. The scenarios envisage that only in LC-90 has very significant investment in new nuclear plant (30GW) taken place by 2035 (this would be equivalent to a new 3GW power station opening every year from about 2025), with 9GW projected in LC, and 4GW in LC-60 by that date. It is probable that the 2035 carbon prices in these scenarios (£37, £97 and £133/tCO$_2$ in LC-60, LC and LC-90 respectively) would provide the kind of price required for these investments, provided that the new generation of nuclear plant is economically and technically proven by about 2015. This cannot be taken for granted, but seems rather more likely than the very challenging timetable for CCS to make its projected contribution in the scenarios.

As discussed in Chapter 3, it is only in the third area of low carbon energy supply, renewables, that the UK Government has firm targets for deployment, in the form of the 15 per cent of final energy demand (probably requiring around 35 per cent of electricity) to come from renewables by 2020 in order to comply with the EU's overall 20 per cent target by that date. This amounts to a ten-fold increase in the share of renewables in UK final energy demand in 2006.

In the MARKAL-MED scenarios, only 15 per cent of electricity is generated from renewable sources by 2020, and this is by assumption (that the 2015 target set by the Renewables Obligation is met *and maintained*), otherwise the model would not choose renewable electricity to this level. More favourable cost assumptions for renewables, as discussed in Chapter 7, result in higher levels of deployment, as might be expected.

To date, Renewables Obligation (RO) targets have not been met – renewable generation (accounted against the RO) in 2007 was 4.9 per cent (BERR, 2008b, p29) against a target for 2007–2008 of 7.9 per cent,[7] a shortfall of 38 per cent. In 2008–2009 the target was 9.1 per cent, but the number of ROCs was still 35 per cent short of the target.[8] While the RO has recently been reviewed and technology 'banding' introduced in order to increase the incentive to install some technologies, the extent to which this will increase installation is still uncertain.

Even with 15 per cent renewable electricity in the MARKAL-MED scenarios, the maximum share of renewables in final energy demand (which also includes non-electricity energy consumed for transport and heat) in the scenarios is 5.77 per cent (in LC-SO), which is obviously well short of 15 per cent. There is therefore a very great policy challenge to increase the deployment of renewables over the next ten years. The UK Government's response to this challenge is the Renewable Energy Strategy (RES), launched in 2009, which envisages 12 per cent renewable heat and 10 per cent renewable transport fuels, as well as 30 per cent renewable electricity (DECC, 2009). The RES consultation document recognized that new policies would need to increase the share of renewables in final energy demand by a factor of three over what current policies (already considered ambitious at the time they were introduced) were designed to achieve (BERR, 2008b, p5).

For the UK the 'realizable mid-term' renewable potential to 2020 totals about 400TWh (IEA 2008, p67), of which the largest components are from onshore wind (28.5TWh), offshore wind (67TWh), biomass for electricity (20.7TWh) and heat (49.5TWh), biogas (16.3TWh), marine (58.9TWh, from tide and wave energy), bio-fuels (domestic, 25.4TWh), solar thermal (56.1TWh) and geothermal heat (53.7TWh). This amounts to about 21 per cent of the UK's projected final energy demand in 2020, so that nearly three-quarters (or about 280TWh) of this will need to exploited by 2020 if the UK is to meet its EU target of 15 per cent of renewables in final energy demand by that date.

The UK Renewable Energy Strategy (DECC, 2009) and Low Carbon Transition Plan (HMG, 2009) suggest a number of policies to seek to meet the 15 per cent renewable target, including the incentivization of renewable heat through a Renewable Heat Incentive, financial support for small-scale power technologies through Feed-in Tariffs, which were introduced in April

2010, reform of the planning system and ensuring grid access for new renewables, making full use of waste for energy and deploying bio-fuels in transport, as well as encouraging the development of electric vehicles. It is still too early to assess the likely success of these policies in delivering their targets, not least because not all of them have yet even been introduced. However, it is worth noting that the slow development of UK renewables to date, especially onshore wind, has been due to such issues as planning and grid access problems, rather than the level of remuneration, which is higher than in some other European countries that have achieved considerably greater deployment (IEA, 2008, p105). These 'non-economic' problems are not likely to be easy to resolve.

If the UK succeeds in meeting the 15 per cent EU renewables target, then it will be very well placed to exceed the renewables projections in the MARKAL scenarios. For example, renewable electricity in LC in 2050 is projected to be only 16 per cent of total electricity. However, if this share was already 30 per cent in 2020, then it is likely that this will at least have been maintained, potentially allowing 380TWh of renewable electricity to substitute for some other low carbon source, for example nuclear or coal-CCS. In LC-90 renewable electricity is 39 per cent of generation in 2050. If 30 per cent had already been achieved by 2020, this seems an eminently feasible projection. In short, while the 2020 EU renewables target is extremely challenging, if it could be achieved, it would make the later carbon reduction targets seem much less daunting.

The policy analysis here has focused on the scenarios with increasing carbon targets. The only areas in which the cumulative emissions scenarios (LC-EA, LC-LC, LC-SO) show a marked difference in technology choice are in respect of vehicle technology and biomass use. LC-SO in 2050 uniquely takes up petrol hybrid and battery cars, and prefers battery and hydrogen LGVs to LGV diesel/biodiesel plug-ins, so that its use of bio-fuels is very small, in contrast to LC-LC, which makes much more use than any other scenario except LC-90 of diesel/biodiesel ICE cars. LC-LC also uses a very large amount of biomass pellets for heating in the service sector, over twice as much as in LC, while LC-SO uses practically none.

Not too much should be read into these specific differences. Perhaps the most significant is that the lower discount and technology hurdle rates in LC-SO result in the uptake of technologies with higher upfront costs and more infrastructure requirements. The policy interpretation of this is government willingness to plan for and enable the deployment of the full range of available technologies in power generation, transport and buildings. Which of these become preferred will depend on how they develop over the next ten years or so. At this stage it is too early to choose between them. Perhaps the most important priority of government policy at present is to ensure that, to the extent possible, the options for the full range of technologies are kept open, so that they all have an opportunity to be commercialized if their development in the meanwhile singles them out as one of the more competitive low carbon technologies of the future.

Annex 5.1: Data for calculation of carbon tax implied by UK Climate Change Levy (CCL)

CCL rates, 2009–2010
Source: www.hmrc.gov.uk/budget2008/bn84.pdf
Electricity: £0.00470 per kWh
Gas: £0.00164 per kWh
Coal: £0.01281 per kg

Carbon emission factors

Energy source	Units	kg CO_2/unit
Grid electricity	kWh	0.537
Natural gas	kWh	0.185
Industrial coal	tonnes	2,457

Source: www.carbontrust.co.uk/resource/conversion_factors/default.htm

Notes

1 Some potential industrial emission reductions measures, notably enhanced energy conservation and CCS from industrial facilities, are not included in the MED model.
2 In these scenarios only nuclear, imported and renewable electricity is counted as primary energy, not the equivalent heat content.
3 These primary energy reductions would be moderated if primary energy in nuclear resources was calculated on a heat content basis (as in some energy statistics publications (e.g. DUKES, 2009)).
4 See IPPR/WWF (2007) for a discussion on sustainable global biomass trade.
5 Because of its intermittent nature, wind capacity needs to be supplemented by extra plant of more reliable output to ensure that peak loads can be met as they arise. Therefore in the scenarios with a high proportion of renewable electricity, overall capacity increases relatively more than generation.
6 This is not the same as the loss of GDP, because GDP is not a welfare measure and because no account is taken of intersectoral interactions and such issues as the macroeconomic impact of wider investment, government spending or trade flows.
7 See Ofgem Press Release 'The Renewables Obligation Buy-Out Fund (2007–2008)', October 7, 2008, www.ofgem.gov.uk/Pages/MoreInformation.aspx?docid=210&refer=Media/PressRel
8 The Obligation also changed from a given percentage of generation to a certain number of ROCs per MWh supplied. See www.ofgem.gov.uk/Sustainability/Environment/RenewablObl/Documents1/Annual%20Report%202008-09.pdf.

References

Anandarajah, G., Strachan, N., Ekins, P., Kannan, R. and Hughes, N. (2009) 'Pathways to a Low-Carbon Economy: Energy systems modelling', UKERC Energy 2050 Research Report 1, Ref. UKERC/RR/ESM/2009/001, UK Energy Research Centre, London, www.ukerc.ac.uk/Downloads/PDF/U/ UKERCEnergy2050/230409UKERC2050CarbonPathwayReviewed.pdf, accessed 8 September 2010

BERR (Department of Business Enterprise and Regulatory Reform) (2008a) *Meeting the Energy Challenge: A White Paper on Nuclear Power*, Cm 7296, Department of Business Enterprise and Regulatory Reform, London

BERR (Department of Business Enterprise and Regulatory Reform) (2008b) *UK Renewable Energy Strategy: Consultation*, Department of Business Enterprise and Regulatory Reform, London

CCC (2008) *Building a low-carbon economy – the UK's contribution to tackling climate change*, UK Committee on Climate Change, London, www. theccc.org.uk

DECC (Department of Energy and Climate Change) (2009) *The UK Renewable Energy Strategy*, Cm7686, The Stationery Office, London, www.decc.gov.uk/ en/content/cms/what_we_do/uk_supply/energy_mix/renewable/res/res.aspx

DUKES (Digest of UK Energy Statistics) (2009) Department of Energy and Climate Change, London

HMG (Her Majesty's Government) (2009) *The UK Low Carbon Transition Plan*, The Stationery Office, Norwich

HMT (Her Majesty's Treasury) (2006) *The Green Book: Appraisal and Evaluation in Central Government*, HM Treasury, London, http://greenbook.treasury.gov.uk/annex06.htm

IEA (International Energy Agency) (2008) *Deploying Renewables: Principles for Effective Policies*, International Energy Agency, Paris

IPPR/WWF (2007) *2050 Vision: How can the UK play its part in avoiding dangerous climate change?*, Institute for Public Policy Research/World Wildlife Foundation UK, London, www.ippr.org.uk/publicationsandreports/ publication.asp?id=572

6
A Resilient Energy System

Jim Skea, Modassar Chaudry, Paul Ekins,
Kannan Ramachandran, Anser Shakoor and Xinxin Wang

Introduction

This chapter returns to the theme of energy system resilience. Chapter 4 defined energy system resilience and noted the types of event or 'shock' which a resilient energy system might be expected to tolerate more effectively. Here, energy system shocks and possible responses are investigated more thoroughly. The *Resilient* and *Low Carbon Resilient* scenarios which are part of the core scenario set are also defined more fully.

The chapter begins with a historical review of shocks and disturbances to energy systems over the last 10–20 years. Building on this analysis, and the policy perspectives on energy security set out in Chapter 3, a small number of 'indicators' of energy system resilience are selected from a long list of possibilities. Some of the indicators refer to the energy economy at the macro level, while others refer to technical aspects of energy infrastructure. Quantitative values are assigned to the macro indicators at this stage to define the *Resilient* and *Low Carbon Resilient* scenarios. The MARKAL-MED model is used to explore the market and technological implications of the two new core scenarios and systematically compare them with the *Reference* and *Low Carbon* scenarios.

The attention then turns to resilience indicators for the energy network industries. These rely on well-established assessment approaches in the electricity and gas sectors and are based on statistical approaches to reliability of supply. The additional levels of investment needed to maintain reliability, especially in a world where intermittent renewable energy plays a larger role, are identified.

The next section of the chapter looks at the capacity of the UK energy system to respond to set of hypothetical shocks under each of the core scenarios. A set of specific possible events, each affecting gas supply, is defined. Using the CGEN model, the impact of these shocks is then assessed in terms of the additional costs associated with re-balancing the energy system to maintain supplies to consumers (e.g. by re-dispatching power stations) and the imputed costs associated with 'lost load' if supply curtailments prove necessary. This analysis has considerable similarities with the 'stress tests' conducted by the UK regulator, Ofgem, through its Project Discovery (Ofgem, 2009).

The final stage is to test a set of measures that would help to mitigate the impacts of the shocks. These measures take the form of investments in physical assets such as gas storage, liquefied natural gas (LNG) terminals or interconnectors additional to those needed to meet the reliability indicators. Without assigning probabilities to the events, this analysis cannot be framed in formal cost–benefit terms. However, if the mitigating investments are regarded as insurance against the shocks taking place, it is possible to calculate how often the shocks would have to recur (the 'return period') before investments can be justified. Note that the 'return period' defined in this way does not refer to the financial rate of return on capital. The chapter concludes with an assessment of the overall cost of building resilience into the energy system.

What can go wrong: shocks to the energy system

Many of the fears around energy security in the UK relate to adequacy of investment in electricity generating capacity (Chapter 2) and fears of politically motivated interruptions to supplies of oil and gas (Wicks, 2009). This section reviews historical experience with supply interruptions and energy system failures in various parts of the world over the last 10–20 years. Disturbances to the electricity system, and to oil and gas supplies, are considered separately.

Two general findings emerge from the review. First, many of the 'shocks' to the gas and electricity systems relate to equipment failures or weather-related events, rather than politically motivated or other deliberate interventions. Second, the duration of impacts differs according to which part of the energy system is affected. Electricity shocks have tended to last for hours or days, gas shocks for weeks or months, and oil shocks for months to years in some cases.

Electricity

Table 6.1 shows a selection of major electricity blackouts and disturbances over the last 20 years. Electricity supply accidents have occurred more frequently in the last ten years, partly reflecting rising energy demand and hot summer conditions.

Electricity blackouts have lasted for several hours up to several days. Compared to oil and gas shocks, electricity shocks have happened more frequently but the durations have been shorter. Almost all of the blackouts were caused by extreme weather and/or the failure of major infrastructure assets.

Gas

Table 6.2 shows that there have been no massive (i.e. global) gas supply disruptions in the last 20 years, although the recent curtailment of supplies from Russia to Ukraine had implications on a continental scale. Most disruptions occurred in specific locations. Some can be attributed to extreme weather events, such as hurricanes in the US in late 2005. Others were caused by weak

Table 6.1 *Major electricity blackouts and disturbances*

Date	Place	Duration	Cause	Loss
23–25 July 2007	Barcelona	More than 3 days	The blackout began when a broken substation cable caused a chain reaction failure in other substations	About 350,000 households and businesses lost power at some point. More than 100,000 had no electricity on the night of 25 July
16–29 July 2006	Queens, New York	About 10 days	Intense heat wave	More than 3 million Americans were affected, some for hours, others for up to ten days
23–24 October 2005	South Florida, Naples, Fort Myers, Miami, Fort Lauderdale, West Palm Beach and Martin county	1 day	Hurricane Wilma	10,000MW capacity loss, affecting over 3 million people
12 July 2004	Athens and southern Greece	70% of the region had no power for an hour. Power was restored to all of Athens in 3 hours. Remote areas were affected for longer	Many planned new upgrades were not integrated in the system until after the yearly peak, which occurred on Monday, 12 July, when high demand due to a heatwave led to a cascading failure	Several million people were affected
28 Sept. 2003	All of Italy with the exception of the island of Sardinia	9 hours	The problem was attributed to a fault on the Swiss power system, which caused the overloading of two Swiss internal lines close to the Italian border	Almost all of the country's 57 million people were affected
23 Sept. 2003	Denmark and southern Sweden	Half a day	The power failure occurred as a consequence of a number of faults in the Swedish power system	4 million businesses and homes were affected. A total loss to the grid of 3000MW or about 20% of Sweden's electricity capacity at the time
14 Aug. 2003	The states of Ohio, Michigan, Pennsylvania, New York, Vermont, Massachusetts, Connecticut, New Jersey and the Canadian province of Ontario	Almost one week	Deficiencies in specific practices, equipment, and human decisions by various organizations	An estimated 50 million people and 61,800MW of capacity were affected
10 Aug. 1996	8 western US states	About 10 hours	A major transmission line failed due to a period of high temperatures and high demand for electricity	4 million people were affected

Sources: BBC (2007); Berizzi, A. (2004); BreakingNews (2007); EIA (2006); Elkraft System (2003); Freeman, M. J. (2006); Northwest Power and Conservation Council (1996); US–Canada Power System Cutage Task Force (2004); Vournas, C (2004); World Energy Council (2003)

Table 6.2 *Gas supply crises and accidents*

Date	Place	Duration	Cause	Loss
Several occasions in 2002, 2006 and January 2009	Ukraine	1 to several days	Russian companies cut off natural gas supplies to the Ukraine and Georgia to force payment of debts	The gas supplies in many European countries which depend on Russian natural gas were threatened or curtailed
16 February 2006	The *Bravo* rig, in Centrica's Rough field, UK	About 4 months	There was a failure of a cooler unit in one of four dehydration units and an explosion occurred in that vicinity	Two people were injured. The whole storage facility was out of service for two months, resulting in higher gas prices for a period
September–December 2005	US	4 months	Hurricanes	US gas production fell by 10% during the last 4 months of 2005
17 and 18 June 2003	Bacton, UK	2 days	On 17 June, in order to address a supply deficit, National Grid Transco (NGT) started a number of localized system balancing actions which did not bring a sufficient physical response. On 17 and 18 June, NGT had to interrupt flows to the Belgian Interconnector and a number of loads on the National Transmission System (NTS) and Local Distribution Zones (LDZs).	In total, National Grid Transco (NGT) interrupted 10.5mcm of NTS loads on 17 June and 11mcm of NTS and LDZ loads on 18 June. NGT restored all NTS and LDZ loads by 1800 hours on 18 June
15 December 1999	Easington terminal, UK		Easington terminal was struck by lightning, limiting the operation of the Rough subterminal.	This reduction in flows contributed to a sharp increase in system average price which by 20 December 1999 had risen to 2.17p/kWh (63.6p/therm);
25 September–14 October 1998	Victoria, Australia	19 days	A vessel in a gas plant fractured, releasing hydrocarbon vapours and liquid	The commercial/industry cost was AUS $1.3bn. Four million people were affected and two people were killed

Sources: AGD (1998); Centrica (2006); IEA (2006); Ofgem (2003); Ofgem (2004);

or ageing gas transmission infrastructure. For example, the gas supply crisis in Australia in 1998 was caused by a fractured vessel in a gas plant (AGD, 1998). There have been three events affecting major UK gas facilities, Bacton, Easington and Rough, in the last decade.

Some gas supply disruptions have arisen for political reasons. The most obvious one in recent years has been the Russia–Ukraine gas supply crisis. Russia cut off natural gas supplies to Ukraine in 2002, 2006 and 2009 in an attempt to increase gas prices and force the payment of debts.

Depending on the reasons for disruption, gas supply disruptions have generally lasted from several days up to a couple of weeks. It took between one and four days to reach an agreement to resolve the various Russia/Ukraine gas crises. However, US gas supply disruptions attributable to hurricanes lasted for four months in 2005.

Oil

The IEA (2007) has identified ten major world oil supply disruptions in the last 50 years. There have been four major disruptions in the last decade. Before that, there had been disruptions roughly once in every ten years.

Among the major disruptions, only one is attributable to severe weather. This was also the shortest disruption, lasting one month. All the others were caused by economic disruptions, political developments at the national/international level, and/or wars in the Middle East. On every occasion, these lasted more than two months. Five out of the ten disruptions lasted for more than six months.

If both the duration of the disruption and the proportion of global production involved are taken into account, the most significant disruption in the last 50 years was the Suez crisis in 1956–1957. The loss in peak supply was about 11.4 per cent of global crude oil production (Table 6.3). Figure 6.1 shows that most disruptions were due to events in the Middle East. The largest loss of supply was associated with the 1979 Iranian revolution. At a national level, shortages of supply have occurred due to industrial action affecting the refining

Table 6.3 Global crude oil production and peak supply loss in each disruption

	1957	1967	1974	1979	1981	1991	2001	2003	2003	2005
Crude oil production (m bbl/d)	17.6	37.1	59.0	66.6	60.6	66.9	77.1	79.5	79.5	84.2
Gross peak supply loss (m bbl/d)	2.0	2.0	4.3	5.6	4.1	4.3	2.1	2.6	2.3	1.5
Ratio %	11.4	5.4	7.3	8.4	6.8	6.4	2.7	3.3	2.9	1.8

Note: m bbl/d = million barrels of oil per day

Source: Earth Policy Institute (2009) and IEA (2007)

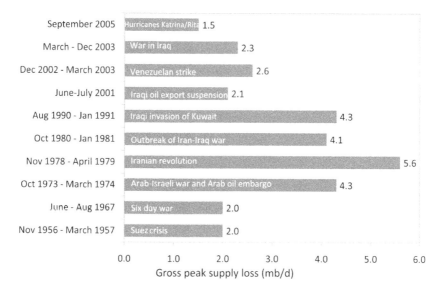

Source: IEA (2007)

Figure 6.1 *Major world oil supply disruptions over the last 50 years*

and distribution of oil products, for example in the UK in 2000 in protest at the level of fuel duty and in France in 2010 in protest at the proposed changes in pension arrangements.

Indicators of resilience

This section turns to indicators that could be associated with energy systems better adapted to meeting the types of shock identified above. The candidate indicators cover three main areas of vulnerability: primary energy supply, energy infrastructure and energy usage. They have been identified by considering the practical concerns of policy-makers discussed in Chapter 3, the theoretical insights into resilience from Chapter 4, and the types of shock that have occurred in reality. Tables 6.4–6.6 note the type of external disturbance to which the resilience indicators apply and, referring back to Chapter 4, the nature of the incertitude (risk, uncertainty or ignorance) which is relevant.

Table 6.4, covering primary energy supply, includes some possible indicators relating to the energy mix. These include novel indicators based on recent work deriving from financial portfolio theory and diversity theory (Bazilian and Roques, 2008) as well as simpler and more conventional indicators such as import dependence and the degree of dependence on the largest single source.

Table 6.5, addressing infrastructure issues, focuses heavily on adequacy of capacity which is central to the UK Government's current approach. It includes standard measures used in statistical assessments of system reliability (loss-of-load

Table 6.4 *Possible resilience indicators for primary energy supply*

Indicator	Type of event	Type of incertitude	Notes
Import dependence	Supply interruption	Uncertainty Ignorance	Could be for the whole economy or specific sectors
Largest single source of supply	Supply interruption	Uncertainty Ignorance	Could be for the whole economy or specific sectors
Diversity/concentration of energy supply, e.g. Herfindahl-Hirschman Index (HHI)	Supply interruption Price shocks	Uncertainty Ignorance	HHI is used by the Office of Fair Trading to assess market concentration
Energy portfolios	Price volatility	Risk	Energy mixes which are efficient in terms of striking a balance between volatility and cost have been explored using financial portfolio theory (Bazilian and Roques, 2008). Could be for the economy as a whole or specific sectors

Table 6.5 *Possible resilience indicators for energy infrastructure*

Indicator	Type of event	Type of incertitude	Notes
Statistical probability of supply interruption in network industries (gas and electricity)	Technical equipment failure Weather-related risks Inadequacy of investment	Risk	Specify as a security standard (i.e. a probability) to allow for the greater use of intermittent supply sources driving up capacity margins
Expected number of annual hours in which energy is unserved	As above	Risk	
Value and/or level of unserved energy	As above	Risk	
Energy storage capacity and/or stocks by fuel and market	Interruption of supply	Uncertainty	Could be measured in hours, days or weeks
Largest single source of supply in an energy market	Interruption of supply	Uncertainty	Expressed as a percentage of supply in any given market
Redundancy in network architecture	Attack on infrastructure Industrial action	Ignorance	The number of pinch points and critical nodes which would need to fail to cause interruption of supply to consumers. The networks could include gas, electricity and distribution of refined oil products

probability; loss-of-load expectation; and value of lost load) but also storage capacity and indicators relating to redundancy in network infrastructure.

Table 6.6 looks at demand side indicators. Most of these address the amount of energy used or financial exposure to energy dependence. They also cover the availability of back-up arrangements or alternative sources of supply for energy sensitive users, e.g. hospitals and banks.

The challenge is to identify a usable set of indicators from the wide range identified in Tables 6.4–6.6 that can be practically applied in energy analysis and, ultimately, in energy policy. For the analysis in this report, indicators that can be easily applied in the available modelling tools are needed. Taking into account these criteria, the small set of selected resilience indicators is as follows. There are three macro-level indicators.

Level of final energy demand

This is used as an operational indicator in EU energy policy (e.g. the Energy Efficiency and Energy Services Directive) serving as a proxy for import dependence and expenditure on energy. This is the easiest demand-related indicator to implement in the models.

Table 6.6 *Possible resilience indicators for energy users*

Indicator	Type of event	Type of incertitude	Notes
Energy demand level	Supply interruption Price shocks	Risk Uncertainty	Often taken as a proxy for exposure to supply interruptions and price shocks in policy formulation. This could be applied to specific sectors, e.g. residential housing, critical transport or types of business
Energy intensity	Supply interruption	Uncertainty	kWh/£ for industry (output) and economy as a whole (GDP). kWh per household in the domestic sector. A lower degree of energy intensity would suggest that an individual sector, or the economy as a whole, would be less vulnerable to supply interruptions
Energy costs	Price volatility Price shocks	Risk Uncertainty	% energy expenditure as proportion of output, expenditure and GDP for industry, households and economy as a whole respectively. This indicator points to the broader economic impact of price shocks
Back-up arrangements for energy sensitive users, e.g. hospitals, banks	Supply interruption	Uncertainty	

Diversity of primary energy supply

This is a constraint on the maximum market share of major supply sources, e.g. coal, oil or gas. It was selected because it is simple to implement in the models and, at an intuitive level, matches well with policy aspirations.

Diversity of generation mix in electricity supply

This is a key indicator because, as shown later in the chapter, this sector turns out to offer some of the lowest cost options for enhancing diversity. A constraint on the maximum market share of major generation types (coal, gas, nuclear) is applied.

There is then a set of indicators relating to reliability in the network industries.

Reliability indicators for electricity

Two indicators are used: value of lost load (VOLL) expressed in £/MWh and loss-of-load expectation (LOLE) expressed in hours per year.

Reliability indicators for gas

Value of lost load (VOLL) is used for industrial gas. Loss of load is not allowed in the household sector, i.e. the assigned value of lost load is implicitly infinite.

Infrastructure investment

The implications of different levels of gas storage and diversity of import options are assessed.

The indicator set is intended to be coherent and not lead to perverse outcomes. A comprehensive set of sensitivity tests, described in Chaudry et al (2010), was conducted using the MARKAL-MED model to ensure that the final definition of the *Resilient* scenario was plausible and consistent.

Quantifying resilience at the macro level

The quantified assumptions made for the macro-level resilience indicators in the *Resilient* core scenario are shown in Table 6.7. The remainder of this section describes how these top-level indicators were derived.

Table 6.7 *Macro-level resilience indicators*

Indicator	Quantified assumption
Final energy demand	Final energy demand falls 3.2% pa relative to GDP from 2010 onwards
Primary energy supply	No single energy source (e.g., gas) accounts for more than 40% of the primary energy mix from 2015 onwards
Electricity generation mix	No single type of electricity generation (e.g., gas, nuclear) accounts for more than 40% of the mix from 2015 onwards

Final energy demand

Reducing energy demand is a key element of, for example, EU energy security strategy. It will reduce vulnerability to all types of insecurity – physical, price and geopolitical. The 3.2 per cent decoupling of final energy demand from GDP that was finally assumed for the *Resilient* scenario is equivalent to an annual reduction of 1.2 per cent in absolute terms (assuming GDP growth of 2 per cent p.a.). Applying this rate of reduction, final energy demand in the *Resilient* scenario is 12 per cent lower than in the *Reference* scenario by 2020, 17 per cent by 2025 and 41 per cent by 2050 (see Table 6.8, from which it can be noted that energy intensity even in the Reference scenario falls by nearly 2% per annum through the projection period after 2015). The demand reduction is implemented in MARKAL-MED by applying a constraint directly on final energy demand. The model determines a portfolio of approaches – energy conservation, fuel switching and price-induced demand reduction – to ensure that the constraint is met.

The 3.2 per cent GDP/final energy de-coupling was based on a modelling sensitivity analysis and a bottom-up assessment of the impact of policy interventions. An ambitious constraint on final energy demand which, at the same time, lay within the bounds of technical and economic feasibility, was sought. Sensitivity tests were conducted using MARKAL-MED with increasingly more ambitious constraints on final energy demand. This tested feasibility from the modelling perspective. These were then compared with the range of uncertainty associated with the delivery of energy efficiency embodied in the UK's Energy Efficiency Action Plan (DEFRA, 2007), as subsequently updated and quantified in the 2008 Updated Energy and Carbon Projections (DECC, 2008). It was also compared with the level of ambition for energy efficiency in the 2008 EU Climate Policy package.

The conclusions of this analysis, described fully in Chaudry et al (2010), are that, out to 2020, the assumptions about final energy demand in the *Resilient* scenario are compatible with DECC assumptions about the 'high' effectiveness of energy efficiency policies put in place after the 2007 Energy White Paper. It is difficult to make a direct comparison between the *Resilient*

Table 6.8 *Final energy demand constraints*

Year	2000	2005	2010	2015	2020	2025	2030	2040	2050
GDP index	100.0	112.9	129.3	143.5	158.4	174.9	193.1	235.4	286.9
Final energy (PJ)									
Reference	6189	6321	6300	6260	6288	6287	6312	6401	6455
Resilient	6189	6318	6291	5933	5567	5224	4902	4316	3801
Final energy/ GDP (% pa)									
Reference %		−1.99	−2.74	−2.18	−1.87	−1.96	−1.88	−1.86	−1.92
Resilient %		−1.99	−2.76	−3.20	−3.20	−3.20	−3.20	−3.20	−3.20

scenario and the EU Climate Policy package because they use a different baseline from which to measure energy demand reductions. However, the levels of ambition are broadly compatible. Therefore the *Resilient* scenario can be interpreted as meaning that ambitious policies, with a relatively high impact, are implemented through to 2020 and that the momentum of energy efficiency policy is maintained afterwards.

Diversity of supply

Diversity constraints relating to primary energy supply, the electricity generation mix and the mix of installed electricity generating capacity have been explored. This exploration is described fully in Chaudry et al (2010). The diversity constraints were formulated in terms of maximum market shares because this is a simple and intuitive characterization and also because such constraints are easily implemented in the MARKAL-MED model. Only a non-linear model could deal with diversity indices such as the Herfindahl-Hirschman Index mentioned in Table 6.4. In each case, a 40 per cent constraint on the market share of the largest source was explored, and interactions with different levels of constraint on final energy demand were tested. The 40 per cent figure was intended to prevent any single energy source from dominating the market. The way that different combinations of constraint affected welfare losses as imputed by the MARKAL-MED model was a key consideration.

The final, and relatively simple, selection of diversity constraints was made for the following reasons

- The constraint on the maximum share of primary energy supply ensured supply diversity in the economy as a whole.
- Generation mix was constrained because the electricity sector was found to play a key role in shifting the primary energy mix. The availability of alternative generating options at similar costs means that diversity can be achieved at a relatively low cost in the electricity sector. The generation mix constraint also helps with security of electricity supply.
- Constraints on installed electricity generation capacity were found to produce perverse outcomes. These drove investment in low capital cost plant (specifically CCGT) which was subsequently used at a low load factor and failed to prevent high market shares (60 per cent and more) for other forms of generation. Constraints on installed capacity, if combined with constraints on generation mix, did not produce a substantively different result from applying a constraint on generation alone.

Resilience: implications for energy markets and technologies

This section describes the consequences of constraining the development of the energy system using the quantified macro-level resilience indicators. The two 'resilient' scenarios – *Resilient* and *Low Carbon Resilient* – are systematically compared with the *Reference* and *Low Carbon* scenarios respectively. In general, the resilience constraints produce quite different patterns of development, both when 80 per cent CO_2 reductions by 2050 are required, and when no carbon constraint is applied. The comparison focuses on 2025 representing the mid-term and 2050 representing the long term. The various 'shocks' that are hypothesized for the energy system later in the chapter are assumed to occur in 2025.

The discussion is grouped around different themes: primary energy demand; final energy demand; electricity supply; CO_2 emissions; marginal CO_2 abatement costs; and demand reduction and welfare costs. However, the discussion also flags important cross-sectoral linkages reflected in overall system change.

Primary energy demand

Figure 6.2 compares how primary energy demand is met in each of the core scenarios. In the *Resilient* scenario, there is a steep reduction in primary energy demand driven by the constraint on final energy. Compared to the *Reference* scenario, primary energy demand is 14 per cent lower in 2025 and 42 per cent lower in 2050. This reduction is achieved by a combination of fuel switching and energy efficiency improvements both on the supply and demand sides.

In 2025, coal demand is 23 per cent lower, gas 13 per cent lower and oil 10 per cent lower. Reduced coal demand reflects a switch from coal to gas in the power sector. However, overall gas demand is down because the residential sector uses 33 per cent less gas. The reduction in residential gas demand is driven by a switch from gas-fired central heating to heat pumps and price-induced demand reduction; the latter is very substantial. Since both coal and gas demand fall, oil demand also has to decline to meet the supply diversity constraint, thereby inducing changes in the transport sector.

In 2050, coal dominates primary energy supply in the *Reference* scenario with a market share of 38 per cent, because of its use in power generation. In the *Resilient* scenario, coal's share declines to 28 per cent and gas has the largest share with 31 per cent. The change in the primary fuel mix is driven mainly by a switch from coal to nuclear in the power sector in order to meet the diversity constraint. Biomass supply is higher in the *Resilient* scenario and meets 6 per cent of primary energy demand, compared to 3 per cent in the *Reference* scenario. However, renewable electricity supply is lower, only just meeting the Renewables Obligation in the power sector. Instead, imported electricity (which was not constrained) is almost three times the level in the *Reference* scenario.

The pattern of primary energy demand in the *Low Carbon Resilient* scenario is similar to that in the *Resilient* scenario. In 2025, demand is 14 per cent lower than in the *Low Carbon* scenario. In terms of its fuel mix, gas demand is 16 per cent lower but coal demand is 35 per cent higher, because of its greater use in power generation (see below). This is because a substantial decline in gas demand (of over 44 per cent) is required to meet the supply diversity constraint.

Primary energy demand in 2050 in the *Low Carbon Resilient* scenario is lower than in the *Low Carbon* scenario. Coal demand falls almost two-thirds. Its market share declines to 14 per cent in the *Low Carbon Resilient* scenario, compared to 32 per cent in the *Low Carbon* scenario, primarily driven by changes in the industrial and power sectors. In the power sector, generation

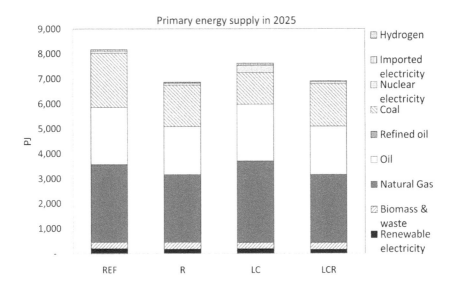

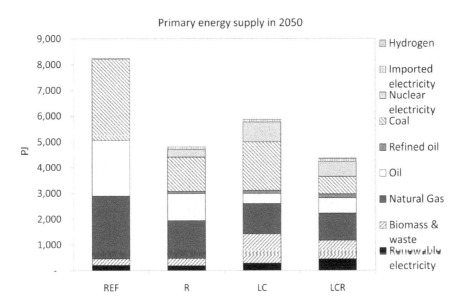

Figure 6.2 *Primary energy supply in the core scenarios*

from coal CCS shrinks by 65 per cent. The share of oil (mainly for transport fuel) in primary energy demand doubles. This is because the transport sector deploys hybrid vehicles rather than the ethanol and plug-in hybrid vehicles seen in the *Low Carbon* scenario. This results in a relatively low use of bio-energy. Overall, the combination of supply diversity, demand reduction and the shift in technology pathways significantly alters the decarbonization pathway.

Final energy demand

Final energy demand in the *Resilient* scenario is 17 per cent lower than in the *Reference* scenario in 2025 (Figures 6.3 and 6.4). Demand falls across all end-use sectors. Residential demand is 27 per cent, industry and service sectors 17 per cent, and transport 7 per cent lower than in the *Reference* scenario. The lower reduction in the transport sector is because of the availability of a wide range of alternative fuel and vehicle technologies (e.g. hydrogen, ethanol, hybrid and plug-in hybrid) that have a lower impact on welfare than demand reduction. There are more limited opportunities for fuel switching in other end use sectors. In 2050, final energy demand is 41 per cent lower than in the *Reference* scenario.

In the *Low Carbon Resilient* scenario, final energy demand is 14 per cent lower than in the *Low Carbon* scenario in 2025. The reduction in residential energy demand is substantial (23 per cent), and is partly delivered by switching from gas-fired central heating to more efficient heat pumps. This enables a large reduction of gas use in the energy system (25 per cent) and thereby lower CO_2 emissions. These lower CO_2 emissions allow other sectors to use more carbon-intensive fuels. For example, electricity generation from coal (without CCS) contributes 23 per cent of total electricity generation in the *Low Carbon Resilient* scenario, whereas the *Low Carbon* scenario has coal only with CCS.

In 2050, final energy demand in the *Low Carbon Resilient* scenario is 15 per cent lower than in the *Low Carbon* scenario. Unlike in the *Low Carbon* scenario, the residential sector continues to use gas because the diversity criterion prevents heat pumps from taking 100 per cent market share. Therefore, gas demand in the residential sector accounts for about 10 per cent of final energy demand, whereas there is no residential gas in the *Low Carbon* scenario. Since residential energy demand is higher than in the *Low Carbon* scenario (due to the restriction on more efficient heat pumps), the model has to work hard to meet the final energy demand constraint. The residential use of gas also means that the power sector has to decarbonize to a greater extent than in the *Low Carbon* scenario.

The transport sector

By 2025, there have been no significant changes in the transport sector. Although total final energy demand declines by 14 per cent in the *Resilient* scenario compared to the *Reference* scenario, transport energy demand declines by around 7 per cent. There are no significant changes in vehicle technology or in the fuel mix.

By 2050, transport fuel demand almost halves in the *Resilient* scenario with respect to the *Reference* scenario (Figure 6.5). This is largely induced by higher prices (Figure 6.6). However, electricity demand more than doubles because car fleets switch from conventional internal combustion engines (ICE) to plug-in hybrid vehicles (Figure 6.7). This change in the car fleet also reduces the demand for bio-fuels. The remaining demand for bio-fuels is forced by the renewable transport fuel obligation (RTFO). The reduction in electricity demand in other end-use sectors exceeds the increase in electricity demand from the transport sector, and therefore total electricity demand declines by 20 per cent (Figure 6.3).

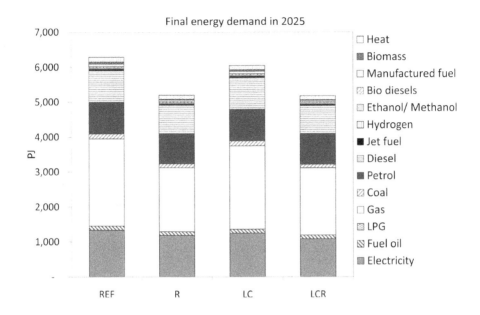

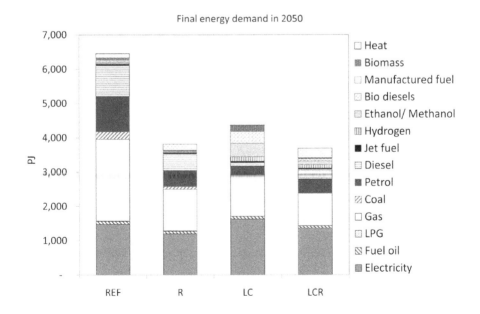

Figure 6.3 *Final energy demand in the core scenarios by fuel type*

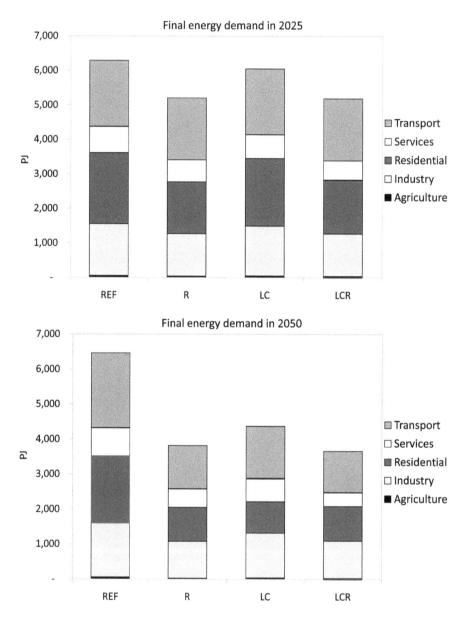

Figure 6.4 *Final energy demand in the core scenarios by sector*

Comparing the *Low Carbon Resilient* and *Low Carbon* scenarios in 2050, total transport fuel demand declines by 22 per cent (Figure 6.5), partly because of price effects. As noted above, unlike other end-use sectors, the transport sector enjoys a wide range of technology and fuel choice. Therefore, the price-induced level of demand reduction is lower than in other sectors, although this varies with vehicle

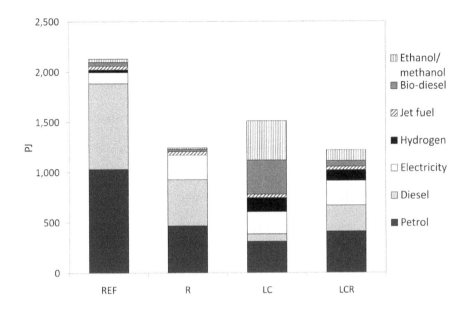

Figure 6.5 *Transport fuel demand in the core scenarios (2050)*

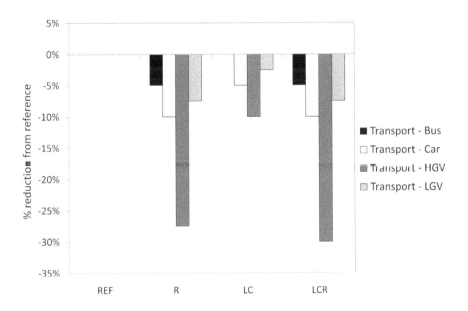

Figure 6.6 *Demand price response in the transport sector (2050)*

technology (Figure 6.6). Transport energy demand also falls because the transport sector uses efficient hybrid vehicles in the *Low Carbon Resilient* scenario compared to the ethanol ICE vehicles seen in the *Low Carbon* scenario (Figure 6.7, see Table 4.8 in Chapter 4 for a description of vehicle technologies).

In the *Low Carbon* scenario, the car fleet relies heavily on ICE (internal combustion engine) ethanol and plug-in hybrid engines while buses and goods vehicles use biodiesel. Bio-fuel ICE vehicles are relatively inefficient, and therefore more efficient petrol-diesel hybrid cars are deployed in the *Low Carbon Resilient* scenario. Although this enables a lower level of transport fuel demand overall, petrol and diesel demand increase by 32 per cent and 172 per cent in 2050 with respect to the *Low Carbon* scenario. Bio-fuel demand (ethanol and biodiesel) declines by three-quarters.

Changes in technology choice result in lower transport fuel demand but higher CO_2 emissions in the *Low Carbon Resilient* scenario compared to *Low Carbon*, indicating a trade-off between final energy and CO_2 constraints.

The residential sector

Up to 2025, imposing the resilience constraints does not lead to significant changes in the residential sector because of the existing stock of heating technologies. However, the residential sector plays a key role in the longer term (Figure 6.8). In the *Low Carbon* scenario, the residential sector contributes to CO_2 emission reduction through the deployment of heat pumps using decarbonized electricity (Figure 6.9). However, in the *Resilient* and *Low Carbon*

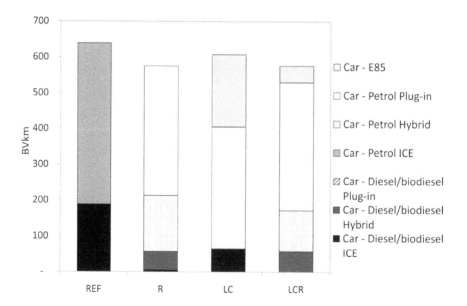

Figure 6.7 *Car fleets in the core scenarios (2050)*

Resilient scenarios, heat pumps are prevented from achieving market dominance, and the sector continues to use gas. Consequently, the level of price-induced reductions in demand for energy services in the residential sector is far higher in the resilience scenarios than in the *Low Carbon* scenario.

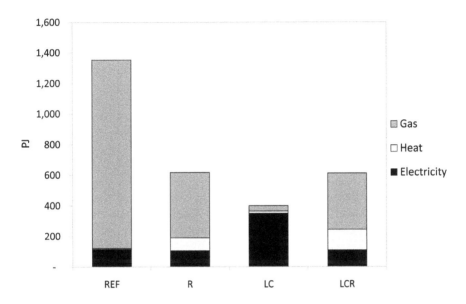

Figure 6.8 *Residential energy system fuel demand in the core scenarios (2050)*

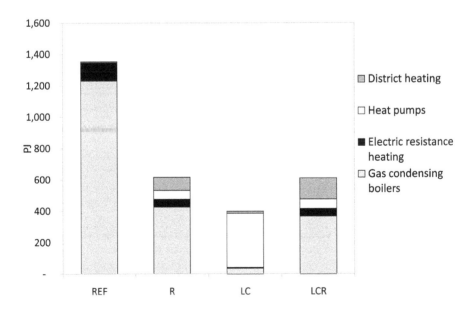

Figure 6.9 *Residential heating in the core scenarios by type (2050)*

Electricity supply

Electricity demand in the *Resilient* scenario is 11 per cent lower in 2025 and 19 per cent lower in 2050 than in the *Reference* scenario (Figure 6.10). In 2025, the biggest reduction in electricity demand is from the industrial sector, followed by the service and residential sectors. The reduction in electricity demand is driven primarily by a price-induced demand response. In 2050, the *Resilient* scenario's electricity demand declines across most end-use sectors, but doubles in the transport sector due to the high penetration of plug-in hybrid cars. Electricity demand in the *Low Carbon Resilient* scenario in 2025 is 14 per cent lower than in the *Low Carbon* scenario.

Figure 6.11 shows the electricity generation mix in the core scenarios. In 2025, the coal-dominated power sector in the *Reference* scenario moves towards gas in the *Resilient* scenario. Coal generation declines by 34 per cent and gas generation almost doubles. However, the reduction in gas demand in the end-use sectors exceeds the increase from the power sector and the net result is lower gas use.

In the *Low Carbon Resilient* scenario in 2025, lower demand reduces electricity generation from gas, nuclear, renewable and imported electricity. Compared to the *Low Carbon* scenario, nuclear generation declines by about 70 per cent while the share of unabated coal increases to 23 per cent from zero in the *Low Carbon* scenario (Figure 6.11). The uptake of coal generation is driven by the power supply diversity constraint. At the same time, the additional CO_2 emissions from coal-based generation are offset by lower CO_2 emissions from end-use sectors due to demand reduction.

In 2050, the power sector diversifies from coal to nuclear in the *Resilient* scenario, compared to the *Reference scenario*. Coal's market share halves to 40 per cent, while nuclear generation contributes 23 per cent compared to zero in the *Reference* scenario. Diversification in the power sector contributes substantially to the diversification of overall primary energy supply. Although total electricity generation declines by 19 per cent, renewable electricity declines by only 8 per cent. Wind generation stays at the same level as in the *Reference* scenario but hydro generation declines to one-third. Imported electricity contributes 8 per cent to total electricity supply compared to 2 per cent in the *Reference* case.

Emissions of CO_2 from the power sector are lower in the *Low Carbon Resilient* scenario in 2050 than they are in the *Low Carbon* scenario. This compensates for emissions associated with the continuing use of gas in the residential sector. Lower emissions are achieved partly via low electricity demand (down 26 per cent) and through the deployment of a combination of nuclear and renewables, especially wind. The market share of nuclear increases to about 40 per cent. Compared to the *Low Carbon* scenario, wind generation almost doubles and accounts for 23 per cent of the generation mix. Coal CCS generation falls by two-thirds and its market share is only 20 per cent compared to 40 per cent in the *Low Carbon* scenario. The shift from coal CCS to renewables is mainly because the electricity system must decarbonize even more in the *Low Carbon Resilient* scenario and cannot accommodate residual emissions from the CCS.

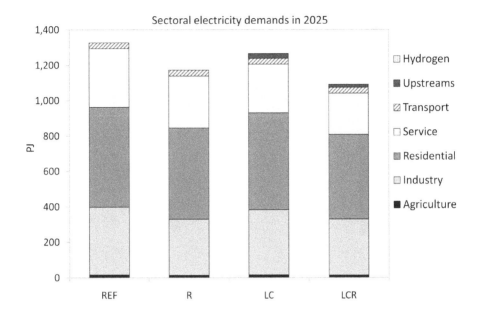

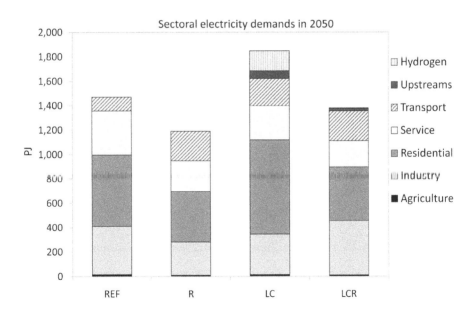

Figure 6.10 *Electricity demand in the core scenarios by sector*

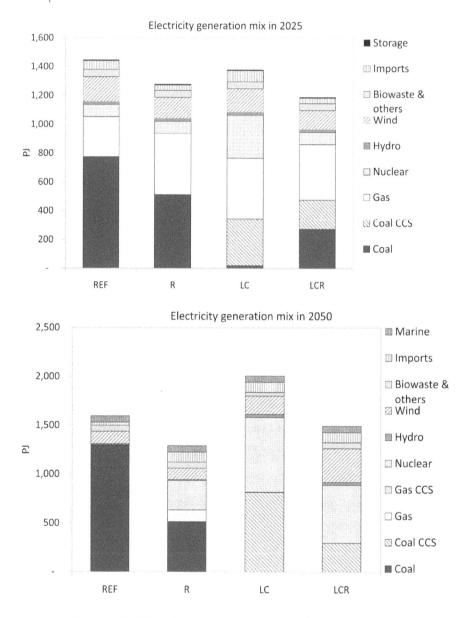

Figure 6.11 *Electricity generation mix in the core scenarios*

Figure 6.12 shows installed generation capacity in the core scenarios. In 2025, coal capacity in the *Resilient* scenario is lower than in the *Reference* scenario while gas capacity is higher. In 2050, coal capacity in the *Resilient* scenario is more than halved compared to the *Reference* scenario while nuclear is deployed. There is no significant change in renewables capacity, but interconnector capacity doubles compared to the *Reference* scenario.

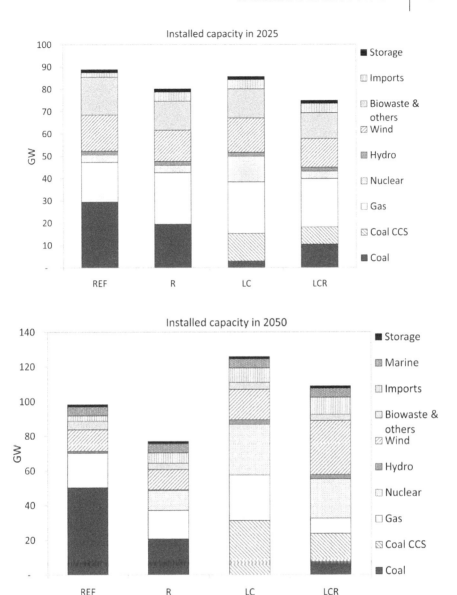

Figure 6.12 *Installed generation capacity in the core scenarios*

There is 10GW unabated coal plant in 2025 in the *Low Carbon Resilient* scenario compared to 3GW in the *Low Carbon* scenario. There is even some unabated coal plant in 2050 in the *Low Carbon Resilient* scenario. There is less coal CCS capacity in the *Low Carbon Resilient* scenario compared to the *Low Carbon* scenario in both 2025 and 2050. In 2025, investment in wind capacity is similar in the two scenarios However, in 2050 wind capacity in the *Low*

Carbon Resilient scenario is almost double that in the *Low Carbon* scenario; nuclear capacity, on the other hand, is slightly lower than in the Low Carbon scenario.

Emissions of CO_2

Due to large reductions in final energy demand, CO_2 emissions are relatively low across all end-use sectors in the resilience scenarios (Figure 6.13). In the *Resilient* scenario, CO_2 emissions in 2025 are 422mt compared to 523mt in the *Reference* scenario. About one-third of this CO_2 reduction is achieved in the residential sector (through demand reduction) while 20 per cent is from the power sector (switching from coal to gas). In the *Resilient* scenario, CO_2 emissions in 2050 are 285mt compared to 583mt in the *Reference* scenario. While all end-use sectors and the power sector contribute to this emissions reduction, around two-thirds comes from the residential sector. The *Resilient* energy system only reduces CO_2 emissions by 52 per cent from the 1990 level, which is still well short of the UK's 80 per cent carbon reduction target.

In 2025, CO_2 emissions from all end-use sectors in the *Low Carbon Resilient* scenario fall from the levels in the *Low Carbon* scenario due to demand reduction and fuel switching. This enables the power sector to retain coal-fired generation, rather than switching to nuclear as in the *Low Carbon* scenario. As a result, CO_2 emissions from the power sector are 65 per cent higher in 2025 in the *Low Carbon Resilient* scenario compared with the *Low Carbon* scenario. In 2050, CO_2 emissions are significantly higher in the transport and residential sectors in the *Low Carbon Resilient* scenario, while the power sector decarbonizes to a greater extent than in the *Low Carbon* scenario, so that the CO_2 intensity of electricity generation in the *Low Carbon Resilient* scenario is 18g/kWh versus 21g/kWh in the *Low Carbon* scenario (Figure 6.14). Higher CO_2 emissions from the residential sector reflect a lesser shift to heat pumps. Also, the transport sector deploys petrol-diesel hybrid cars, rather than plug-in hybrids and ethanol ICE as seen in the *Low Carbon* scenario.

Marginal cost of CO_2 abatement

Figure 6.15 shows the marginal cost of CO_2 abatement in 2025 and 2050. In 2025, the marginal cost of CO_2 abatement is £5/tonne in the *Low Carbon Resilient* scenario, half of that in the *Low Carbon* scenario. Since the resilience energy constraints themselves reduce CO_2 emissions, the marginal cost of the extra CO_2 abatement required to meet the target declines. However, in 2050 the marginal abatement cost of £320/tonne is double that in the *Low Carbon* scenario. This high marginal cost is a result of limited decarbonization in the residential sector forcing the power sector to decarbonize through the deployment of more expensive renewables.

Energy demand response and welfare costs

It is clear that the energy systems in the *Resilient* and *Low Carbon Resilient* scenarios are much smaller than those in the *Reference* scenario and therefore

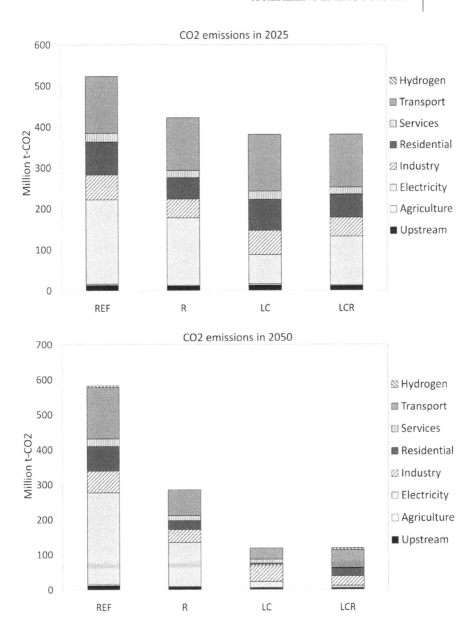

Figure 6.13 *CO₂ emissions in the core scenarios by sector*

incur a lower energy system cost (Figure 6.16). Undiscounted energy system costs in the *Resilient* scenario in 2025 are about £2.5 billion lower than in the *Reference* scenario as opposed to a £2.3bn increase in the *Low Carbon* scenario. However, in the resilience scenarios, price-induced demand reduction in the end-use sectors results in a loss of consumer surplus, leading to a higher

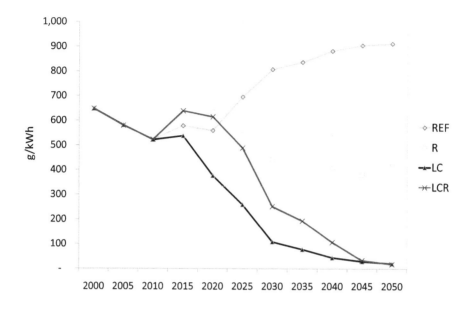

Figure 6.14 *CO_2 intensity of electricity generation in the core scenarios*

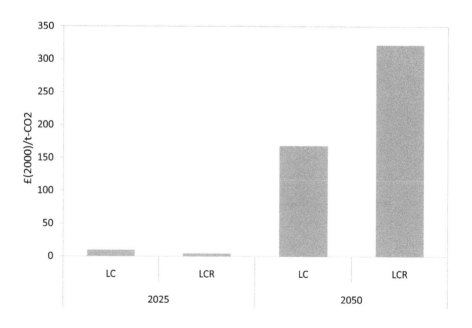

Figure 6.15 *Marginal cost of CO_2 abatement*

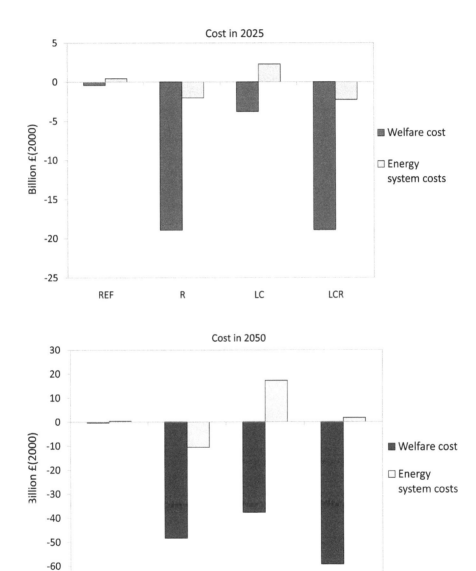

Figure 6.16 *Welfare costs and changes in energy system costs in the core scenarios with respect to the Reference scenario*

welfare cost. As shown in Figure 6.16, in the *Resilient* scenario, the welfare costs are about £19bn in 2025 and increase to £48bn in 2050.

The *Low Carbon Resilient* scenario requires a substantial demand response to meet the combined low carbon and resilience constraints. As already noted in Figures 6.2 and 6.3, by 2050, demand reductions in the *Resilient* and *Low Carbon Resilient* scenarios are far higher than those in the *Low Carbon* scenario. The £19bn welfare cost in 2025 is the same in the two resilience scenarios, but far higher than in the *Low Carbon* scenario. The welfare cost in 2050 increases to £60bn in the *Low Carbon Resilient* scenario compared to £37bn in the *Low Carbon* scenario.

Table 6.9 compares the range of price-induced reductions in demand for energy services in the core scenarios with respect to the *Reference* scenario. The ranges refer to different types of energy use, e.g. heating versus lighting and appliances. Because of the final energy demand constraint, the reductions in demand for energy services are far higher in the *Resilient* and *Low Carbon Resilient* scenarios than they are in the *Low Carbon* scenario. For example, residential demand in 2025 is only marginally lower in the *Low Carbon* scenario, whereas it is up to 27 per cent lower in the *Resilient* scenario. The lower demand response from the transport sector reflects the availability of a wide range of alternative vehicle technologies. The welfare costs arising from demand reduction are discussed in more detail in the final section of this chapter.

Reliability in the network industries

This section is concerned with the adequacy of investment in supply capacity and infrastructure. It covers risks, as defined in Chapter 4, to which probabilities can be assigned. These include variability in demand, including those related to weather, and the probabilities of plant outages through technical failure. Reliability standards are set so as to ensure that the risk of outages is kept to acceptable levels. There can be no guarantee that demand can be met at all times.

MARKAL-MED uses a simple 'capacity margin' approach to determine the adequacy of generation capacity in the electricity sector. Capacity margins are 'rules of thumb' based on more fundamental statistical analyses. The loss-of-load probability (LOLP) – the number of winters per century in which demand

Table 6.9 *Demand reduction in the core scenarios with respect to the Reference scenario*

End-use sectors	R	LC	LCR	R	LC	LCR
	2025			2050		
Industry	10%–30%	2%–7%	10%–30%	22%–50%	10%–32%	21%–50%
Residential	13%–27%	2%–5%	15%–27%	35%–50%	10%–30%	30%–50%
Services	2%–20%	2%–3%	5%–18%	13%–36%	+3%–18%	8%–35%
Transport	2%–13%	0%–2%	2%–13%	5%–28%	0%–10%	5%–30%

will not be fully met – is another reliability indicator. A capacity margin of around 20 per cent and a LOLP of nine winters per century were the standards used by the former Central Electricity Generating Board (CEGB).

If new types of plant, particularly intermittent renewables, take a large share of the electricity market, then the old assumptions about LOLP and capacity margin may break down. Periods of interruption may be longer and more load may be lost. There is a need to adopt a more fundamental statistically-based approach to reliability.

The basic approach is to assess the value of gas or electricity to customers and multiply by the probabilities of outages occurring to obtain the expected welfare cost of unserved demand. Both the WASP model of the electricity system and the CGEN model of gas and electricity infrastructure use the concept of value-of-lost-load (VOLL) to balance the cost of outages against the cost of investment in additional capacity to improve reliability. 'Lost load' can also be known as 'energy unserved'. In addition, the WASP model applies a constraint on the loss-of-load expectation (LOLE): the number of hours per year in which demand is expected not to be met. Where there is substantial investment in intermittent wind generation, using the LOLE and VOLL constraints will increase the capacity needed to meet reliability standards in the electricity sector.

The input assumptions about VOLL and LOLE are shown in Table 6.10. These assumptions come from a variety of sources, including Kariuki and Allan (1996) and from reviewing practice in a number of different countries. This section first considers the approach to reliability in the electricity sector and then goes on to look at gas.

Reliability of electricity supply

Applying the reliability indicators in Table 6.10 to the electricity system using the WASP model shows that the conventional capacity margin approach used in MARKAL-MED will lead to increasingly unreliable supply if significant amounts of intermittent renewable capacity comes onto the system. Figure 6.17 shows how the required system capacity margin (the fraction by which installed capacity exceeds peak demand) changes between 2005 and 2050 under the *Low Carbon* scenario if the more formal reliability approach based on VOLL and LOLE is applied. The difference relates mainly to the degree of renewables on the system. With intermittent renewables electricity shortages tend to become longer, involving deeper load cuts, even if the frequency of such events remains the same.

Table 6.10 *Reliability indicators for gas and electricity*

Value of lost load	£5/kWh (residential electricity)
	£40/kWh (industrial electricity)
	£5/therm (industrial gas)
	Lost load not allowed (residential gas)
Loss-of-load expectation (LOLE)	4 hours per year (0.05% of year) for electricity

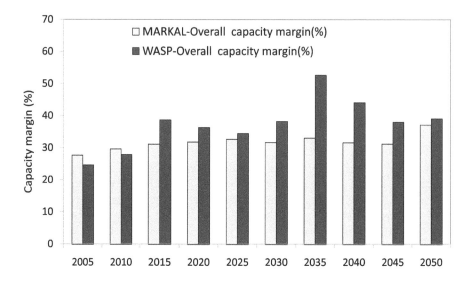

Figure 6.17 *Capacity margin using different reliability approaches,*
Low Carbon scenario

Figure 6.18 shows how, in the same *Low Carbon* scenario, LOLE exceeds accepted norms under the conventional capacity margin approach (as used in MARKAL-MED); accepted loss-of-load expectations under current conventional systems range between two and eight hours per year. With more intermittent renewables on the system, the conventional approach could lead to loss of load as high as 150 hours per year by 2040. In the later years of the *Low Carbon* scenario, nuclear forms an increasing part of the mix and LOLE falls off.

The additional capacity and electricity system costs associated with applying the more formal reliability approach are shown in Table 6.11. Beyond 2020, the additional capacity required on the system to maintain reliability is in the range 5–12GW depending on the scenario and the precise point in the projection. The cost of maintaining this capacity, which would be seldom used, is largely associated with capital costs and could run into several hundreds of millions of pounds per year. As an indication of these, the £354m incurred in the *Low Carbon Resilient* scenario in 2020 is equivalent to £1.03/MWh of all electricity generated and £9.85/MWh of wind energy generated. The modelling suggests just over 12GW of wind on the system at this point.

Reliability of gas supply

When the UK was self-sufficient in natural gas, unanticipated fluctuations in supply could be accommodated by stepping up production from the North Sea. With the prospect of the UK becoming largely dependent on imports, other measures are required to ensure reliability of gas supplies. These include

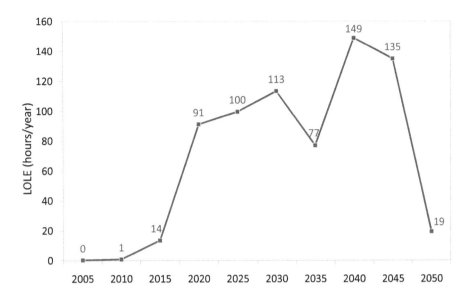

Figure 6.18 *Loss-of-load expectation using capacity margin approach,
Low Carbon scenario*

greater interconnection with Europe, opening up to global LNG markets and
investing in storage.

The CGEN model has been run using the reliability indicators from
Table 6.10 to assess the investments needed to ensure reliable gas supply
through to 2050 under each of the four core scenarios. After taking account
of current and committed projects, the CGEN model chooses between additional pipeline interconnectors, LNG terminals and gas storage facilities.
Table 6.12 shows the investments selected under the four core scenarios
in addition to current and committed capacity. New interconnectors are
not selected, but there is considerable investment in new LNG terminals to
compensate for declining domestic supply. This is largely driven by assumptions about the relative cost of continental gas and gas available through

Table 6.11 *Additional capacity and system costs to ensure reliability*

	Additional system capacity (GW)			Additional system cost (£m pa)		
	2020	*2035*	*2050*	*2020*	*2035*	*2050*
Reference	1.3	5.5	5.5	67	274	277
Low Carbon	3.7	11.5	4.4	187	575	219
Resilience	6.8	5.9	9.1	341	296	457
Low Carbon Resilient	7.1	5.4	6.2	354	269	312

Table 6.12 *Gas infrastructure investments*

	Reference (REF)	Low Carbon (LC)	Resilient (R)	Low Carbon Resilient (LCR)
Interconnectors	No additional	No additional	No additional	No additional
LNG terminals	40mcm/d 2015	40mcm/d 2015		
	20mcm/d 2020	20mcm/d 2020	40mcm/d 2020	40mcm/d 2020
	60mcm/d 2025	60mcm/d 2025	40mcm/d 2025	40mcm/d 2025
	40mcm/d 2030	40mcm/d 2030	60mcm/d 2030	60mcm/d 2030
Storage	2000mcm 2015	1000 mcm 2015	No additional	No additional

LNG markets. New, additional storage is selected in the *Reference* and *Low Carbon* scenarios but not in the resilience scenarios, where final gas demand is much lower.

Figure 6.19 shows the gas market balance out to 2030 under each of the four core scenarios. The shaded areas indicate the maximum supply capacity for UK domestic production, LNG imports and pipeline imports as determined by the CGEN model. This shows clearly the declining role of UK gas production and the growth of LNG capacity in particular. The bars indicate the annual

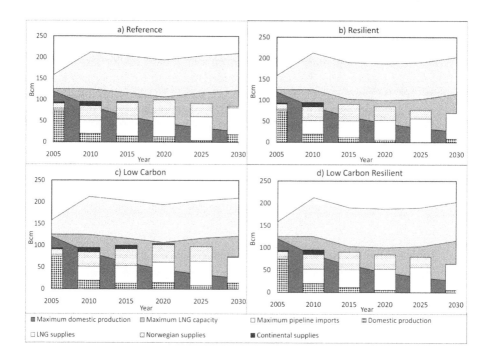

Figure 6.19 *Gas supply/demand balance in the four core scenarios*

supply from these sources required to meet UK demand. This illustrates starkly the degree to which the UK will become import dependent. The broad pattern across all scenarios is that LNG capacity substitutes for UK domestic production and, in the 2020s, for Norwegian imports. In the two resilience scenarios, where gas demand is lower, Norwegian imports are reduced more quickly.

Hypothetical system shocks

Building on the earlier historical review of 'shocks' in the energy sector, a set of hypothetical shocks to the UK energy system is now defined. These all relate to natural gas supply. Gas was prioritized because:

- there are well-established requirements for major consuming countries to maintain significant quantities of oil stocks under International Energy Agency (IEA) and EU rules; and
- shocks to the electricity system tend to be of a much shorter duration and can be mitigated by addressing the reliability issues considered in the previous section.

The shocks are assumed to occur in mid-winter, in the year 2025, under each of the four core scenarios.

The three possible 'shocks' result in the loss of the gas supply facilities described in Table 6.13. It has been assumed that the impact of each shock is experienced over three different durations – 5, 40 and 90 days. In each case it has been assumed that the shock occurs in mid-winter, nominally 1 January 2025, during a period of 'average cold spell' demand. These are deliberately severe events. However, they are within the range of recent experience. As shown in Table 6.2, an explosion at the *Bravo* rig in the Rough field took the storage facility out of service for two months in 2006. Gas supplies through Easington were interrupted for five days after it was struck by lightning in 1999.

Table 6.13 *Description of facilities lost in the hypothetical shocks*

Facility	Description	Size
Easington gas terminal	Connects the UK gas system to the Rough storage facility and the Langeled pipeline from Norway	Can deliver 120mcm/day (equivalent to 35% of UK winter demand). Rough can store 3.3bcm of gas, equivalent to 10 days' average winter demand
Bacton gas terminal	Connects the UK to continental Europe via Zeebrugge and Balgzand. Also links to some domestic production	Can deliver 144mcm/day (equivalent to ~40% of UK winter demand). Also used for export
Milford Haven LNG terminal	Two terminals being commissioned in 2009	In 2009, Milford Haven could deliver 75mcm/day (equivalent to >20% of winter demand)

Table 6.14 shows the impact of the three events under each of the core scenarios. The key messages are as follows.

- The loss of the largest terminal, Bacton, has the largest impact; it affects both imported and domestic gas supplies.
- The energy system can 'ride through' the loss of Easington or Milford Haven under the *Resilient* and *Low Carbon Resilient* scenarios – and the impact of losing Bacton is also much diminished. This is because these two scenarios are characterized by lower levels of residential gas demand (which is strongly seasonal). The system can cope better when demand is less 'peaky'. Demand reduction therefore demonstrably contributes to energy system resilience.
- The imputed value of unserved energy is an order of magnitude larger than the changed system costs. System costs generally rise as more expensive gas is sourced and coal substitutes for gas in electricity generation. However, this does not take account of the likely response in energy spot markets following such events, which would tend to increase costs further.
- The patterns of response are complex because the facilities play different roles in the gas network. In none of the scenarios is it necessary to curtail electricity supplies. Rather, response involves exercising interruptible gas contracts, re-dispatch of the electricity system, use of distillate oil at certain CCGTs and non-contracted industrial gas interruptions.

Table 6.14 *Impact of 40-day shocks in the four core scenarios*

	Energy unserved (mcm)	Value of energy unserved[1] (£m)	Change in system operating costs[2] (£m)
Reference (REF)			
Bacton	1839	3404	−7
Easington	1049	1942	+137
Milford Haven	866	1604	+104
Low Carbon (LC)			
Bacton	1718	3179	+29
Easington	1155	2138	+144
Milford Haven	1015	1878	+89
Resilient (R)			
Bacton	244	452	+203
Easington	–	–	–
Milford Haven	–	–	–
Low Carbon Resilient (LCR)			
Bacton	704	1303	+135
Easington	–	–	–
Milford Haven	–	–	–

Notes: (1) using the values of lost load in residential and industry from Table 6.10; (2) this does not allow for the likely rise in spot prices for gas

Table 6.15 assesses how different lengths of interruption could affect outcomes. The loss of Easington is taken as an example. The clear message is that shorter periods of interruption, of the order of a few days, can be accommodated through system adjustments with very little loss of load. However, at some point, costs increase rapidly, though not linearly with the length of the interruption.

Mitigating the shocks

The analysis above is based on the assumption that investment takes place to meet the reliability standards set out earlier. It is also possible to undertake additional infrastructure investment that would help to mitigate the impacts of major shocks. This section investigates the benefits that such investment might bring, and sets them against the costs. Table 6.16 shows seven possible projects which would increase the resilience of the gas supply network.

Table 6.15 *Impact of the loss of Easington for different periods*

	Energy unserved (mcm)	Value of energy unserved (£m)	Change in system operating costs (£m)
Reference			
5 day	14	26	+29
40 day	1049	1942	+137
90 day	1857	3438	+294
Low Carbon			
5 day	12	23	+32
40 day	1155	2138	+144
90 day	2127	3937	+242

Table 6.16 *Gas infrastructure projects*

Project	Capacity	Capital cost (£m)
Storage facility similar to Rough	3000mcm delivering 40mcm/d, located near St Fergus	475
Two storage facilities	Salt cavities each with a capacity of 500mcm delivering 40mcm/d	550
Expansion at one LNG terminal	40mcm/d at Teesside	400
Expansion at two LNG terminals	20mcm/d at each of Teesside and Isle of Grain	405
New gas interconnector	40mcm/d through Theddlethorpe	340
Backup distillate storage[1] at CCGTs	5 days' storage at 6GW plant	15
Major distillate storage at CCGTs	40 days' storage at 6GW plant	215

Note: (1) 'distillate storage' refers to storage of light distillate oil which can be used as a substitute for natural gas to fire a CCGT. Distillate oil is more expensive than natural gas but its use is acceptable for a short period.

The loss of Bacton, the most severe shock, for a 40-day period under the *Low Carbon* scenario is considered in detail. The pay-off from the mitigating investments will manifestly be less under the *Resilient* and *Low Carbon Resilient* scenarios under which the combined electricity and gas system is much more able to ride through shocks. The 40-day Bacton shock stands as a good proxy for other shocks, periods and scenarios as the pattern of impacts is similar.

Table 6.17 shows the degree to which each individual mitigating investment would reduce the volume of energy unserved and associated costs. The biggest reduction in the cost of the shock comes from the expansion of import facilities, be they new LNG terminals or a new interconnector. Five days' distillate storage has little impact (as might be expected for a 40-day outage) but dedicated gas storage has half of the impact of more import facilities.

The impact of dedicated new gas storage is critically dependent on how much gas is in store at the time of the shock. For the major storage facility, two options have been considered:

1 that the facility is half full in mid-winter; and
2 that it is kept completely full for emergencies.

Full storage has a significant effect on the conclusions, reducing the volume of energy unserved to the same level as for import facilities. Note also that the analysis does not take into account changes in spot market prices that might be expected to take place.

Making a mitigating investment can be regarded as taking out insurance against the eventuality of adverse events. If the event is expected to occur regularly, taking out the investment could reduce costs in the longer term. If it were extremely rare it might be better to forgo the insurance costs and accept the consequences. The 'return period' in Table 6.17 refers to the frequency with which the event would need to take place for each of the mitigating investments to pay off in the long run. It is based on a simple calculation that involves dividing the expected loss as a result of the adverse event by the annualized capital cost of the mitigating investment. Table 6.17 shows, for example, that if the investor requires a market rate of return of 10 per cent real, investing in two new LNG terminals might be expected to pay off in the long run if a 40-day outage at Bacton were to occur more frequently than once every 35 years. Given the improbability of its happening as frequently as this, it is almost impossible to conceive of this as a good investment in a market context.

On the other hand, investment in these mitigating measures could be regarded as being in the public interest for strategic reasons. At a rate of return on investment of only 3.5 per cent real, the Treasury 'social' discount rate, the investment might still 'pay off' if the event were to occur as infrequently as once in 100 years. There might therefore be a case for the regulator to allow (or mandate) the costs of such an investment to be passed on to consumers, whereby companies would be prepared (or required) to accept a lower risk-free rate of return. The potential difficulty with this solution is that such assets could, without careful restrictions, be used for arbitrage in the gas markets and inhibit investments undertaken on purely market grounds.

With lengthy return periods, the question of the changing structure of the energy system in the long term becomes an issue. If an event is likely to occur only once every 50 years or so, it may not be desirable to invest in a project that appears to pay off using the insurance analogy, if gas use is virtually eliminated from the energy system over that time period. Further research would be needed to elucidate this question.

Adding up the costs of resilience

This chapter has developed scenarios describing resilient energy futures for the UK with and without the achievement of decarbonization policy goals – the *Resilient* and *Low Carbon Resilient* scenarios. It has shown that the resilience scenarios not only reduce exposure to uncertain global markets through reduced energy demand and hence imports, but also result in a greater ability to withstand 'shocks' to gas supply.

However, there are costs associated with building in resilience. Ultimately, deciding how much to invest in order to promote resilience in the energy system is a political decision that must be informed by evidence, albeit in the light of deep uncertainty. In this section the evidence from the preceding analysis is drawn together to assess the overall economic impact and desirability of investing in resilience.

Table 6.17 *Impact of mitigating investments: Bacton out for 40 days*

	Energy unserved (mcm)	Reduction in cost of shock (£m)	10% investor rate of return		3.5% regulated rate of return	
			Additional annual costs (£m)	Return period[3] (years)	Additional annual costs (£m)	Return period (years)
Baseline investment Additional:	1718	–	–	–	–	–
Storage facility[1]	1104	1093	52	21	17	63
Storage facility[2]	832	1598	52	31	17	93
Two storage facilities	1246	832	60	14	20	42
One LNG terminal	786	1572	44	36	15	108
Two LNG terminals	789	1569	45	35	15	105
Gas interconnector	790	1600	37	43	12	129
Distillate storage	1685	53	2	32	1	96
Major distillate storage	1246	767	24	32	8	96

Notes: (1) facility half full in mid-winter; (2) facility completely full; (3) the 'return period' refers to the frequency with which an event would need to take place before an investment was justified. It does not refer to the financial rate of return on capital.

Table 6.18 summarizes how system costs and welfare costs and the value of energy unserved change in moving from the *Reference* and *Low Carbon* scenarios to the *Resilient* and *Low Carbon Resilient* scenarios respectively. Note that this omits any consideration of the additional costs associated with the failure to mitigate climate change that is implicit in the higher emissions in the *Reference* scenarios. The high-level goals in Table 6.18 refer to the resilience constraints imposed on final energy demand and the energy mix. Electricity reliability costs refer to the additional system costs of maintaining capacity sufficient to meet reliability criteria over and above maintaining a conventional capacity margin. These costs are essentially associated with the greater use of intermittent renewables. Finally, infrastructure costs are those associated with mitigating the effects of major disruptions to the gas system. Note that these costs are not additive and should be considered only as illustrative of broad orders of magnitude.

Table 6.18 *Estimated costs associated with different aspects of resilience in 2025*

	Reference[1] ↓ Resilient	Low Carbon[2] ↓ Low Carbon Resilient
High-level goals		
Reduction in annual system cost[3]	£2.5bn	£4.6bn
Loss of welfare[4]	Up to £19bn	Up to £15bn
Mitigation of welfare loss through 25% conservation[5]	£3.1bn	£2.3bn
Reduced value of energy unserved by 40-day loss of Bacton[6]	£3.0bn	£1.9bn
Electricity reliability (applies to all scenarios)		
Cost of higher capacity margins[7]	~£300m	
Infrastructure (applies to all scenarios)		
Enhanced gas import or storage capacity[8]	£45m	

Notes: (1) the changes associated with moving from the *Reference* to the *Resilient* scenario; (2) the changes associated with moving from the *Low Carbon* to the *Low Carbon Resilient* scenario; (3) from MARKAL runs; (4) change in consumer and producer surplus assuming a price-induced response in the residential sector (see Figure 6.16); (5) mitigation of welfare loss when conservation measures deliver about 25 per cent of residential demand reduction (see text for explanation); (6) calculated from Table 6.14 by comparing energy unserved in, for example, the *Reference* and *Resilient* scenarios; (7) the cost of additional capacity determined by WASP as opposed to MARKAL (see Table 6.11). These additional costs vary considerably from one year to another; (8) annualized cost of two LNG terminals (calculated from Table 6.16)

The high level of welfare costs imputed in the *Resilient* and *Low Carbon Resilient* scenarios need to be treated with caution because of some of the detailed modelling assumptions. As discussed in Chapter 4, the uptake of residential conservation measures has been kept constant across the core scenarios by 'turning off' MARKAL-MED's endogenous selection of conservation measures, because energy service demand assumptions for the residential sector have been taken from the Environmental Change Institute's UK Domestic Carbon Model (UKDCM). The UKDCM has its own methodology and assumptions on the uptake of residential conservation measures based on historical uptake, behavioural change and affordability. This means that reductions in demand in MARKAL-MED are induced by prices only and not through the selection of conservation technologies. Price-induced demand responses are responsible for the high welfare costs shown in Figure 6.16 and reproduced in Table 6.18.

In practice however, additional high-cost conservation measures could be a more cost-effective way of achieving the energy demand reductions required by the resilience scenarios. Therefore additional sensitivity analysis has been carried out by allowing such conservation measures to be taken up in the residential and service sectors. This sensitivity analysis is aimed at investigating the impact on welfare costs. The results show that the welfare loss in the *Resilient* scenario compared to the *Reference* scenario in 2025 is reduced by £3bn, falling to £16bn (compared to £19bn without conservation measures), and in 2050 falling to £36bn compared to £48bn. In fact, the additional conservation measures turn out not to increase demand (and therefore consumer surplus) in the residential and service sector. Instead, demand adjusts in other end-use sectors. Energy savings are effectively passed on to substitute for more expensive measures in other sectors. For example, industrial energy demand is reduced by 18–47 per cent, depending on the fuel and subsector, if residential and service sector conservation is allowed, compared to 23–50 per cent in the *Resilient* scenario without such conservation. The additional conservation measures reduce the welfare loss in the *Low Carbon Resilient* scenario by £2.3bn in 2025, and from £60bn to £50bn in 2050. However, there is no significant change in the energy pathways.

The costs of ensuring electricity system reliability (£300m) and reinforcing gas infrastructure (£45m) are one or more orders of magnitude less than the costs associated with welfare changes at the system level associated with decarbonization (£3bn) or improving the macro-indicators of resilience (up to £16bn). Enhanced electricity reliability appears to cost around £10–15 per household per year while additional investment in gas storage or import facilities appears to run at around £2 per household.

A key point is that high-level resilience goals, particularly those that involve significant demand reduction, have potential costs that are more than an order of magnitude higher than those associated with guaranteeing reliability of supply or insuring against infrastructure loss. The benefits from these goals in terms of energy system resilience would need to be at least as great to justify pursuing them. However, as shown above, if the high-level goals are pursued then the case for infrastructure measures is considerably reduced.

Policy implications

Achieving the high-level goals of reduced energy demand, imports and greater supply diversity can be achieved through the vigorous pursuit of fairly conventional policy instruments. The key is a very strong emphasis on policies to improve energy efficiency in buildings and transport. The emphasis on the demand side needs to be much stronger than in a pure *Low Carbon* scenario. Keeping up the pace of investment in renewables and nuclear will also contribute.

As regards reliability and redundancy in the electricity system, current UK Government policy is to deliver an adequate capacity margin by relying on markets, through price signals, to deliver capacity that may only be rarely used. There is now a widespread view that current market arrangements may not be sufficient to guarantee reliable energy supply while ambitious low carbon targets and renewable energy goals are pursued. During 2010, market arrangements are being reviewed by the Government, Ofgem and the Committee on Climate Change.

This chapter suggests that there is potentially a case for investment in further 'strategic' gas infrastructure beyond that which the market would deliver if the supply-led energy strategy embodied in the *Low Carbon* scenario is pursued. By itself, that investment would be relatively modest and would add little to consumer costs.

There are three possible models for stimulating such investment: Government provides the appropriate framework for the market to make the investment; the regulator permits or mandates the investment through price reviews, but the investment is provided by the regulated companies; or Government carries out the investment itself. The latter model appears unlikely. The key policy question is whether the benefits of driving this investment through rate of return regulation outweigh the disadvantages of driving out investment made on a purely market basis.

Finally, policy is not the only factor that could drive down energy demand and enhance resilience. In Chapter 9, a 'lifestyles' scenario resulting in demand reductions even greater than those in the resilience scenarios is considered. This results in a much reduced energy system cost. The detailed assumptions and conclusions are set out in Chapter 9. What may be noted here is that the crucial issue for welfare is whether the demand reduction is as a result of voluntary lifestyle choice (as in Chapter 9), or is imposed in order to achieve other goals, as in the *Low Carbon* and resilience scenarios. This chapter, and Chapter 5, have shown that, in these cases, the welfare costs associated with these goals may be high.

References

AGD (Australian Government Attorney-General's Department) (1998) EMA Disasters Database, Canberra

Bazilian, M. and Roques, F. (2008) *Analytical Methods for Energy Diversity and Security*, Elsevier, Oxford

BBC (2007) 'Barcelona blackout into third day', BBC News website, accessed 20 June 2010, http://news.bbc.co.uk/1/hi/world/europe/6915728.stm

Berizzi, A. (2004) 'The Italian 2003 blackout', *IEEE Power Engineering Society General Meeting*, 2004, vol 2, pp1–7, http://ieeexplore.ieee.org/stamp/stamp.jsp?arnumber=01373159&tag=1

BreakingNews (2007) 'Barcelona blackout leaves thousands without power', www.breakingnews.ie/archives/?c=WORLD&jp=mhcwaucwkfql&d=2007-07-23

Centrica (2006) *Interim Report on Rough Incident*, Centrica Storage Ltd., Staines, Middlesex

Chaudry, M., Ekins, P., Ramachandran K., Shakoor, A., Skea, J., Strbac, G., Wang, X. and Whitaker, J. (2010) *Building a Resilient UK Energy System*, UK Energy Research Centre Working Paper, London, www.ukerc.ac.uk/support/tiki-index.php?page=ResiliencePaper&structure=Energy+2050+Overview

DECC (Department of Energy and Climate Change) (2008) *Updated Energy and Carbon Projections,* URN 08/1358, Department of Energy and Climate Change, London

DEFRA (Department for Environment, Food and Rural Affairs) (2007) *UK Energy Efficiency Action Plan*, Department for Environment, Food and Rural Affairs, London

Earth Policy Institute (2009) *World Oil Production, 1950–2007*, www.earth-policy.org/index.php?/plan_b_updates/2007/update67#table1

EIA (Energy Information Administration) (2006) *Electric Power Monthly March 2006*, http://tonto.eia.doe.gov/ftproot/electricity/epm/02260603.pdf

Elkraft System (2003) *Power failure in Eastern Denmark and Southern Sweden on 23 September 2003 – Final report on the course of events*, Energinet, Fredericia, Denmark, www.energinet.dk/NR/rdonlyres/BC99F243-304D-4ADD-B769-46351A959C85/0/Powerfailurereportsept2003.pdf

Freeman, M.J. (2006) 'Power Outages Hit U.S. Grid; Utility Deregulation to Blame', *Executive Intelligence Review*, vol 33 (31), pp60–62, www.larouchepub.com/eiw/public/2006/2006_30-39/2006_30-39/2006-31/pdf/eirv33n31.pdf

IEA (International Energy Agency) (2006) *Natural Gas Market Review 2006*, IEA, Paris, www.iea.org/textbase/nppdf/free/2006/gasmarket2006.pdf

IEA (International Energy Agency) (2007) *IEA Response System for Oil Supply Emergencies,* IEA, Paris

Kariuki, K. and Allan, R. (1996) 'Evaluation of reliability worth and value of lost load', *IEE Proceedings – Generation, Transmission and Distribution*, vol 143 (2), pp171–180

Northwest Power and Conservation Council (1996) *Blackout of 1996*, Portland, OR, www.nwcouncil.org/history/Blackout.asp

Ofgem (Office of Gas and Electricity Markets) (2003) *Summer Interruptions 17 and 18 June 2003: Conclusions*, Ofgem, London, www.ofgem.gov.uk/Markets/WhlMkts/CompandEff/Archive/4263-Summer_interruptions_conclusions_Aug03.pdf

Ofgem (Office of Gas and Electricity Markets) (2004) *The review of top up arrangements in gas: Conclusions document*, Ofgem, London, www.ofgem. gov.uk/Networks/Archive/The%20Review%20of%20Top%20Up%20 Arrangements%20in%20Gas%20-%20Conclusions%20Document.pdf

Ofgem (Office of Gas and Electricity Markets) (2009) *Project Discovery: Energy Market Scenarios*, Ofgem, London, www.ofgem.gov.uk/Markets/ WhlMkts/Discovery/Documents1/Discovery_Scenarios_ConDoc_FINAL. pdf

US-Canada Power System Outage Task Force (2004) *Final Report on the August 14, 2003 Blackout in the United States and Canada: Causes and Recommendations,* US Department of Energy, Washington DC, https:// reports.energy.gov/

Vournas, C. (2004) *Technical Summary on the Athens and Southern Greece Blackout of July 12, 2004*, Regulatory Authority for Energy, Athens, Greece

Wicks, M. (2009) *Energy Security: A national challenge in a changing world*, Department of Energy and Climate Change, London

World Energy Council (2003) *Blackouts: Southern Sweden and Denmark*, 23 September, website news report, www.worldenergy.org

7
Accelerating the Development of Energy Supply Technologies: The Role of Research and Innovation

*Mark Winskel, Gabrial Anandarajah, Jim Skea
and Brighid Jay*

Other contributors:
*Chiara Candelise, Henry Jeffrey, Nils Markusson, Hannah Chalmers,
Donna Clarke, Gail Taylor, Sophie Jablonski, Geoff Dutton,
Christos Kalyvas, Paul Howarth, David Ward and Nick Hughes*

Introduction

This chapter considers the potential for accelerating the development of a number of emerging low carbon energy supply technologies, and the possible impact of accelerated development on UK energy system decarbonization pathways. The technologies covered here include a number of large scale renewables (wind power, marine energy, solar PV and bio-energy) and also nuclear power, carbon capture and storage and hydrogen fuel cells. Most of these are electricity generation technologies, as discussed in Chapter 5, the electricity sector may have an increasingly important role in the UK's future energy system.

The analysis suggests that accelerated development could open up more affordable and more diverse decarbonization pathways over the longer term, with currently emerging technologies assuming a significant role in the future energy supply mix. This implies that, in mapping out desirable decarbonization strategies for the UK, it is important to take accelerated development potential into account.

Realizing the benefits of accelerated development will require raised and sustained efforts on low carbon technology innovation – investment which promises significant return in the longer term. In response to rising concerns about climate change and energy security, international and domestic spending on energy research, development and demonstration (RD&D) has risen significantly in recent years, from very low levels over the past two decades. The scenarios and system modelling presented here suggest that this can be economically justified in the context of energy system decarbonization over the next 40 years, especially if the investment costs of accelerated development are assumed to be shareable internationally. Within these broader international efforts, UK public and private RD&D can make important contributions, and under a long-term view of investment, the analysis indicates there is an economic case for a step change increase in UK annual public spending on energy RD&D.

However, although it carries shorter-term implications for energy policy and RD&D support, supply-side technology acceleration – as represented here – only introduces major changes to UK energy system deployment patterns over the longer term. Over the shorter term, the most affordable opportunities for decarbonization lie elsewhere, such as with improved energy efficiency and conservation, lifestyle changes (see Chapter 9), and more mature supply-side technologies.

Given the illustrative nature of the scenarios presented here, and the many uncertainties involved in long-term energy system change, no firm messages should be drawn in terms of identifying technology winners and losers. Most of the technologies analysed here, and many others not included, could play a significant role in UK decarbonization pathways over the longer term. Rather than a premature attempt at selecting particular emerging technologies, the prescription that emerges is for expanded support of a diverse range of designs *across* different technology fields (e.g. different types of renewables, carbon capture technologies and advanced nuclear designs) and *within* fields (such as different types of solar PV cells, advanced bio-energy technologies and fuel cell designs). While the specific outcomes of such investment cannot be predicted with any confidence, it promises substantial overall benefit to society in addressing the long-term challenge of decarbonization.

The chapter proceeds as follows: the next section considers the role of technological innovation in energy system change, in terms of the current position in 2010, forward-looking ambitions, and a brief review of historical trends. Then, the Accelerated Technology Development (ATD) scenarios are described; this is followed by a discussion of the role of scenarios and system modelling. The ATD scenario results are then presented, in terms of their impact on UK decarbonization pathways. The last section considers the implications and challenges of the ATD scenarios.

Technological innovation and energy system change

The ATD scenarios consider the prospects for the accelerated development of a range of emerging low carbon energy supply technologies – and the possible impact of such acceleration on UK energy system decarbonization pathways. As discussed in Chapter 3, highly ambitious targets for UK energy system change by 2020 and 2050 – in terms of both decarbonization and low carbon technology deployment – have been set at UK and European levels. Most ambitiously, the European Renewable Energy Directive sets an EU-wide target of 20 per cent of all energy consumed to be provided by renewable energy sources by 2020 (CEC, 2009). Within this, the UK national target is 15 per cent – a remarkably ambitious figure given the UK's very modest track record of renewables deployment – the equivalent figure in 2008 was 2.25 per cent (HMG, 2009a).

Because renewable technologies are technically and economically more readily deployable at scale in electricity generation than transport or heating, the UK Government's 'lead scenario' for complying with the Renewable Energy Directive envisages renewable supply technologies providing more than 30 per cent of all electricity produced in the UK by 2020 – requiring an unprecedented programme of renewables build over the next decade (HMG, 2009a). The lead scenario also involves renewable technologies meeting 12 per cent of heat demand and 10 per cent of transport demand by 2020. From a longer-term view of energy system change to 2050, the huge level of ambition for change over the next decade in the Renewable Energy Directive raises some concerns, in terms of the possibility of locking-in to pathways which may make achieving deeper change after 2020 more difficult, or more expensive. This is a key issue for the present analysis, and it is returned to later in the chapter.

A number of studies have devised technology deployment scenarios for meeting 2020 targets for decarbonization and renewables deployment (see for example, Pöyry, 2008; RAB, 2008; SKM, 2008; CCC, 2009; Redpoint/Trilemma, 2009). By contrast, the primary concern in this chapter is decarbonization pathways over longer timescales, and in particular, the possibly important role of emerging technologies in the UK's energy supply system from now to 2050. Realizing policy ambitions for deep decarbonization will involve deep social and economic changes to energy production and consumption, with significantly improved efficiencies of energy use, and lifestyle changes to enable reduced energy demand. At the same time, any full response to this challenge should also consider how to best support the accelerated development of emerging low carbon energy supply technologies.

For the UK (and many other countries), however, these technology development capacities have been severely eroded over recent history – levels of funding for energy RD&D and support for national research facilities declined sharply after the mid-1980s, under the combined influences of cheap oil and gas and the liberalization and privatization of the energy sector (see Figure 7.1). Only recently have growing concerns about climate change and energy security prompted a reversal of this trend, and increased spending on energy RD&D. Looking ahead, this upward trend is not assured – governments may be tempted to scale back RD&D efforts if economic growth falters or if fossil fuel prices fall – but climate change and security concerns are likely to provide long-term drivers for raised public investment. One of the key implications of the research presented here is that such investment, if sustained over time, promises significant rewards, in opening up more affordable ways to achieve deep cuts in carbon emissions.

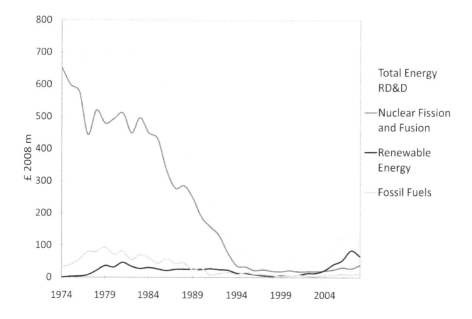

Source: IEA Statistics, 2009 database

Figure 7.1 *UK Public spending on energy RD&D, 1974–2008*

The recent resurgence in energy-related innovation activity globally and in the UK is encouraging the development of a large number of low carbon energy supply technologies, including the technologies analysed here – offshore windpower, solar PV, marine energy, bio-energy, carbon capture and storage, advanced nuclear power and hydrogen fuel cells. There are many others: the Carbon Trust has identified around 50 low carbon technology 'families', and many hundreds of specific 'variants' (Carbon Trust, 2009, p10). Many of these are supported by dedicated policy initiatives, investment programmes, developer firms and support agencies. Assessing the effectiveness of this activity is a formidable challenge, with many possible pitfalls. Technologies which are routinely compared in scenario or modelling exercises, or debates on energy futures, may be quite dissimilar in terms of their technical or economic maturity, fuel or resource requirements and implications for the wider energy system.

Another important factor shaping long-term energy system change – again, something often neglected in policy debates and scenario exercises – is the relative inertia and resistance to change in energy systems. To a greater extent than other parts of the economy, such as IT or biotechnology, energy systems reflect and rely on embedded technical and social commitments, and tend to lock-in around established technologies and institutions (Unruh, 2000). This general resistance to more radical change (a property also described as 'path dependence') means that, for energy systems, the future is often shaped by the past to a significant degree, so that responses to the global challenges of decarbonization and energy security are likely to be conditioned by national and regional histories.

It is useful, in this context, to consider recent historical patterns of energy production in the UK; a more detailed review is presented in Chapter 2. UK energy production has historically been highly dependent on fossil fuels across power, heating and transport, and remains so today. While coal use has declined steadily since 1970, gas and electricity consumption have substantially increased. Coal-fired generation made up two-thirds of electricity generated in the UK for many decades up to 1990. Since then, natural gas has become a key fuel for electricity production, but coal-fired generation also remains important, despite high levels of associated CO_2 emissions. Nuclear power has been a significant contributor to UK electricity generation for decades, contributing around 15–20 per cent of all power supplied. Until very recently, renewable energy technologies were almost insignificant contributors to overall energy production in the UK (other than hydropower installations in Scotland). The UK failed to sustain its programmes of renewable energy technology development initiated in the 1970s, and since then, environmental imperatives have been a relatively weak and intermittent driver of change in the UK energy system. The introduction of the Renewable Obligation in 2002 has supported growth in renewable generation, which rose to 6.9 per cent of all electricity supplied in 2009; output from wind power alone rose by 20 per cent between 2008 and 2009 (DECC, 2009; 2010).

Institutionally, the major change over the past 20–30 years was the liberalization and privatization of the energy industries in the 1980s. These changes, combined with a freeing-up of oil and gas reserves internationally, were associated with a major shift away from national systems of energy production, to international flows of capital, technology, fuel supplies, and international ownership of power companies and equipment suppliers. Economic liberalization also led to the running-down of long-term investment programmes, a shift away from capital intensive technology, and an emphasis on 'asset sweating' and short-term investment horizons. Some other longstanding features of the UK energy system survived liberalization – for example, a relative neglect of energy efficiency and conservation, and of regional and local interests (Smith, 2006).

These technical and social characteristics, often built up over decades, are now coming under pressure from a host of new policy initiatives being introduced to help meet ambitions for decarbonization and enhanced security. These reforms are increasing the possibility of more radical and more rapid system change, so that history may well become a less reliable guide to the future. Nevertheless, the present-day UK energy system still reflects an embedded orientation to large scale supply technologies, fossil fuels, and national level infrastructures, and a relative neglect of energy demand management, regional or local interests and environmental policy imperatives. In the absence of strong and persistent external pressures or system shocks, these features will tend to privilege certain pathways for system change above others. Of particular concern for this chapter is that the recent imposition of very ambitious policy targets for decarbonization and renewables deployment – onto a system with inherited short-termism from the liberal market era – may lead to a systematic myopia, and a locking out of longer-term opportunities for change. This concern is returned to later.

The accelerated technology development scenarios

The pace and direction of technological change is a major uncertainty in understanding, anticipating and managing the development of energy systems over time. This uncertainty, and also the rather hidden nature of much technological change in energy systems, means that despite their importance, innovation dynamics often have rather limited, superficial representation in energy system scenario and modelling exercises, and in wider debates on energy and climate change.

Explicitly addressing this issue here, for a selected number of emerging energy supply technologies, has involved devising scenarios of *accelerated technology development* (ATD) which assume high levels of technological progress over time from now to 2050, and comparing these to 'non-accelerated' scenarios which follow the same decarbonization trajectory (the Low Carbon (LC) and Medium Carbon (LC-60) scenarios described in Chapter 5). 'Accelerated development' is interpreted here as the effect of raised innovation efforts on lowering technology cost and improving technology performance (i.e. the emphasis is on the left-hand side of the innovation chain model shown in Figure 7.2). There is less emphasis on market-pull effects, although these may also be crucial drivers of technology development. In the ATD scenarios, additional innovation efforts drive the commercialization of emerging technologies from now to 2050. Technology adoption in the ATD scenarios is decided on a least cost basis, using the MARKAL-MED energy systems model (as described in Chapter 4).

Source: Foxon, 2003; Grubb, 2004

Figure 7.2 *The innovation chain*

The accelerated technologies in the ATD scenarios include a number of renewables: wind power, marine energy, solar PV and bio-energy, and some other emerging low carbon technologies: carbon capture and storage (CCS), hydrogen/fuel cells, and advanced designs of nuclear power. This is a limited selection of the many low carbon supply technologies now emerging, and within this selection, there was some further focusing down on particular technology systems and designs from among many possible configurations (see Table 7.2, below, and Winskel et al, 2009, for more details). The ATD scenarios offer a limited analysis of prospective technological acceleration in energy supply, and as such, they are an illustrative exercise – they explore the possible impact of technology progress under particular assumptions for particular technologies. Given the partial nature of this exercise, and the large uncertainties involved in technological innovation and wider aspects of energy system change, only rather tentative messages can be drawn from the scenarios presented here.

Energy supply technologies are now mostly developed by international networks of private firms and public sector organizations. In analysing the prospects for accelerated technology development, therefore, the primary references have been to long-term international trends and prospective break-throughs in cost and performance from now to 2050. In essence, the ATD scenarios explore the effect of *global* innovation trends on *UK* energy system responses to climate change. The working assumption is that the UK is a 'buyer' of technologies developed internationally. In general, therefore, the impact of UK deployment support policies on technology cost and performance have not been explicitly taken into account. In one case – marine energy – UK 'niche' support policies are considered capable of fostering a significant degree of overall technological improvement internationally, and this was explicitly taken into account in the marine energy ATD scenario; for more information, see chapter 3 in Winskel et al, 2009.[1]

As is appropriate for a study of the possible impact of global innovation trends on the UK energy system, the ATD scenarios are supported by energy system modelling which uses exogenous data representations of technology learning, rather than endogenous learning rates. Because of this, the ATD scenarios are oriented toward *learning-by-development* for emergent technologies over longer timescales, rather than *learning-by-deployment* for more mature technologies over the shorter term (Grubb and Ulph, 2002). As is discussed later, this way of representing innovation has some consequence for the outcomes of the analysis, and its possible implications for policy.

The prospects for accelerated development were considered for each technology by developing 'narratives' of development, taking into account possible incremental trends and step-change breakthroughs from now to 2050, drawing on a wide range of sources, including UK and international roadmaps, other scenario exercises and expert consultation. Each of these technology specific narrative accounts includes an overview of the current status of the technology field under investigation, an assessment of its prospects for accelerated development, the specific RD&D challenges involved, the relative status of UK RD&D

capacities in the international context, and the role of policy in supporting acceleration (see Winskel et al, 2009).

These narrative accounts were then used as the basis for devising quantitative datasets of accelerated technology development in terms, for example, of reduced cost, improved performance, earlier availability of advanced designs, or the introduction of novel designs. These datasets provided the inputs for UK energy system scenarios from now to 2050 using MARKAL. An example of the capital cost assumptions associated with the ATD scenarios, for solar PV, and how these compare to data in the non-accelerated LC scenario, are shown in Figure 7.3.

In this way, a series of technology specific acceleration scenarios were devised, and then a set of aggregated ATD scenarios were assembled in which emerging technologies are accelerated in parallel. The ATD scenario set is listed in Table 7.1. The *Low Carbon* (LC) scenario provides the 'non-accelerated' baseline scenario here because of the focus on the potential contribution of accelerated development to decarbonization of the UK energy system. As well as the scenarios listed in Table 7.1, a number of additional variant ATD scenarios were also devised to consider, for example, the failure of certain supply technologies to be developed (such as CCS or hydrogen/fuel cells), or to have delayed availability or higher or lower levels of performance.

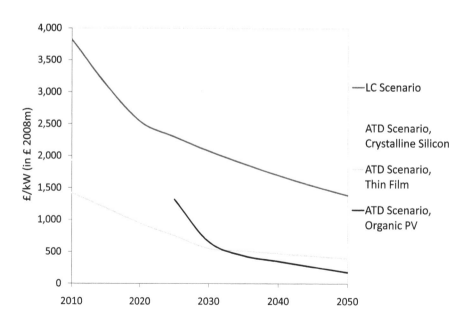

Figure 7.3 *Capital cost assumptions for the ATD solar PV scenarios*

Table 7.1 *Accelerated technology development (ATD) scenario set*

	Scenario name	Description
Non-accelerated low carbon scenarios	Low Carbon (LC)	Low carbon pathway with 80% CO_2 reduction by 2050
	Medium Carbon (LC-60)	Low carbon pathway with 60% CO_2 reduction by 2050
Single technology ATD scenarios	LC-60-Wind LC-60-Marine LC-60-Solar LC-60-Bio-energy LC-60-Nuclear LC-60-CCS LC-60-Fuel cells	Single technology accelerated development scenarios, low carbon pathway with 60% CO_2 reduction by 2050
Aggregated ATD scenarios	LC-Renew	All four renewable technologies accelerated, low carbon pathway with 80% CO_2 reduction by 2050
	LC-Acctech	All seven low carbon technologies accelerated, low carbon pathway with 80% CO_2 reduction by 2050

The UK energy system model used here, the UK MARKAL elastic demand model (MARKAL-MED), is described in Chapter 4. As Chapter 4 explains, the energy supply portfolios selected using MARKAL-MED from now to 2050 are based on maximizing total social welfare (the sum of consumer and producer surplus) from many different available supply and conversion pathways, consistent with UK policy ambitions for decarbonization to 2050. UK CO_2 emission reduction trajectories are imposed here as a constraint: the single technology ATD scenarios impose a 'straight line' decarbonization pathway from now to 2050 consistent with 60 per cent CO_2 emission reductions by 2050 (compared to 1990 levels), reflecting the level of UK policy ambition at the time they were constructed in 2008. The aggregated ATD scenarios also impose a 'straight line' decarbonization pathway, but consistent with 80 per cent CO_2 emission reductions by 2050 (branching away from the 60 per cent pathway in 2020), reflecting the raised policy ambition introduced in late 2008 in the UK Climate Change Act (HMG, 2008).[2]

The same decarbonization trajectory is imposed for both accelerated and non-accelerated scenarios, so that they illustrate alternative ways of meeting the same overall policy objective under different assumptions of supply-side technology progress. In reality, of course, accelerated technology development could allow for deeper (or earlier) carbon reductions to be made for the same overall cost to society, but this is not formally considered here. (The impact of higher and lower decarbonization ambitions to 2020 and 2050 is explored in Chapter 5.)

The UK Climate Change Act 2008 incorporated the Climate Change Committee's recommendation that emission reduction targets should be applied to the basket of six greenhouse gases (GHGs) covered by the Kyoto Protocol (CCC, 2008). Given the focus here on energy sector emissions, the scenarios only constrain CO_2 emissions. (Non-CO_2 emissions associated with decarbonization and energy security are considered in Chapter 10.) The Committee also recommended adoption of an interim target of 34 per cent GHG reductions to 2020, subsequently embedded in the UK's legislated GHG 'budget' to 2020 (a target which, it suggested, should be raised to 42 per cent upon the agreeing of a successor international framework to the Kyoto Protocol). Because non-CO_2 GHG emissions have declined substantially since 1990 – an overall decline of 50 per cent is projected from 1990 to 2020 – the equivalent 'CO_2 only' reduction ambition for 2020 corresponding to a 34 per cent reduction in GHG emissions is 29 per cent; for a 42 per cent GHG reduction target, the equivalent CO_2-only target is 40 per cent (CCC, 2008). In all the scenarios reported here, the CO_2 reduction constraint imposed at 2020 is just over 26 per cent, i.e. below the UK's CO_2 target of 29 per cent. In the LC, Renew and Acctech 80 per cent scenarios, 29 per cent CO_2 reduction is achieved in 2021; 40 per cent in late 2027.

Scenarios, system modelling and the real world

The ATD scenarios illustrate the possible role of accelerated technology progress in meeting UK decarbonization ambitions. They are *not* forecasts of UK energy system development; there are far too many uncertainties and unknowns involved – technological, economic and political (including ongoing changes to policy frameworks for low carbon technology development) – to make reliable predictions of this kind, even over relatively short timescales. Rather than forecasting, the intention is to offer structured insights into key enablers of change (Strachan et al, 2009), and more specifically, to consider possible responses to decarbonization imperatives over the next 40 years, under a series of distinctive assumptions.

The system level scenarios discussed below are generated using MARKAL-MED, based on maximizing total social welfare over the entire time period, assuming 'perfect foresight' about the cost, performance and availability of technology and other inputs. The modelling results are therefore based on an implicit representation of decision-making in the energy sector characterized by well operating competitive markets with full understanding of present and future costs. In reality, of course, the energy sector is characterized by highly imperfect markets, and high levels of uncertainty about technology cost and performance. Also, as well as the comparative costs of energy supply technologies, many other factors also affect investment in energy systems, including perceived investment risks (Gross et al, 2007), changes to energy networks and storage technologies, human behaviour and lifestyle, energy efficiency and many other regulatory, organizational and political factors. While a number of these wider factors are explored in other chapters in this book, system modelling remains a highly simplifying and abstract way to represent energy system change. Despite these limitations, however, the scenarios offer useful insights into the dynamics of change – such as the possible long-term role of innovation – which may not be readily accessible by other means.

Scenario exercises based on present-day knowledge and perceptions may under-represent the potential of emerging innovations to contribute to system change over the longer term. In many energy system modelling exercises, a small number of relatively well understood, more mature options dominate the market over the short term and crowd out more radical or more emergent technologies over the longer term (ERP, 2010). In the LC scenario, for example, nuclear power and coal with CCS provide the major contributions to power sector decarbonization from now to 2050, with other low carbon supply technologies making much smaller contributions. Nuclear power delivers virtually all of the expansion of low carbon electricity production after 2035, capacity which is used to enable the electrification of transport and heating. In essence, the power generation mix in this scenario reflects the characterization of nuclear power as a relatively established and economic low carbon supply technology; it also reflects optimistic assumptions about the cost and performance of CCS. By contrast, as is discussed below, a number of other emerging technologies play significant roles in the ATD scenarios.

The early stage lock-ins seen in many scenario exercises have real world correspondences – energy supply investments are long lasting and require a host of interests to support them, so that, as discussed earlier, energy systems tend to lock around established technologies. This is an inevitable and necessary feature of system development (Walker, 2000). It becomes a problem when the drivers and expectations around a system undergo significant change, and decarbonization – effectively, an imperative to escape from *carbon lock-in* built up over many decades – presents a particularly systemic challenge (Unruh, 2000). In this context, it is important to step back from present-day conditions and perceptions and explore a number of alternative possible system futures. For this chapter, the particular concern is that the potential role of emerging supply options be acknowledged, and explored in a structured way within overall system change.

However, the ATD scenarios are concerned not only with finding appropriate ways to represent acceleration in system level scenarios, but also with the 'real world' expectations, enablers and barriers involved in accelerated technology development. This means interpreting the general definition of 'accelerated development', described above, terms of the specific circumstances of each technology. Each of the broad technology categories in the ATD scenarios spans a number of different systems, typically referred to as different design generations, such as third or fourth generation nuclear fission, or first, second or third generation solar PV. For each technology, devising scenarios of accelerated development involved addressing both possible incremental improvements within generations, and more radical or disruptive shifts between generations. This highlighted discernible differences in the feasibility and desirability of accelerated development for different technologies, and the relative emphasis on incremental or radical change.

For example, for relatively mature and large, capital intensive technologies such as nuclear power, more radical inter-generational technology acceleration may be less attractive, because it is associated with increased technical and economic risk, possibly incurring higher costs for investors and project developers. As a result, much of the focus of the nuclear power community in the UK is on supporting current generation plant, rather than more ambitious and radical designs some decades into the future. For less mature and more modular technologies, by contrast – such as solar PV – the emphasis may be much more on accelerated development of future generations, and RD&D efforts to capture step-changes in cost and performance.

The technical and financial risk associated with more radical innovation is a general dilemma for technology development in liberalized economies, and is especially significant in the context of long-term system change. Over the shorter term, the primary focus is inevitably on *deployment* rather than *development* – that is, on established technologies which carry reduced technical and economic risk, and also on non-technical institutional factors affecting deployment, such as licensing and planning processes. Over longer timescales, however, as the ATD scenarios illustrate, more radical or disruptive intergenerational innovation becomes potentially very significant.

As discussed above, the ATD scenarios emphasize the *learning-by-RD&D* potential of emerging and more radical options, rather more than the *learning-by-deployment* potential of more mature technologies. While this provides a useful corrective to short-termism and lock-in, it may presume an unrealistic degree of responsiveness and far-sightedness, with investors and others responding quickly to changing technology costs. It may also underplay the learning potential of more mature technologies; there is evidence that some established technologies in the energy sector have been able to steadily reduce their costs over time (McVeigh et al, 1999). Once technologies assume a central position in the energy system, resources are directed to support their gradual improvement and this can become very significant over long time periods.

There are high levels of technical, economic and political uncertainty involved here, making it impossible to identify a optimum decarbonization pathway from now to 2050, or to strike an ideal balance between short and long timescales or incremental and radical innovation. In addition, what appear as 'sub-optimal' paths from a techno-economic perspective may offer significant wider benefits, such as enhanced security or new industry creation. These are pressing matters for UK energy policy, given, for example, the very large scale deployment of renewables technologies anticipated over the next decade. As discussed later, the very substantial commitments required here, in supply chains, installation capacities and project finance, will build up significant momentum behind some technology pathways, and condition the response to imperatives for deeper change after 2020.

ATD scenarios and UK decarbonization pathways

Accelerated technology development opens up more affordable paths for achieving deep long-term reductions in carbon emissions in the ATD energy system scenarios. The impact of accelerated development becomes more significant after 2030, as emerging technologies become affordable, and as overall decarbonization ambitions increase. For example, the suggested economic benefit of technology acceleration is much greater under an overall decarbonization ambition of 80 per cent rather than 60 per cent. A number of emerging technologies provide important contributions to UK energy system change in the ATD scenarios (see Table 7.2 and Figure 7.5, below). *Bio-energy* technologies play an important role from 2020 onwards, in helping to decarbonize different energy services (power, heat and transport). *Offshore wind*, *marine energy* and *solar PV* are deployed much more significantly in accelerated development scenarios, although mostly after 2030. Accelerated *hydrogen fuel cells* development has a major role in transport sector decarbonization after 2030. Coal-fired generation using *carbon capture and storage (CCS)* plays a significant part in the UK energy supply mix as early as 2020 in both non-accelerated and accelerated scenarios, reflecting relatively aggressive assumptions about the pace of CCS development in the LC scenario. (Additional scenarios were therefore produced to illustrate ATD decarbonization pathways in the absence of CCS, or the delayed availability of CCS.) The accelerated development of *nuclear power* has a more modest impact, reflecting its relative maturity, as discussed above. However, nuclear power plays a much greater role if CCS is assumed to be unavailable, or to have delayed availability.

Electricity supply sector

The ATD scenarios feature alternative pathways for decarbonizing the UK power system in the longer term, with significantly increased contributions from emerging technologies, especially offshore wind power, but also marine, solar PV, bio-energy and in some scenarios, nuclear power (see Figure 7.5, below). For example, in the LC Acctech scenario, there is a major rise in offshore wind exploitation over the longer term, with about 70GW installed by 2050, generating around one-third of all power supplied. Similarly, Figure 7.4 illustrates the impact of accelerated development of marine energy in different scenarios, with earlier initial deployments, and much greater longer-term deployment, as compared to non-accelerated scenarios. Overall, by making a number of emerging technologies more commercially deployable before 2050, technology acceleration enables more diverse low carbon supply portfolios to emerge.

However, the ATD scenarios have much less impact on deployment over the shorter term, up to 2020. Combined cycle gas turbine technology remains the most significant power generation technology over the next decade in all scenarios, but with emerging contributions from renewables (wind power and bio-electricity) and CCS-abated coal in a number of scenarios. However, if CCS technology is assumed not to be available, there is a significant increase in wind power deployment, with 21GW deployed by 2020.

Over the medium term, to 2035, the power sector is substantially decarbonized, and the supply portfolio is transformed in all scenarios. Significant differences have now emerged between non-accelerated and accelerated model runs. Coal-CCS is a key supply technology in most scenarios, but bio-energy and marine energy deploy more significantly over the medium term in LC Acctech,

Table 7.2 ATD scenarios: summary of assumptions and impacts[3]

	Cost and performance assumptions in ATD scenario	Impact of ATD on deployment in UK decarbonization pathways
Wind power	ATD cost and performance assumptions were applied to offshore wind: an aggressive capital cost reduction of 10% to 2020, based on a global learning rate of 10% to 2020. After 2020 a lower but sustained learning rate to 2050. Onshore wind resource capacity limits were raised for ATD scenario, but no cost or performance gains were applied.	The accelerated development of offshore wind power has a major long-term impact on deployment (especially after 2030) but a more moderate impact over the short–medium term. Offshore wind has an expanded short-term role if CCS technology is assumed to be unavailable before 2030. More onshore capacity is also deployed over the next decade under ATD assumptions.
Marine energy	ATD scenario involves 'niche learning' driven by UK and global supported deployment to 2015, with an aggressive global learning rate of 10% after 2015 out to 2050. ATD scenario covers wave and tidal flow; it doesn't include tidal barrages or lagoons.	Marine acceleration has significant impact on deployment. First deployments are seen much earlier than in non-accelerated scenarios (after 2010 in ATD Marine, and after 2020 in LC, see Figure 7.4). Initial deployment is tidal flow, with wave becoming dominant after 2030. Long-term deployment is constrained by resource availability assumptions.
Solar PV	The ATD solar scenario involves accelerated global learning for first, second and third generation PV (see Figure 7.3 for unit cost assumptions). UK-specific deployment support policies for solar PV are not included in the analysis. The ATD scenario does not cover concentrating solar power or solar thermal technologies.	Solar PV acceleration has a significant long-term impact, with low cost third generation solar cells deploying in ATD scenarios. The lack of earlier deployment possibly reflects the absence of UK deployment support policies in the ATD scenarios. Long-term deployment is also constrained by installable capacity assumptions. A higher long-term contribution is seen in scenarios where CCS has limited availability.
Bio-energy	Bio-energy scenario focuses on 5 areas offering potential for accelerated development: gasification for heat and power; plant biotechnology; agro-machinery; ligno-cellulosic ethanol and fast pyrolysis. ATD assumptions were applied as reduced capital and O&M cost, improved crop yields and increased conversion efficiency. To restrict importing, ATD feedstock productivity assumptions were only applied to domestic resources.	Bio-energy acceleration has a significant impact on UK decarbonization pathways over short, medium and long timescales. Domestic biomass feedstocks are deployed in power, heating and transport sectors at different times in the ATD scenarios (see discussion of 'system level effects' in main text, below). The most significant technology changes are bio-engineering improvements to crops yields, improved gasification technology and development of second generation ligno-cellulosic ethanol.

	Cost and performance assumptions in ATD scenario	Impact of ATD on deployment in UK decarbonization pathways
Nuclear power	ATD assumptions were applied to Gen III, III+ and IV nuclear fission technologies, and also nuclear fusion, in terms of reduced cost, improved efficiency and earlier availability. ATD nuclear assumptions are relatively modest, reflecting the relative maturity of nuclear fission and the long-term timescales needed for fusion development.	Nuclear acceleration has moderate medium–long impact on deployment, but with much greater medium- and long-term impact if CCS is assumed to be unavailable, or to develop later. Only Gen III fission reactors are deployed; more advanced fission technologies are not deployed, reflecting assumptions of their relatively limited advantages. Fusion deployment is not seen in any ATD scenarios before 2050.
Carbon Capture and Storage (CCS)	Assumptions about the availability, cost, and performance of CCS are optimistic in the LC (non-accelerated) scenarios, and were left essentially unchanged for ATD scenario. First availability is just before 2020. Scenario assumptions do not explicitly distinguish between different types of CCS technology.	Coal CCS has major medium- and long-term role in most scenarios, although its longer-term role is limited by residual emission assumptions in 80% decarbonization pathways. Delayed availability also limits its long-term role, as other technologies become commercially available. Gas-fired plant with CCS is not deployed in the ATD scenarios, reflecting relative fuel price assumptions.
Hydrogen fuel cells	The ATD scenarios developed for stationary power and transport applications are based on aggressive global learning and capital cost reduction assumptions. The transport FC scenario data relate primarily to polymer electrolyte membrane (PEM) fuel cells, although other types are also included in the analysis.	Hydrogen fuel cells (HFC) acceleration has a major long-term impact on transport sector decarbonization, with the mass conversion of transport fleets to HFC technology after 2030. HFCs have only a minor role in power generation sector. HFC acceleration also has significant effects on the wider energy system (see main text, under 'system level effects').

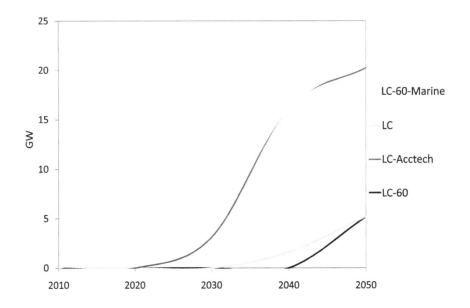

Figure 7.4 *Marine energy installed capacity under accelerated and non-accelerated scenarios (selected data, smoothed)*

compared to LC. If CCS technology is not available during this period, there are much greater contributions from nuclear power (with 21GW of nuclear capacity by 2035) and renewables – with bio-electricity (22GW) and marine energy (14GW) making significant medium-term contributions; wind power deployment remains relatively modest, at 20GW.

By 2050, the size and make-up of the power supply sector has again been transformed as the energy system decarbonizes more deeply. Low carbon electricity now provides an important means of decarbonizing the wider energy system, but accelerated and non-accelerated scenarios are associated with distinctive long-term decarbonization pathways. The role of coal-fired technology is now limited as residual CO_2 emissions associated with carbon capture technology become significant under higher decarbonization ambitions as the 80 per cent target approaches.[4] Instead, *zero* emission carbon technologies deploy very significantly in the longer term – mostly nuclear power in the non-accelerated LC scenario, and mostly renewables technologies in ATD scenarios, especially offshore wind power. Other renewables also find significant deployment, but at levels restricted by either assumed resource limits (for marine or biomass), or cost-effective installable capacity limits in the UK context (for solar PV). The contribution of bio-electricity declines in the longer term, as limited domestic biomass resources are used to decarbonize transport and heating sectors, rather than power. Gas-fired generation has an important longer-term role in ATD scenarios, but as a reserve plant technology, rather than bulk supply.

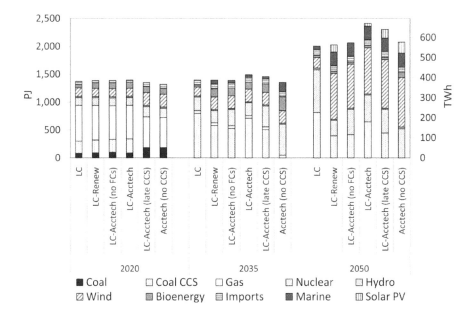

Figure 7.5 *Power generation portfolios under different ATD scenarios*

The scenarios also suggest that achieving deep emission reductions may involve the development of a much larger UK power supply industry over the long term. While this expansion is seen with or without accelerated development, it is more pronounced in some accelerated development scenarios, with, for example, installed generating capacity doubling between 2030 and 2050 in the LC Acctech scenario. Fuel cell technology acceleration has a distinctive influence here – hydrogen production by electrolysis creates substantially raised demands for low carbon electricity. Combined together, the accelerated development of renewables and fuel cells technologies is associated with a particularly large long-term expansion of the UK power sector.

The proportion of electricity supplied from renewables in the LC Acctech scenario increases steadily over short, medium and longer timescales, from 20 per cent in 2020, to 30 per cent in 2035 and 55 per cent by 2050. The equivalent proportions in the LC Renew scenario are only marginally higher, but if CCS technology is assumed to be unavailable before 2030, there is a significant increase in the renewables contribution, rising to 27 per cent by 2020, and 58 per cent by 2050 (Figure 7.6).

System level effects of technology acceleration

Energy system modelling enables a structured exploration of the possible dynamics of energy system change under accelerated development assumptions. As well as combining together the technology-specific accounts of accelerated development, so as to consider cross-technology competition and

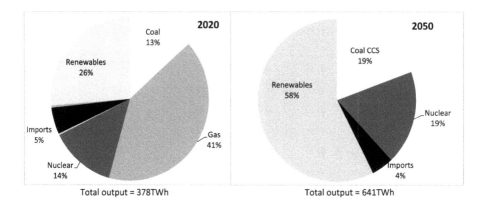

Figure 7.6 *Share of electricity generated, LC Acctech (late CCS) scenario*

synergies (as discussed above for the power sector) this also draws attention to some of the wider possible impacts of technology acceleration on preferred decarbonization pathways. These wider effects, such as the relative emphasis on decarbonizing different energy services (power, transport or heating) or on demand-side management versus supply-side transformation, may be non-intuitive or unexpected.

For example, biomass lends the energy system an important degree of flexibility of response to the decarbonization imperative, because it can be used in different forms across heating, transport or power generation. However, domestic biomass resources are limited, and given concerns about the environmental sustainability of imported biomass, ATD feedstock productivity assumptions were only applied to domestic resources. In the ATD scenarios, the preferred use of limited domestic feedstocks depends on the overall level of permitted carbon emissions, and the cost and availability of other low carbon supply options. In the single technology ATD bio-energy scenario, for example, biomass resources are used mostly for electricity generation in the medium term, but mostly for residential heating in the longer term (see Figure 7.7, below).

In aggregated scenarios, biomass resources are used mainly for electricity generation where there is a reduced overall requirement for decarbonization (such as in scenarios with a 60 per cent reduction ambition to 2050); in mid-term supply portfolios (in the 2020s and 2030s) in 80 per cent reduction pathways; or when other technologies for decarbonizing electricity (such as CCS) are assumed to be unavailable. By contrast, biomass resources are used mainly for decarbonizing the heat sector over the longer term in 80 per cent scenarios where there are other affordable ways to decarbonize the power sector (such as in LC Acctech), or in the transport sector, if fuel cell technology acceleration is excluded.

Similarly, hydrogen fuel cell technology acceleration provides an alternative, affordable way of decarbonizing the transport sector, with significant

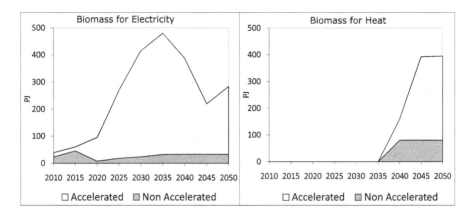

Figure 7.7 *Use of biomass for power and residential heating, non-accelerated (LC-60) and accelerated (LC-60-Bio-energy) scenarios*

consequences for the wider energy system. For example, it is associated with a distinctive decarbonization pathway for residential heating, with much less reliance on heat pumps than in LC, and a significant uptake of biomass boilers using pellets. In the LC scenario, residential heat pumps provide 842PJ of 'useful energy' for space and water heating in 2050, but only half as much (422PJ) in the LC Acctech scenario; wood pellet boilers provide 344PJ of energy for heating in LC Acctech, but are not used at all in LC. While there is a very significant growth in the long-term size of the power sector in both LC and LC Acctech scenarios, this has distinctive drivers: heat pumps and electric vehicles' adoption in LC, and hydrogen production for transport decarbonization in LC Acctech.

Because low carbon electricity is an important enabler of system-wide decarbonization, the absence or presence of CCS has significant effects across the system. ATD scenarios without CCS feature less overall demand for electricity (as electricity is generated from more expensive sources), greater take-up of solar PV, and a switching of biomass resources from residential heating to power and transport. The delayed commercialization of CCS (to the post-2025 period) reduces its long-term role, as other low carbon supply technologies mature, and as residual emissions from CCS become more significant as overall decarbonization ambitions increase.

In terms of broader decarbonization pathways for different energy services, the ATD scenarios follow broadly the same pattern as non-accelerated scenarios. The electricity supply sector decarbonizes first and most thoroughly, and is substantially decarbonized by 2030. Other energy sectors (transport and heating) then decarbonize after 2030. However, by creating alternative low carbon supply options, technology acceleration changes the types of technologies enabling these changes. The same overall pattern of falling primary energy demand from now to 2050, as the energy system decarbonizes, is followed with or without accelerated technology development. However, electricity from

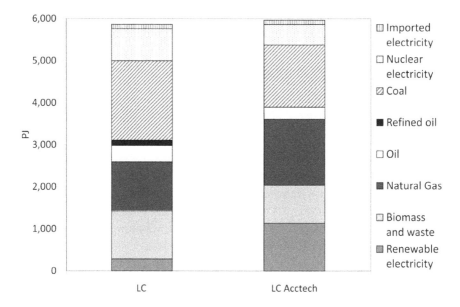

Figure 7.8 *Primary energy demand by fuel in 2050, LC and LC Acctech scenarios*

renewable sources provides a much greater proportion of primary demand by 2050 under accelerated development assumptions: almost 20 per cent in LC Acctech, compared to under 5 per cent in LC (Figure 7.8). In terms of final energy demand for different fuels, LC Acctech features higher demand for hydrogen, and reduced demand for electricity and bio-fuels in transport; by 2050, hydrogen is the dominant transport fuel in LC Acctech scenarios which include accelerated fuel cells development.

Long-term benefits of technology acceleration

The ATD scenarios offer some indication of the overall economic benefits of accelerated technology development. These advantages accrue mostly in the long term, as accelerated development creates more affordable ways to achieve deep carbon emission reductions. Across all services (transport, industry, residential, commercial services and agriculture), energy demands reduce over time as energy becomes more expensive as the system decarbonizes. (This reflects the price elasticity of demand, as represented in MARKAL-MED.) In LC Acctech, however, because there is greater availability of affordable low carbon supply options over the longer term, the energy cost increases and associated demand reductions are significantly reduced, compared to the LC scenario.

Two parameters – the marginal cost of CO_2 abatement, and the overall 'welfare cost' of decarbonization – allow for some measure of this benefit. It is important to reiterate here that given the many uncertainties regarding technology cost and performance, and other uncertainties in long-term energy system change, these figures can only offer an illustration of the possible benefits of accelerated development, rather than a robust cost–benefit analysis.

In all scenarios consistent with 80 per cent decarbonization by 2050, the marginal cost of carbon abatement increases over time as progressively more expensive carbon abatement options are deployed. In the accelerated development scenarios, however, this increase is significantly less than in non-accelerated equivalents; by 2050, the marginal cost of CO_2 abatement is just over £130/tonne in LC Acctech, compared to just under £170/tonne in LC. Around half of this difference is associated with fuel cells acceleration: in the LC Acctech variant which excludes fuel cells acceleration, the suggested marginal cost of CO_2 abatement in 2050 is just over £150/tonne.

The modelling results also suggest that technology acceleration has the potential to substantially reduce the overall welfare cost of decarbonization, especially as carbon emission reduction ambitions increase. As described in Chapter 4, MARKAL-MED expresses welfare savings as the sum of reduced energy system costs and increased consumer surplus. In 2050, for example, annual welfare savings associated with technology acceleration increase by around two-thirds (from £5.7 billion to £9.5 billion) if the overall decarbonization ambition is raised from 60 per cent to 80 per cent.

Between 2010 and 2050, accelerated technology development provides a total saving in the welfare costs of achieving 80 per cent decarbonization of £36bn (Figure 7.9).[5] On a UK 'break-even' basis, and assuming accelerated development is driven mainly by RD&D, this figure could be translated into a annual budget for *additional* UK RD&D investment in low carbon technology development of just under £1bn per annum (indicated as a dashed line in Figure 7.9). This compares to *total* public spending levels of around £160m in 2008 (see Figure 7.1). Given the long timescales involved, and the early stage of development of the technologies covered here, much of the investment would need to be provided by public funding programmes. A long-term view of investment and return is assumed here: although there are some shorter-term benefits from accelerated development, the major benefits only start appearing after 2030.

However, the accelerated development of low carbon energy supply technologies is more appropriately seen as a *global* rather than national challenge, with RD&D investments being made internationally, and UK commercial firms and publicly funded organizations making greater or lesser relative contributions across different technology fields. This means that the costs of technology acceleration are, at least in principle, shareable internationally. (In practice, there are substantial political, institutional and organizational barriers to developing a globally shared programme of low carbon technology development.)

Estimates of the possible costs of a global programme of accelerated development of low carbon supply technologies have been provided by the International Energy Agency (IEA). Although the IEA and ATD scenarios differ in their detailed assumptions, IEA global scenarios of energy system change distinguish between lower and higher rates of supply-side technology development in a broadly similar manner to the LC and LC Acctech scenarios (IEA, 2008). Considering only a similar basket of technologies as those analysed

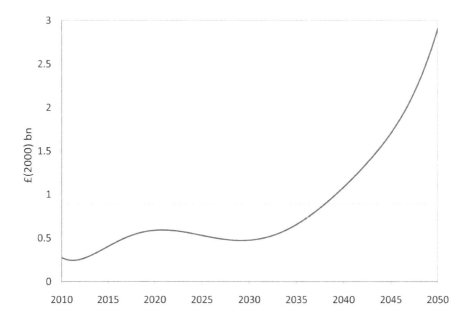

Figure 7.9 *Welfare increases associated with technology acceleration (2010–2050)*

in the ATD scenarios, and allocating the UK 'share' of RD&D costs by UK share of global GDP, the additional investment costs to the UK associated with a global programme of accelerated development are around US$8bn (about £5.5bn).[6] Put alongside the much larger suggested national benefit to the UK arising from accelerated development (£36bn), the overall message here is that the benefits of acceleration are likely to significantly outweigh the investment costs, in terms of additional RD&D spending.

Implications and challenges

Innovation strategy and research priorities

A summary of the innovation and research priorities associated with accelerated development of the technologies considered here is presented in Table 7.3. The table shows areas where enhanced innovation efforts and RD&D investments are needed to generate the improvements in technology availability, cost and performance associated with the ATD scenarios. Given the unpredictabilities of technological change, especially over long time periods, the impact of RD&D investment on technology cost is inevitably uncertain. In addition, many other factors influence the cost (more properly, the price) of technology. However, while there can be no 'hard-wired' relationships between inputs, processes and outcomes, the ATD scenarios *do* identify particular areas where improved innovation strategies and increased RD&D efforts have the potential to stimulate accelerated learning effects, improved technical performance and reduced cost.

The themes and priorities summarized in Table 7.3 draw on a wide range of sources, as detailed in Winskel et al, 2009. This list extends beyond those technologies which are deployed in the ATD scenarios of UK energy system change; the supply portfolios in the ATD scenarios omit a number of emerging technologies which may well be developed and deployed internationally and in the UK over the next 40 years, such as second generation thin-film PV, advanced bio-energy technologies such as algal bio-fuels and advanced designs of nuclear fission reactors. Given their limitations and qualifications, the ATD scenarios should not be used to identify technology winners and losers and associated research priorities. Rather, Table 7.3 draws on the wider 'off-model' analysis of innovation themes and research priorities in the ATD scenarios. For some of these, but not all, the ATD scenarios illustrate their significant potential role in UK decarbonization pathways over the longer term.

Even so, this coverage is far from exhaustive. As well as the emerging supply technologies analysed here, there are a host of other emerging low carbon supply options not covered, and there is also a need for innovation efforts and RD&D support to be directed to 'enabling' technologies also not analysed here, such as emerging types of power storage, smart grids and demand-side management. The omission of these enabling technologies means that the ATD scenarios do not represent the possibility of a more radical transition to highly distributed energy systems. As discussed earlier, the ATD scenarios are intended to illustrate an important dynamic in overall system change – the potential role of supply-side accelerated technology development – rather than to provide a comprehensive analysis of the myriad possible low carbon innovation opportunities.

Shorter-term policy: from now to 2020

The ATD scenarios should be considered in the context of the changing policy drivers for energy system change. As discussed in Chapter 3, the policy and regulatory framework for the UK energy system is now being substantially remade, and ambitious targets have been established for both decarbonization and renewables deployment, especially over the next decade to 2020. In particular, the UK Government is reforming renewables deployment support measures to encourage 30 per cent or more of electricity generation to be produced from renewable sources by 2020 (HMG, 2009), enabling the UK to meet its commitments under the EC's Renewable Energy Directive (CEC, 2009).

In the LC Acctech scenario, the proportion of electricity supplied from renewable technologies in 2020 is 20 per cent, and does not rise above 30 per cent until after 2035. Clearly, there is a significant disparity here between the deployment of renewables presented in the LC Acctech scenario, and UK policy targets. In part, this reflects highly aggressive assumptions about the availability and cost of CCS in the ATD scenarios, such that it plays a significant role in many modelled supply portfolios as early as 2020. If CCS is assumed not

Table 7.3 *Innovation themes and research priorities for ATD scenarios*

	Innovation themes	Specific research priorities, examples
Wind power	• Improved wind turbine efficiency for low speed (onshore) locations • Improved condition monitoring and enhanced reliability of offshore turbines • Expanded offshore electricity transmission infrastructure • Electricity storage technologies and/or demand-side management	• Blade materials technology • Control algorithms • Generator design • Aerodynamic design (limited scope for improvement) • Offshore balance of system costs • Offshore resource characterization
Marine energy	• Consensus on designs (system concepts and components) • Greater collaboration on generic technologies and components • Improved operational data on prototype performance in real operating conditions • Analyse feasibility of more radical designs • Promote knowledge transfer from other sectors	• Resource modelling and measurement • Device modelling • Moorings and seabed attachments • Power take-off and control • Installation and O&M costs • Environmental impact assessment • System simulation
Solar PV	• Low cost, stable and efficient cells • Reduced production costs • Improved control systems and storage technologies • Reduced balance of system costs for all module types • Bringing together materials researchers and plant designers • Improved performance prediction tools	*For crystalline silicon* • Increasing cell efficiency • Improved materials utilization (in particular silicon) • Higher yielding processing *For thin film cells* • Increasing cell efficiency • Improved device structure and substrates • Large area deposition techniques • Improved manufacturing processes (including roll-to-roll) *For organic cells* • Increased cell efficiencies and device lifetimes
Bio-energy	• Improved 'carbon efficiency' of feedstocks • Improved conversion technologies • System level research on optimal use of limited biomass resources • Lifecycle and environmental impact analysis • Establishing standards for biomass trade and use	• Improved efficiency, cost, flexibility of existing conversion technologies • High yielding second generation biomass with minimal land and water requirements • 'Third generation' novel feedstocks such as algae and artificial photosynthetic systems • Novel conversion technologies, e.g. pyrolysis

	Innovation themes	Specific research priorities, examples
Nuclear power	*Fission* • Supporting existing plant operations • Enabling deployment of advanced reactor systems • Solutions for waste management (including legacy waste) and plant decommissioning *Fusion* • Plasma performance • Enabling technologies • Materials, component performance and lifetime	*For generation III and III+ reactors* • Long-term materials irradiation and structural integrity • Control, instrumentation, monitoring • Lifetime prediction • High temperature materials • Fuel burn-up and long-life fuel cores *For generation IV reactors* • Advanced materials • Fuel fabrication and high burn-up fuel • Thermal hydraulics • Spent fuel reprocessing and recycling *For nuclear fusion* • Steady-state operation and divertor performance • Superconducting machine • Power plant diagnostics and control
Carbon capture and storage	• Demonstrating existing component technologies in an integrated CCS system • Planning and building transport infrastructure • Optimizing the retrofitting of capture technology onto power plants • R&D for potential future improvements, e.g. efficient, low cost capture and integrity and capacity of storage, especially aquifers	*Capture* • *Post-combustion:* resistant amine solvents or alternatives • *Pre-combustion:* improved membrane or pressure swing separation of CO_2 from H_2 and improved O_2 separation • *Oxyfuel combustion:* lower cost O_2 separation from air, better membranes for CO_2 separation, chemical looping *Storage* • Assessing aquifer storage potentials • Evaluating CO_2 sealing and leakage • Monitoring and verification technology from existing applications
Hydrogen and fuel cells	• Cost reduction of the hydrogen drivetrain • Cost reduction of hydrogen production chains • System integration for hydrogen • Safety and reliability of hydrogen applications • Compliance with long-term sustainability needs	*Drivetrain* • Cell components (membrane, catalyst, materials) • Periphery components (air supply, humidification, valves, power and control); onboard storage *Hydrogen production chains* • Electrolysers, biomass gasification systems, CCS and standard components and instruments *System integration* • Integration of drivetrain, onboard storage and auxiliary safety equipment, valves, and electronics • Integration with renewables in island/remote systems • Harmonized regulations, codes and standards • Hydrogen from renewables, fossil fuel + CCS, or nuclear

to be available, significantly higher levels of renewables penetration are seen, with 27 per cent of generation by 2020 in LC Acctech (no CCS), rising to 40 per cent by 2030.

Even with CCS excluded, however, overall levels of renewable electricity generation in the ATD scenarios are lower than those in the 'lead scenario' in the UK Government's *Renewable Energy Strategy* (RES) (HMG, 2009). In LC Acctech (no CCS), renewables provide just under 100TWh of electrical energy in 2020, compared to 117TWh in the RES lead scenario. Within this, wind power is a more dominant technology in the RES, with around 27GW installed by 2020 in the lead scenario, roughly evenly split between onshore and offshore (around 13GW offshore is deployed in the RES lead scenario). In LC Acctech (no CCS), 21GW of wind power is installed by 2020, with only around 8GW of this from offshore. However, the RES also makes clear that the Government's ambitions for offshore deployment to 2020 are 'much greater' than those in the lead scenario, with 20GW described as 'achievable' and 33GW as 'feasible' (HMG, 2009, p41).

Realizing these much greater ambitions will require much greater levels of financial support for offshore wind deployment over the next decade than those assumed in accelerated development scenarios. Looked at differently, the UKERC scenarios suggest that the UK Government's highly ambitious aspirations for offshore wind deployment over the next decade may direct the UK energy system into a more expensive decarbonization pathway from now to 2020. While the UK has a vast offshore wind resource, the ATD scenarios portray offshore wind as an emerging technology, much of whose potential is most economically realized well after 2020. By contrast, bio-energy provides a somewhat greater short-term contribution in the ATD scenarios, with 30TWh of bio-electricity generated in 2020 in LC Acctech (no CCS), compared to 26TWh in the RES lead scenario. This said, an ambitious UK offshore wind deployment programme over the next decade may lead to significant cost reductions through learning-by-deployment. Although the ATD wind power scenario includes aggressive cost reduction assumptions for offshore wind technology, based on global deployment projections, the possible learning effects from UK supported deployment were not explicitly represented.

The scenarios are also relatively modest in overall decarbonization ambition to 2020, compared to the upper end of current UK policy ambitions. In the LC and LC Acctech decarbonization pathway, CO_2 emissions reductions are just over 26 per cent in 2020; 29 per cent reduction (the established UK target) is achieved very shortly after, in 2021, but 40 per cent reduction (the UK's declared target upon wider international agreements being struck) is not achieved until 2027.

Meeting a 40 per cent target for CO_2 emission reductions by 2020 cannot easily be achieved by earlier deployments of the low carbon supply technologies seen in the ATD scenarios. For example, there are substantial engineering, financial and planning challenges involved in increasing the contribution of nuclear power by 2020, although a combination of extended lifetimes of existing plant, and a programme of fleet build of new reactors could provide for a more sustained nuclear contribution than seen in the ATD model runs (this is discussed in Chapter 6 of Winskel et al, 2009). It is also unlikely that CCS technology will be commercially available at significant scale before 2020 – indeed, a somewhat later impact from CCS is seen as more credible by some observers. In the ATD scenario, if CCS is assumed to be unavailable throughout the 2020s, other technologies deploy to a much greater extent, so that 40 per cent decarbonization by 2030 is associated with further expansion of bio-electricity and an emerging role for marine energy; offshore wind expansion is seen only after 2030.

This suggests that ambitious targets for renewables deployment and decarbonization over the next decade are unlikely to be most affordably delivered by a dominant reliance on offshore wind, but are more likely to feature enhanced contributions from bio-energy and onshore wind (where barriers are mostly non-technical), and also, an emphasis on demand-side measures, such as improved energy conservation and efficiency, and lifestyle changes (see Chapter 9 for a discussion of the latter). In sum, accelerated renewables *development* has more powerful effect on UK *deployment* patterns in the medium to longer term than over the next decade, especially for offshore wind.

Longer-term policy: from 2020 to 2050

Current UK energy policy is being driven largely by relatively short-term targets for decarbonization and renewables deployment by 2020. Reflecting this, the UK Government's *Low Carbon Transition Plan* (HMG, 2009b) is aimed at providing a 'route-map' for the UK's transition to 2020, with post-2020 change understood as an essentially follow-on problem. While the *Plan* refers to the need to 'consider the nature and timing of future key decisions in view of the necessary trajectory beyond 2020' (HMG 2009b, p169), the bulk of the plan is concerned with setting out a detailed set of measures to take effect over the next decade. The main policy concern is identifying and supporting an optimal pathway over the next decade – with less consideration given to longer-term pathways. By contrast, the ATD scenarios are primarily oriented towards the longer term. The scenarios were devised to explore the dynamics of long-term system change, and the system level scenarios are generated by maximizing welfare over the entire time period from now to 2050.

Given their longer-term outlook, the ATD scenarios highlight some of the risks of policy focus on 2020 targets, in terms of the possible neglect of longer-term opportunities for change, and trade-offs involved over different timescales. For example, the ATD scenarios suggest that the accelerated development of hydrogen fuel cells may offer an economically attractive means of transport decarbonization after 2030. However, one already emerging response to decarbonization is the deployment of electric vehicles – highlighted, for example, in the Renewable Energy Strategy (HMG, 2009a). Given the need for substantial financial, manufacturing and infrastructure commitments, it is possible that the development of electric vehicles over the next two decades will greatly limit the prospects for hydrogen fuel cells technologies thereafter. The specific emphasis on fuel cells in LC Acctech should be treated with some caution here, given the omission of accelerated development of battery storage technologies, making the relative costs of hydrogen vehicles more attractive.

Nevertheless, there are similarly substantial and growing financial, organizational and institutional commitments around, for example, offshore wind (discussed above) and micro-renewables. To an extent, this is an unavoidable consequence of system change, under the decarbonization imperative. As stated earlier, lock-in is a prerequisite for change. Nevertheless, the capacities developed over the next decade to meet 2020 targets – in terms of financial capital, physical infrastructure, supply chains and corporate strategy – will shape the direction and pace of change thereafter. From a longer-term perspective, the danger is of new types of technological and social lock-ins being created around those technologies which attract high levels of deployment support over the shorter term, and a locking-out of others which have the potential to become attractive options for deeper decarbonization over the longer term.

There are important overall implications for policy here, in terms of the need for more fully taking into account the interconnectedness of short- and long-term policy targets, and a more explicit justification of allocating support across shorter-term deployment and longer-term development. There are indications that these longer-term issues are being addressed. For example, the Carbon Trust has called for a more systematic and transparent 'technology focused approach' to low carbon innovation policy, involving a distinction between *focused support* for deployment over the shorter term, and *option creation* for RD&D over the longer term (Carbon Trust, 2009). It is important that these recommendations find a response in policy. In a context of increasingly urgent change, and a crowded landscape of organizational interests and policy initiatives, there is a danger that longer-term opportunities for change will be marginalized.

Summary and conclusions

This chapter has considered the prospects for the accelerated development of a range of emerging low carbon energy supply technologies – and the possible impact of technology acceleration on UK energy system decarbonization pathways from now to 2050. The analysis suggests that technology acceleration has the potential to create new decarbonization pathways, with significantly increased contributions from emerging low carbon supply technologies over the longer term. These pathways promise a number of advantages, in terms of reduced cost, greater diversity of supply, and less reliance on demand reductions over the longer term. In devising decarbonization policies for the UK, therefore, it is important that the role of accelerated technology development is taken into account.

As well as a general consideration of the opportunities and challenges associated with technology acceleration, the analysis has involved devising scenarios of accelerated technology development for several emerging technologies, and also the construction of a series of scenarios of UK energy system evolution from now to 2050 under assumptions of accelerated technological progress. The system level scenarios suggest that accelerated technology development could open up more affordable ways to achieve deep decarbonization of the UK energy system over the longer term (after 2030) as emerging technologies become commercially attractive and as overall decarbonization ambitions increase. The suggested benefits of technology acceleration are much greater in scenarios with an 80 per cent decarbonization ambition by 2050 compared to those with a 60 per cent ambition.

In the shorter term accelerated technology development has a more modest impact. Even under accelerated development assumptions, the suggested level of renewable energy deployment by 2020 is lower than that envisaged in UK policy targets, especially for offshore wind. This indicates that achieving shorter-term policy targets will require high levels of financial support for *deployment*, and cannot rely on improvements in technology cost and performance delivered by accelerated *development*. It also suggests that current UK policy, oriented towards very high ambitions for renewables deployment over the next decade, risks directing the energy system into relatively expensive decarbonization pathways in the short term, and possibly locking the system into more expensive pathways after 2020.

The long-term economic benefits of technology acceleration appear to significantly outweigh the added costs in terms of additional RD&D spending, especially if the investment costs of accelerated development are assumed to be shareable internationally. There is a strong case for much greater efforts globally on low carbon energy supply RD&D, to which the UK should make a proportionate contribution, especially in areas of particular strength. Under a long-term view of investment, a step-change increase in UK public spending on energy supply RD&D is economically justified.

A number of the emerging technologies analysed here (and others not considered) have the potential to contribute significantly to UK energy system decarbonization over the longer term. Given the illustrative nature of the scenarios presented here, and the many uncertainties involved in long-term change, the prescription that emerges is for an increase in RD&D efforts across a broad portfolio of emerging technologies, rather than an early selection of preferred technologies, or a narrowing down of RD&D spending onto a few areas. Indeed, there is some evidence that maintaining design diversity in early stage technology development is a precondition for later mass diffusion (e.g. Jacobsson and Bergek, 2004).

UK energy policy targets for large scale low carbon technology deployment over the next decade effectively represent a focus on *learning-by-experience* to drive improved cost and performance. By comparison, the accelerated development scenarios emphasize the role of expanded and sustained international investment in RD&D, and illustrate its potentially powerful long-term impact on the cost and performance of renewables and other emerging technologies, so that their eventual deployment is less dependent on market subsidies. Other evidence suggests that for emerging energy supply technologies such as offshore wind, R&D spending generates significantly greater cost reductions than equivalent spending on deployment (Frontier Economics, 2009). In practice, innovation involves an interplay between research-based and experience-based learning, for different technologies.

Without strong and sustained interventions, energy systems tend to lock-in around established technologies and practices, so that the possible benefits of emerging opportunities for change may be missed. In a context of increasing urgency, and under highly ambitious short-term targets for decarbonization and low carbon technology deployment, there is a need to recognize, and support, less immediate but potentially very significant opportunities for change over the longer term.

Notes

1   UK policy support for offshore wind technology development and deployment may also have a significant impact on the costs and performance of the technology internationally (Carbon Trust, 2008; 2009), but this was not explicitly taken into account in the ATD wind power scenario.
2   Total cumulative CO_2 emissions between 2000 and 2050 in the UKERC 80 per cent pathway are 20.39GT. Unless otherwise stated, a discount rate of 10 per cent is applied to all costs and benefits associated with the scenarios presented here.
3   See Winskel et al, 2009, for more details on the cost and performance assumptions in the ATD scenarios.
4   The assumed CCS CO_2 capture rate is 90 per cent for the UKERC scenarios discussed here.
5   Calculated as the net present value of the cumulative difference in welfare costs between non-accelerated (LC) and accelerated development (LC Acctech) scenarios over the period 2010–2050, discounted at UK Government recommended long-term social discount rates of 3.5 per cent for Years 1–30, and 3.0 per cent for Years 31–40 (HM Treasury, 2008). Because most of the benefits of technology acceleration accrue in the long term, the non-discounted saving is significantly higher, at around £88bn.
6   Calculated as the difference in aggregated RD&D investment costs from now to 2050 in the IEA's 'ACT' and 'BLUE' scenarios, as specified for comparable supply technologies in Chapter 8 of the IEA's report on *Energy Technology Perspectives*, 2008 (IEA, 2008). Note that the IEA analysis excludes marine energy.

References

Carbon Trust (2008) *Offshore Wind Power: Big Challenge, Big Opportunity*, Carbon Trust, London

Carbon Trust (2009) *Focus for success: A new approach to commercialising low carbon technologies*, Carbon Trust, London

CCC (Committee on Climate Change) (2008) *Building a low-carbon economy – the UK's contribution to tackling climate change*, TSO, London

CCC (Committee on Climate Change) (2009) *Meeting Carbon Budgets – the need for a step change*, TSO, London

CEC (Commission of the European Communities) (2009) *Directive 2009/28/ EC of the European Parliament and of the Council of 23 April 2009 on the promotion of the use of energy from renewable sources* (Renewable Energy Directive), Commission of the European Communities, Brussels

DECC (2009) *Digest of UK Energy Statistics: Long-term Trends*, Department of Energy and Climate Change, London

DECC (2010) *Energy Trends – March 2010*, Department of Energy and Climate Change, London

ERP (Energy Research Partnership) (2010) *Energy Innovation Milestones to 2050*, Energy Research Partnership, London

Foxon, T. (2003) *Inducing Innovation for a low-carbon future: drivers, barriers and policies*, Carbon Trust, London

Frontier Economics (2009) *Alternative policies for promoting low carbon innovation: Report for Department of Energy and Climate Change*, Frontier Economics, London

Gross, R., Heptonstall, P. and Blyth, W. (2007) *Investment in electricity generation: the role of costs, incentives and risks*, UK Energy Research Centre, London

Grubb, M. (2004) *Technology Innovation and Climate Change Policy: an overview of issues and options*, Keio Economic Studies, Keio University, Tokyo

Grubb, M. and Ulph, D. (2002) 'Energy, the Environment and Innovation', *Oxford Review of Economic Policy*, vol 18 (1), pp9–106

IEA (International Energy Agency) (2008) *Energy Technology Perspectives: Scenarios and Strategies to 2050*, IEA, Paris

Jacobsson, S. and Bergek, A. (2004) 'Transforming the energy sector: the evolution of technological systems in renewable energy technology', *Industrial and Corporate Change,* vol 13 (5), pp815–849

HMG (Her Majesty's Government) (2008) *Climate Change Act 2008*, TSO, London

HMG (Her Majesty's Government) (2009a) *The UK Renewable Energy Strategy*, TSO, London

HMG (Her Majesty's Government) (2009b) *The UK Low Carbon Transition Plan: National strategy for climate and energy*, TSO, London

McVeigh, J. Burtraw, D., Darmstadter, J. and Palmer, K. (1999) *Winner, Loser, or Innocent Victim? Has Renewable Energy Performed as Expected?* Discussion Paper 99-28, Resources for the Future, Washington DC

Pöyry (2008) *Compliance Costs for Meeting the 20% Renewable Energy Target in 2020*, Pöyry Consulting, Oxford

RAB (Renewables Advisory Board) (2008) *2020 Vision – How the UK can meet its target of 15% renewable energy,* Renewables Advisory Board, London

Redpoint/Trilemma (2009) *Implementation of the EU 2020 Renewables Target in the UK Electricity Sector: RO Reform*, Redpoint Energy Ltd, London

SKM (2008) *Growth Scenarios for UK Renewables Generation and Implications for Future Developments and Operation of Electricity Networks*, SKM, Newcastle

Smith, A. (2006). *Multi-level governance: Towards an analysis of renewable energy governance in the English regions*, SPRU Electronic Working Paper Series Paper No. 153, University of Sussex, Brighton

Strachan, N., Pye, S. and Kannan, R. (2009) 'The iterative contribution and relevance of modelling to UK energy policy', *Energy Policy*, vol 37 (3), pp850–860

Unruh, G.C. (2000) 'Understanding carbon lock-in', *Energy Policy*, vol 28 (12), pp817–830

Walker, W. (2000) 'Entrapment in Large Technology Systems: Institutional Commitment and Power Relations', *Research Policy*, vol 29 (7–8), pp833–846

Winskel, M., Markusson, N., Moran, B., Jeffrey, H., Anandarajah, G., Hughes, N., Candelise, C., Clarke, D., Taylor, G., Chalmers, H., Dutton, G., Howarth, P., Jablonski, S., Kalyvas, C. and Ward, D. (2009) *Decarbonizing the UK Energy System: Accelerated Development of Low-carbon Energy Supply Technologies*, UKERC Energy 2050 Research Report No. 2, March 2009, UK Energy Research Centre, London

8

A Change of Scale? Prospects for Distributed Energy Resources

Adam Hawkes, Noam Bergman, Chris Jardine, Iain Staffell,
Dan Brett and Nigel Brandon

Introduction

This chapter presents a more in-depth consideration of the opportunities and challenges associated with distributed energy resources. Specifically, it examines very small-scale energy generation, that of residential micro-generation and associated technologies, which could in the future form part of a much more active, integrated and ultimately valuable demand-side. However, such systems cannot be classified as standard generation technologies with definite economic, energy security and environmental performance benefits. This is because their uptake and application are framed by a diverse array of technical, institutional, economic, behavioural and policy-related factors.

The discussion in this chapter links closely with that of Chapter 9, and is written in the context of rapid system change as described in the book as a whole. This is the key overarching theme of the chapter: the (potential) importance of micro-generation in the context of energy system transformation.

The following sections examine the technologies themselves, behavioural, social and institutional factors that influence their successful uptake and appropriate utilization, and the policy frameworks that can support them.

Challenges in the residential sector

The residential sector is responsible for just under one-third of UK national energy consumption, and just over one-quarter of greenhouse gas emissions. It is important in terms of economically-rational climate change mitigation because it has one of the largest cost-effective potentials for emissions reduction (IPCC, 2007). Despite these qualities, and the sector being a focus of mitigation policy, it has historically been very difficult to make inroads there due to an array of factors including the diversity of social, institutional, technical and economic circumstances of people and buildings, along with the non-traded nature of emissions within the sector.

Notwithstanding lack of significant progress in terms of achieving the technical potential in energy efficiency, it can be shown that the sector has changed radically over past decades. This is particularly apparent regarding changes in the energy mix. As shown in Figure 8.1, coal and other solid fuels dominated heating applications in 1970, with oil and gas playing a much smaller role. The situation by 2000 was remarkably different, with coal virtually eliminated, and a vast increase in the utilization of gas. This change was driven by an *en masse* switch to central heating systems with natural gas-fuelled boilers, with an ensuing increase in average comfort level via higher internal temperatures that are more consistent across the dwelling (Utley and Shorrock, 2008). Of course there was also an increase in the number of dwellings, and subsequently an aggregate increase in total energy consumption for the sector.

In spite of the increase in final energy consumption, greenhouse gas emissions for the sector have steadily fallen since 1970, largely due to the switch from coal to natural gas. This is shown in Figure 8.2, which presents the relative contribution of each energy type consumed to aggregate emissions. Whilst this may seem promising in terms of long-term emissions reductions, it should

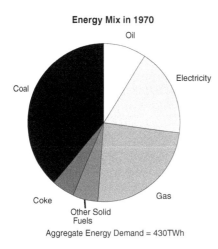

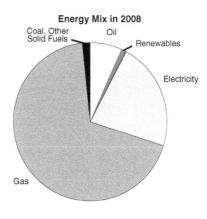

Source: DECC, 2009a

Figure 8.1 *Change in mix of energy consumed in the UK residential sector between 1970 and 2008*

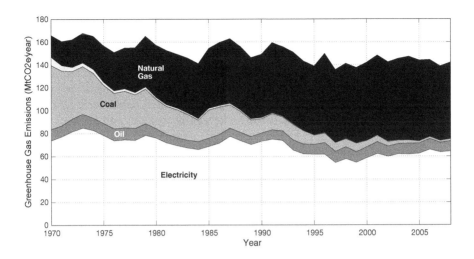

Source: DECC, 2009a

Figure 8.2 *Greenhouse gas contribution by source for the UK residential sector from 1970 to 2008*

be noted that even if this trend could be maintained, by 2050 it would result in aggregate annual emissions of approximately 115MtCO$_2$e, which is only 23 per cent below 1990 levels and well short of what may be expected from this sector given its potential for economic mitigation actions. To compound this observation, the lack of further opportunities for switching heating fuel from coal to natural gas implies that the underlying drivers of this broad historical trend cannot be sustained in the future.

It follows that in order to meet medium- and long-term emissions targets, alternative means of decarbonization must be identified and pursued. Potential options in this regard can be broadly categorized as combinations of the following:

- alternative low carbon heating fuels becoming mainstream (e.g. biomass and waste, hydrogen) along with the demand-side technologies to utilize them;
- long-term trends in aggregate energy demand being reversed; or
- thermal demand in buildings being shifted onto the electricity sector, whilst simultaneously decarbonizing electricity supply.

Clearly these propositions are tremendously challenging, and fundamental shifts in demand-side interventions, supply-side technology, and supporting infrastructure are required. Encouragingly, however, the sweeping change seen in the sector since 1970 suggests that this may be possible. One potential element of such a shift is increasing use of micro-generation technologies, which are defined for the purposes of this chapter as small-scale electricity and/or heat generation suitable for installation and use in single residential dwellings. These systems are interesting because they can provide a source of low carbon energy supply, a range of opportunities for much more intelligent integration of the demand-side into the broader energy system, and perhaps most importantly the opportunity to bring about awareness and behavioural change regarding people's interaction with their energy system. Of course, actually realizing such developments is far from assured, with each technology exhibiting specific pros and cons, and broader market, institutional and behavioural issues bearing deeply upon outcomes.

The following section provides an overview of the technologies, including a description of their basic mode of operation, representative costs and efficiencies, as well as the practical issues encountered with installing and operating them.

Technology characteristics, performance and suitability

The 'micro-generation' label refers to a highly diverse set of technologies which show marked differences in output, cost effectiveness and suitability. The different technologies can be grouped into three categories:

1   low carbon heating: *condensing boilers, biomass heating* and *heat pumps*;
2   renewables: *solar photovoltaics, solar thermal* and *micro-wind*; and
3   combined heat and power (CHP): *Stirling engines, internal combustion engines* and *fuel cells*.

Figure 8.3 depicts typical installations of these technologies within the home, showing how they interact with the electricity and heat distribution systems.

These technologies improve on the traditional means of heating and powering houses either by using carbon-free renewable energy or by utilizing existing fuels with far greater efficiency. Low carbon heating has significant potential for reducing both domestic fuel bills and environmental impacts, as around 80 per cent of energy demand from UK homes is for heat (Utley and Shorrock, 2008). Renewable micro-generation contributes to decarbonization in much the same way as larger centralized equivalents, reducing reliance on fuel imports and providing zero carbon electricity from the wind or sun. Micro-CHP on the other hand increases the efficiency of natural gas consumption by producing power at the point of use and capturing the otherwise-wasted heat.

The performance of these technologies in people's homes is not widely understood as they are new and evolving, and highly sensitive to their operating conditions. Not every technology is suitable for every property, and in some cases improper installation results in disappointing performance.

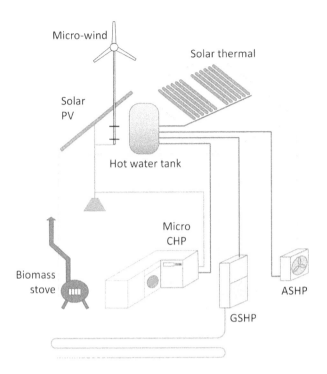

Source: Staffell et al, 2010

Figure 8.3 *Diagram depicting the typical size and position of different micro-generation technologies within the home. Note that the figure is illustrative only. In particular, no house should contain all the technologies, because some of them perform the same function and are substitutes, not all homes are suitable for all technologies, and solar PV and solar thermal should be located on the same (south-facing) roof of the house.*

Technology descriptions

Low carbon heating

Condensing boilers

Condensing gas boilers are the standard heating technology in the UK and became mandatory for new installations in 2004 due to the efficiency improvements made over regular furnaces and boilers. By condensing water vapour in the flue gases and extracting its latent energy, leading boilers offer efficiencies of 91.5 per cent (note that all efficiencies in this section are measured against the Higher Heating Value (HHV)) in laboratory tests (DEFRA, 2008), compared to a maximum of 82 per cent for older non-condensing models (BRE Ltd et al, 2002).

Two independent trials of 'A-rated' boilers, however, show that efficiency is 4–5 per cent lower in people's homes than in the laboratory, averaging 82–89 per cent (Carbon Trust, 2007; Orr et al, 2009). This is due to losses in the storage tank and pipework, and from cycling boilers on and off several times a day; both of which result in drastically lower efficiencies for hot water production during summer. Modern boilers also consume a significant amount of electricity, averaging 225kWh per year (Orr et al, 2009).

Due to their relative simplicity and scale of manufacture, condensing boilers are relatively cheap; typical 24kW$_{th}$ models retail for £650–800 (Staffell et al, 2010). As with other micro-generation, labour-intensive installation adds significantly to this, raising total costs to about £1,500 in new build and £2,000–2,500 in existing properties.

Despite initial scepticism about the durability of condensing boilers, current systems are believed to reliably operate for 10–15 years provided they are serviced annually.

Biomass heating

Biomass for heating typically refers to wood produced from dedicated short rotation coppice (SRC) and timber industry wastes. This can be a sustainable and low carbon fuel provided that only well managed sources are used. Logs are the cheapest form available, but require substantial storage space due to low energy density and a drying time of at least a year. Compressed sawdust pellets are a more convenient alternative, which can be automatically fed into a burner and combust with higher efficiencies.

Biomass micro-generation options range from traditional open fires, to wood burning stoves (single-room space heaters) and fully integrated boilers. Stand-alone room heaters of around 5kW$_{th}$ cost £2,000–4,000, whereas 10–20kW$_{th}$ boilers are significantly more expensive at £5,000–15,000 (Bergman and Jardine, 2009). Running costs are dominated by the price of wood-fuel: logs bought in bulk are cheaper than natural gas (1.8–2.6p/kWh), but pellets are more expensive (3.0–4.2p/kWh) (Staffell et al, 2010).

Traditional fireplaces are inefficient and give only 10–30 per cent useful heat; however, more modern systems attain 70–90 per cent efficiency depending on the fuel used (Horio et al, 2009). Simpler log-based systems also require regular human intervention for daily refuelling and ash removal, fuel delivery, servicing and cleaning. People who are accustomed to mains gas and electricity perceive refuelling and storage to be substantial inconveniences, even though modern boilers can operate unattended for 6–12 months due to automated fuel feeding and ash collection (SolarFocus GmbH, 2008).

Initial problems when switching to biomass are often related to fuel quality and moisture content (South West of England Regional Development Agency, 2008). However, reliability and lifetime are generally thought to be the same or better than for conventional gas boilers, so long as fuel quality and the boiler itself are properly maintained.

Heat pumps

Heat pumps are a form of heating that operate via a refrigerant absorbing thermal energy in a relatively cold zone (outside the home), and then releasing that energy into a relatively hot zone (inside the home). This is achieved by capitalizing on the physical properties of a working fluid (a refrigerant) and a pressure difference between the hot and cold side of the cycle. A heat pump can be seen as a 'refrigerator working in reverse'. Both gas and electric heat pumps exist, although only electric types are commercially available at the residential scale. In an electric heat pump, a cold (liquid) refrigerant is pumped around an external heat exchanger to extract heat from its surroundings, which it does by evaporating. The refrigerant is then compressed, which has the effect of increasing its pressure and thus its temperature. Heat can then be extracted for space and water heating, which causes the gas to liquify again and the cycle is repeated. Two types of electric heat pump are commercially available for the sector: air source (ASHP) which collects heat from the air via a unit outside the dwelling; and ground source (GSHP) which uses plastic tubes laid underground.

Figure 8.4 shows the annual energy balance for an average UK home using a heat pump. The average home uses 3.3MWh of electricity and 15MWh of heat. If the heat is supplied via a standard condensing boiler, then the home will need 17.4MWh of gas and 9.2MWh of energy input to a central power station, giving 26.6MWh of primary energy demand overall. Using a heat pump, no gas would be required but there would be an additional 4.7MWh of electricity demand. The total electricity demand of 8.0MWh would require 22.1MWh of energy to power stations, a 4.5MWh (17%) saving on the conventional system. The greater overall efficiency associated with a heat pump is due to the fact that it has extracted 14.1MWh of 'renewable' heat from the environment.

Small ASHP single-room heaters (up to $4kW_{th}$) can be purchased for as little as £250. Full systems that integrate with hot water and central heating cost in the region of £2,500–5,000 for both types over the range of $5–20kW_{th}$ (Staffell et al, 2010). ASHP installation can be relatively standard (although at present planning permission is required), resulting in total installed costs around £3,500–7,000. Laying the ground pipes for a GSHP adds around £800 per kW_{th} to installation, giving a total cost of around £10,000–12,500. Despite this high capital cost, the overall cost of domestic heating can be reduced due to the high operating efficiency and reduced maintenance requirements.

Efficiency is represented by the coefficient of performance (COP), which is the ratio of heat output to electricity input. A COP of over 2.5 will typically give lower CO_2 emissions than a condensing boiler. COP is highly dependent on the temperature difference between the external heat exchanger and the internal heat distributor. A warm external loop and cool internal loop will maximize efficiency, so it is beneficial to use low-temperature radiators or direct air heating. During winter when there is greatest demand for heating, UK air temperatures average 4°C. Ground temperatures stay around 10°C all year round, allowing GSHPs to offer higher COP than ASHPs (Met Office, 2008). Annual average COP values for ASHP lie around 2.9–3.5 in UK

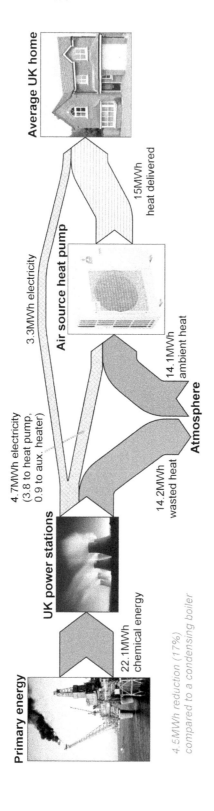

Primary energy

22.1MWh
chemical energy

*4.5MWh reduction (17%)
compared to a condensing boiler*

UK power stations

14.2MWh
wasted heat

4.7MWh electricity
(3.8 to heat pump,
0.9 to aux. heater)

Atmosphere

14.1MWh
ambient heat

Air source heat pump

3.3MWh electricity

15MWh
heat delivered

Average UK home

Source: Staffell et al, 2010

Figure 8.4 *A Sankey diagram showing the flows of energy required to heat and power a typical house for a year using a heat pump. Arrows denote the transfer of energy from one system to another, with thickness that is proportional to the amount of energy*

conditions, and 3.9–4.8 for GSHPs, if using a low temperature heating system (Staffell, 2009b) and accepting the associated higher installation costs. High-temperature ASHPs, which can be applied in a retrofit that does not involve low-temperature underfloor heating, have recently become available but these models are generally more expensive than their low-temperature counterparts.

Reliability is good, with compressors expected to last 15–25 years and heat collectors up to 50 years (Energy Saving Trust, 2000). Maintenance requirements are low, and there is no risk from natural gas leakage or explosion.

Renewables
Solar photovoltaics

Solar photovoltaics (PV) are semiconducting materials that convert energy from sunlight directly into electricity. Glass coated panels made from 'first generation' mono- and multi-crystalline silicon cells are the most common form; however a variety of 'second generation' thin films are now available integrated into roof tiles, shingle and foils (Ekins-Daukes, 2009). A typical installation of $2kW_e$ peak output ($kW_{e,pk}$) consists of 10–20 modules each of around $1m^2$ connected to an inverter which determines the maximum power and converts the DC output into grid-compatible AC.

Prices fell sharply during the global recession of 2009 which eased demand for silicon. Average prices in December 2009 were £2,600–3,800 per $kW_{e,pk}$, although building-integrated panels are nearly twice as expensive (Munzinger et al, 2006; Staffell et al, 2010). Complete system prices are also higher as they include the cost of an inverter and other system components, plus installation costs which vary significantly between sites. The total cost of retrofitting $2kW_{e,pk}$ panels onto existing houses is therefore £10,000–13,000, reducing to £8,000–11,000 for new-build installations (Munzinger et al, 2006; Bergman and Jardine, 2009).

PV performance is measured in terms of the annual energy yield (expressed as kWh per $kW_{e,pk}$) rather than efficiency; this accounts for the varied levels of insolation, spectral quality and cell temperature. Academic studies have shown that silicon panels produce 700–800kWh/kW_{pk} in the UK, equivalent to a capacity factor of 8–9 per cent (Staffell et al, 2010). Despite having lower laboratory efficiency, second generation CIS (Copper Indium (Gallium) (Di) selenide) and amorphous silicon appear to be the most effective technologies in cold and overcast UK conditions. Extensive trials of PV systems in the UK have demonstrated slightly lower yields of 600–800kWh/kW_{pk} (Munzinger et al, 2006).

The durability and reliability of PV are extremely good; guaranteed lifetimes of 25 years are standard, and advanced ageing techniques suggest that today's first-generation modules could last 40–50 years (Wohlgemuth et al, 2005). The inverter is the most fragile component and will typically require replacement after 10–15 years, however, even with zero maintenance, the long-term rate of power loss is only 0.3–0.6 per cent per year (Staffell et al, 2010).

Solar thermal

The basic concept of solar water heating has been employed for many decades, predating the widespread use of electric and gas heating. The cold and overcast UK winter means it provides a poor match with space heating requirements, and is most often used only to supplement hot water production.

A roof-mounted collector is connected to a hot water storage tank by pipes which carry a mix of water and antifreeze. The collector consists of a flat metal plate or partially evacuated glass tubes, selectively coated with a heat absorber. Evacuated tube designs are slightly more expensive, but offer greater efficiency and can deliver water at a higher temperature. A pump is used to circulate the working fluid which consumes a small amount of electricity (30–90kWh/year), although some systems incorporate a small PV panel to remove all external dependency (Martin and Watson, 2001; Forward et al, 2008).

Solar thermal systems can be very simple and inexpensive, as high-tech components and manufacturing are not required. However, bare systems in the UK retail for around £1,500, and fully installed systems cost anywhere from £2,000 up to £8,000 (Bergman and Jardine, 2009; Ekins-Daukes, 2009). There is little technical justification for the observed price differential; however, a new hot water tank and condensing boiler are often sold with the panel (Fisher et al, 2008; Ekins-Daukes, 2009), which would account for some of the observed variations.

A well run system can transfer 40–50 per cent of the incident sunlight to the hot water tank, meaning a typical 4m² system can produce around 1400–1700kWh of hot water per annum (Martin and Watson, 2001; Forward et al, 2008). Not all of this energy goes to use in smaller households though due to their limited demand for hot water (Hill, 2009), and there is evidence that behavioural factors (e.g. showering in the morning) can further degrade observed performance. Very little has been published on the durability of solar thermal panels, but it is expected that like solar PV, systems should last 30 years or more with minimal maintenance (Fisher et al, 2008).

Micro-wind

Micro-wind turbine designs are varied, including 2- to 6-bladed horizontal, vertical, and cross-flow designs plus building-augmented turbines. Micro-turbines are designed to minimize noise and vibration, and must be able to operate in more turbulent conditions with rapid changes in wind speed and direction. Horizontal axis wind turbines are the most common and well developed design, using about 2m-diameter blades mounted at 2–3 metres above the roof-line.

Popular building mounted models in the UK cost £2,000–3,000, while larger free-standing turbines above 5kW (suitable for rural locations) are significantly more expensive (Staffell et al, 2010).

Despite the range of designs, most turbines exhibit the same relationship between wind speed and power output: below 3–4ms⁻¹ no power can be produced; output then increases with the cube of wind speed until the rated design speed at around 12ms⁻¹ (force 6 on the Beaufort scale) when the generator is running at full capacity. This strong dependence means that consistent and high wind speeds are necessary for a turbine to give good annual energy yields. Simulations of localized wind speeds suggest that utilization (note that utilization of 10 per cent is equivalent to 876kWh/kW$_{pk}$) of only a few per cent should be expected from urban areas, with up to 10 per cent in suburban and 15–20 per cent in rural locations (Phillips et al, 2007; Clark et al, 2008). Two separate trials in the UK demonstrated an average output of just 53kWh per year (0.85 per cent), with no urban or suburban site generating more than 200kWh (Encraft, 2009; Energy Saving Trust, 2009).

There is little long-term experience with building mounted micro-turbines due to their relative immaturity; lifetimes are therefore estimated rather than proven in the field. Manufacturers and organizations expect lifetimes to be 15–22.5 years (Staffell et al, 2010). However, there have been reports of reliability problems with early models, with catastrophic failures such as blades and tail fins being ripped from the turbine (Encraft, 2009).

Combined heat and power (CHP)

The benefits of generating power on-site and capturing the by-product heat are illustrated in Figure 8.5 overleaf.

Internal combustion engines

Reciprocating internal combustion (IC) engines are similar to the well established automobile engine and usually operate on natural gas for micro-CHP applications. The engine crankshaft is connected to an AC generator, and heat is extracted from the exhaust gases and cooling fluid pumped around the engine (Onovwiona and Ugursal, 2006). A fixed optimal running speed is used to give minimum fuel consumption, which means power output cannot be changed. Instead of reducing output, the engine cycles on and off according to the amount of heat demand (Aki, 2007). Noise levels are a common concern, however modern units are no louder than a refrigerator (about 45dB) (Kayahara, 2002).

The only engine suitable for single-family domestic properties is the 1kW$_e$ Honda ECOWILL, which produces 2.8kW$_{th}$ of heat plus 36kW$_{th}$ from a supplementary boiler (Aki, 2007). In Japan and the USA the ECOWILL costs £4,000–7,000 installed, although a number of larger (4–5 kW$_e$) engines are available in Europe for slightly more (Staffell et al, 2010).

Manufacturers of larger engines quote efficiencies of 25–30 per cent electrical and 85–90 per cent total. Several field trials have shown that real-world efficiency lies around 21–25 per cent electrical and 79–84 per cent total (Teekaram, 2005; Onovwiona and Ugursal, 2006; Carbon Trust, 2007; Thomas, 2008). Results from demonstrations of the ECOWILL have not been published, so real-world performance can only be assumed from the slightly lower quoted efficiencies of 20.5 per cent electrical, 77.5 per cent total (Aki, 2007). Emissions of nitrogen oxides and carbon monoxide from older models are around double those of a gas boiler (Teekaram, 2005); however, newer engines such as the ECOWILL use catalytic converters to reduce these to zero CO and around 80mg NO_x per kWh of fuel combusted (Thomas, 2008).

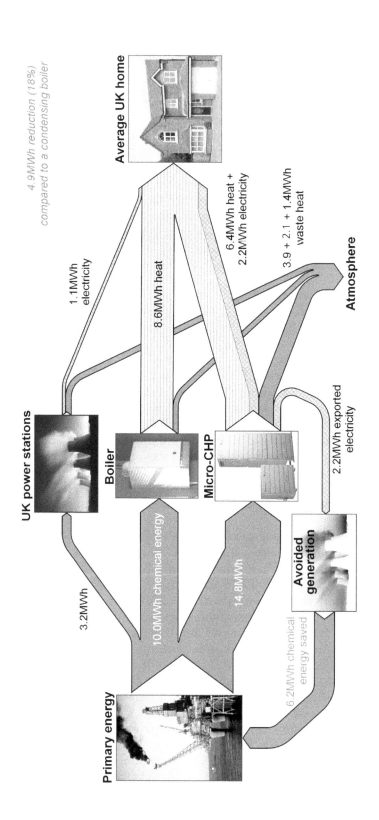

Figure 8.5 *A Sankey diagram showing the flow of energy required to heat and power a typical house for a year using fuel cell based micro-CHP. Electricity exported from the fuel cell is credited with generation in UK power stations which can be avoided, which gives a net reduction in primary energy consumption*

Durability is good due to the maturity of the technology, so a minimum of 10–20 years of intermittent operation is expected (Kayahara, 2002). Maintenance requirements are often overstated, as changes of oil, coolant and spark plugs are only required every 6,000 hours (3 years) for the ECOWILL (Kayahara, 2002).

Stirling engines

Stirling engines differ from IC engines in that fuel combustion occurs outside of the cylinders. Rather than being explosive, combustion is continuous and more tightly controlled, removing the need for catalytic converters and reducing engine noise (Onovwiona and Ugursal, 2006). There are several engine–generator configurations: notably kinematic designs using a traditional crankshaft, and newer free-piston designs with a more efficient linear alternator. Stirling engines produce significantly more heat than electricity, and so are operated to follow heat demand, only running intermittently during summer.

As with IC engines there is currently only one domestic-scale system, the $1kW_e$ $8–13kW_{th}$ WhisperGen. Free-piston engines were due to be launched in 2010 by both Baxi and Enatec, offering a lower heat-to-power ratio (HPR) of 5:1 from smaller wall-hung units (Enatec Micro-cogen B.V., 2007; Baxi, 2008). Stirling engines are expected to cost more than IC engines due to the precision engineering required; however, at present the WhisperGen is the cheapest micro-CHP system available. This appears to be due to heavy subsidies provided in the UK, where limited numbers have been made available for £3,000, compared to £11,000 in Germany (Staffell et al, 2009).

Larger Stirling engines have electrical efficiencies of 15–30 per cent; however, this falls sharply with capacity, so micro-CHP systems offer approximately half this: 6 per cent for the WhisperGen and 12 per cent for the Ecogen (Baxi, 2008; Whisper Tech, 2003). Large UK field trials of the WhisperGen demonstrated 4–8 per cent electrical and 75–80 per cent total efficiency (Carbon Trust, 2007), and other studies have shown 5.5–11 per cent electrical and 80–88 per cent total efficiency from different models (Staffell et al, 2009). Efficiency is strongly influenced by operating pattern and running time, and can halve during summer months or in smaller homes (Carbon Trust, 2007).

Stirling engines are expected to be more robust than IC engines due to reduced mechanical wear. System lifetimes have not yet been widely verified, but ten years' operation is expected (Onovwiona and Ugursal, 2006). As with other emerging technologies, reliability has proven a major issue, with 25 per cent of the WhisperGens installed in UK trials requiring repairs within their first year (Carbon Trust, 2007).

Fuel cells

Fuel cells operate differently from engines, electrochemically converting fuel directly into electricity rather than combusting it. Natural gas is reformed into hydrogen in a fuel processor which is then catalytically split into ions and electrons within the fuel cell, producing a DC current and water. CO_2 is also produced when hydrocarbon fuels are reformed, although emissions of other air pollutants are one-tenth those from engine based systems (Pehnt, 2003). Two types of fuel cell are predominantly used for micro-CHP: polymer electrolyte membrane (PEMFC) and solid oxide (SOFC). Low temperature (about 80°C) PEMFCs use hydrated fluoropolymer composites and small quantities of platinum catalyst, whereas SOFCs use thin ceramic composites and chromium alloys that can withstand 500–900°C operating temperatures (Hawkes et al, 2009a). Heat is extracted from the fuel processor and exhaust gases, giving between 0.8 and 1.6kW$_{th}$ per kW$_e$.

Commercialization has been protracted over the last decade; however, PEMFC systems are now on sale throughout Japan and several manufacturers are looking to launch systems in Asia and Europe in the next three years. These are the most expensive form of micro-generation, with a 1kW PEMFC costing £15,500 in Japan, and SOFC costing at least £30,000 (Staffell and Green, 2009). However, after economies of scale and system optimization it has been estimated that unsubsidized prices of £5,000 per kW$_e$ could be offered within ten years (Staffell and Green, 2009).

The electrical efficiency of fuel cells is the highest of all CHP technologies and can rival that of modern combined cycle gas turbine (CCGT) systems. SOFC have demonstrated 50 per cent HHV efficiency operating on natural gas, and leading PEMFC systems offer 31–33 per cent (Staffell, 2009a). Total efficiencies are, however, lower than for CHP engines due to difficulties in capturing low-grade heat. Field trials in Japan show that efficiency is reduced in people's homes, due to lengthy system start-up routines and degradation of performance over time. In four years of trials, PEMFC therefore attained 27–31 per cent electrical and 68–75 per cent total efficiency (Staffell, 2009a).

Improving durability is a top research priority (Hawkes et al, 2009b), as only five years of intermittent operation has been demonstrated by current PEMFC systems, and three years from SOFC (Hawkes et al, 2009a). Reliability has also been a concern in early trials, as 90 per cent of systems required repairs in their first year of operation (Omata et al, 2007). The commercial models now being sold are expected to be substantially improved, offering the industry-wide target of ten years' (40,000 hours') operation.

Table 8.1 summarizes the data presented in this section and reference (Staffell et al, 2009), giving the general range of cost and performance that can be expected for each technology.

Suitability and practical issues

The incumbent domestic energy system in the UK is incredibly convenient; an unobtrusive boiler provides heat and the electricity grid is effectively invisible to the customer. Micro-generation technologies therefore face tough competition and cannot always offer the convenience that customers are used to.

As well as issues with practicality, performance in the field is found in many cases to differ widely from manufacturers' specifications, laboratory studies and consumer expectations, often due to avoidable problems with installation and operation.

Table 8.1 *A comparison of metrics for micro-generation technologies*

	Typical capacity (kW)	Efficiency/COP/ annual yield	Current installed cost (£,000)	Expected lifetime (years)
Condensing boiler	25_{th}	82–89%	1.5–2.5	10–20
Biomass heater	$<5_{th}$	70–80%	2–4	20–30
Biomass boiler	$10–20_{th}$	80–90%	5–15	
ASHP	$5–15_{th}$	2.9–3.5	1–6	15–25
GSHP		3.9–4.8	10–15	(up to 50)
Solar photovoltaic	1.5_e	1000–1300kWh	6–10	25–50
Solar thermal	3_{th}	800–1700kWh	2–8	25+
Micro-wind	$0.6–1.2_e$	50–150kWh urban 500–1000kWh rural	*2+*	*15*
IC engine	1_e 3_{th}	$20\%_e$ $75–80\%_{total}$	*4–6*	*10–20*
Stirling engine	1_e $5–13_{th}$	$4–8\%_e$ $75–80\%_{total}$	*3 (subsidized)*	*10*
Fuel cell	$0.7–1_e$ $0.5–3_{th}$	$30–35\%_e$ $70–75\%_{total}$	*15+*	*3–10*

Note: Figures in italics indicate limited certainty. Efficiencies are given as higher heating value (HHV), except for biomass technologies. Installed costs are for the chosen typical capacity, and will fall over time.

Installation location

Installers must carefully consider where to locate micro-generation technologies, as a surprising number of installations perform well below specification due to poor siting (Munzinger et al, 2006; Encraft, 2009). Solar panels can end up shaded by other parts of the building (usually satellite dishes), and wind turbines are routinely installed in areas with insufficient wind speeds to meet the manufacturer's recommendations.

While the yield from solar panels can be estimated with reasonable accuracy, micro-wind output depends strongly on local wind conditions and surrounding geography, which are difficult to quantify. Obstructions such as buildings and trees severely disrupt airflow, slowing it down and creating turbulence (Watson et al, 2007). There is no simple method for estimating wind speeds at a particular site, so ideally they should be measured onsite, although time and cost prevents this from being commonplace. Installers perform a basic site survey using national databases of high-altitude wind speeds with simple correction factors for rural, urban and high-rise areas (BWEA, 2009). This method systematically overestimates wind speeds and thus predicts energy yields that are around double those achieved in practice (Encraft, 2009).

Due to the visible protrusion of a wind turbine, planning permission is required in the UK which adds £200 to the installation costs and places a regulatory obstacle in the way of owners. Retrofitting any renewable technology on a roof is a relatively costly procedure, although they can be incorporated into new-build constructions for very little additional cost.

Equipment size

The installer must also choose a suitable capacity of micro-generation to match the heat and power loads of the individual house, which is not usually a worry with traditional heating systems. Due to the complexity of determining system size, a site survey is usually performed to determine the property's heat losses and identify actions to reduce these.

The obvious consideration is that smaller systems will have lower capital costs, but these will meet a smaller portion of demand meaning greater reliance on less efficient backup systems, or an inability to meet peak demands if no backup is present. Under-sized heat pumps suffer during the winter, as continuous running leads the outside unit to freeze over and repeated defrosting is required to keep it operating effectively.

Most technologies have a lower efficiency when run at partial load or frequently cycled on and off, so heating systems that are over-sized for the property consume more fuel (Thomas, 2008). This is most notable with Stirling engines due to their high ratio of heat to power output. Whereas excess electricity can be exported, heat production is constrained during summer, meaning these units can only operate close to rated efficiency in large houses with high demand for hot water (Carbon Trust, 2007; Hawkes et al, 2009a). Similarly, solar thermal panels are oversized for the one- or two-person households they are often installed into, meaning excess hot water is produced during summer that goes to waste (The Energy Monitoring Company Ltd, 2001).

Property size

Property size is also an important consideration, as some technologies are particularly large. Current fuel cell systems are the size of a fridge-freezer, and so are installed outdoors or in a basement or utility areas. Biomass boilers are similarly large, and also require a substantial amount of space outdoors for fuel storage. The average home requires 3–5 tonnes of wood per year, meaning a hopper of around 1×2×3 metres is needed to store a year's worth of pellets. Customers wanting to use logs for heating would need several times this amount of space for storage, or must accept regular deliveries throughout the heating season.

The main unit of a heat pump can be relatively small; however, GSHPs need an extensive heat exchanger to be installed below ground. This can consist of shallow trenches laid horizontally over an area of 400–800m², requiring a large garden that can be dug up (Staffell et al, 2009). The inconvenience of doing this in existing (rather than new-build) properties that have the available space is a major factor in a consumer's decision to purchase. The other option is to use two or more boreholes dug to around 100m depth; however, this requires planning permission and a geological survey in the UK (Staffell et al, 2009), and entails substantial additional cost.

Operation

The way in which a micro-generation system is used will ultimately govern how well it performs. Sub-optimal conditions can be devastating to performance, and often result from a lack of basic information given to the owners.

Simple actions such as lowering space heating temperatures can greatly improve the efficiency of any heating technology with no perceivable loss in comfort. A performance improvement can therefore be gained by using large surface area radiators or under-floor heating which use water at 30–50°C. However, the cost of retrofitting a new heating system into existing properties means they are more viable for new-build properties.

Monitoring how well the technology is running is particularly important, as instability in the UK electricity grid causes frequent shutdown of PV, CHP and micro-wind inverters. Better use of self-diagnostics or system information displays, combined with user education can minimize the impact of these all-too-common events.

Lifestyle changes can also help micro-generation technologies to operate more effectively, as discussed further in Chapter 9. The majority of solar thermal panels in the UK achieve only a quarter of their full potential as customers' demand for hot water in the morning means the backup boiler has heated the water tank before the sun is given a chance (Hill, 2009). The typical manufacturers' claim of providing up to 70 per cent of hot water demand is therefore not widely experienced in the field (The Energy Monitoring Company Ltd, 2001; Forward et al, 2008; Hill, 2009).

Micro-generation cannot be treated simply as a black-box that will automatically lower cost and emissions – it must be installed and used appropriately. This can require an improved awareness of one's own energy habits and sometimes changes in behaviour. However, by engaging consumers with their energy usage habits, micro-generation can create the 'win–win' of providing more sustainable energy generation and inspiring behavioural change that lowers domestic energy demands. This and further institutional and social issues are explored in the following section.

The human dimension: installers and householders

Micro-generation is often thought of in terms of technological innovation and grid decentralization, looking at potential energy and emissions savings in the residential sector with blanket assumptions about uptake and use of the different technologies by householders, usually thought of simply as 'consumers'. However, the human dimension is critical; householders' choices in purchase and use of micro-generation are non-trivial to predict, and the behaviour and policies of installers, retailers and manufacturers also play an important part in both uptake and use of micro-generation. This section reviews some of the behavioural aspects and how they affect the micro-generation side of the energy system, focusing on residential energy. Also, further analysis of relevance to this topic is available in Chapter 9.

Installers and the skill base

In 2008 there were approximately 100,000 micro-generation installations in the UK, the large majority of which were solar thermal (Element Energy, 2008). The potential is for millions of installations, and this order of magnitude is needed if these technologies are to play a major part in the UK energy sector.

The exact number of installers is not known, but is probably several hundred strong. There are at least 370 companies who have installed micro-generation under the Low Carbon Buildings Programme (LCBP, Bergman and Jardine, 2009a). Most of these companies (313) install only one of the common micro-generation technologies, although the smaller markets of the newer technologies seem to be 'piggy-backing' on the more established solar thermal industry. Less than 10 per cent of the companies installing solar thermal are also offering another technology, but a quarter to a third of the companies offering ground source heat pumps, PV, micro-wind or wood-fuelled systems offer solar thermal (Bergman and Jardine, 2009a). This is consistent with a trend of market maturity: companies installing the less mature technologies are less likely to concentrate on that technology alone, instead offering a diversity of products to remove economic and technological risk. Many of these installation companies only have a handful of domestic installation jobs a year (although there are non-domestic installations and those that fall outside the LCBP), not enough for full time employment. Micro-generation is usually not the main business of the company, for example, companies which specialize in plumbing could also install solar thermal or ground source heat pumps.

This small market size has several consequences which need addressing if it is to be increased to the point of millions of installations. First, a much larger skill base is required, including on the one hand more installers for individual technologies, project managers and others for increasing installation numbers, and on the other more holistic expertise looking at the house as a whole system. As described, in 2009, most installers usually offered only one technology. Moreover, many installers might be familiar with only one brand – because courses run by manufacturers of system components are more attractive than courses for best practice, as they are shorter and cheaper (Hill, 2009). At present, there is no sign of installers considering the house as a whole system and installing the most appropriate technology (or technologies) for that property, and there is no incentive for them to do so (Bergman and Jardine, 2009a). Beyond installers, the industry as a whole does not focus on whole-system thinking, e.g. solar thermal installers might not know how best to install a system in line with the existing heating system, and manufacturers might not be producing the best products for minimizing energy in existing houses (Hill, 2009).

Another issue to consider is the effect of the small, uncertain micro-generation market on prices and installers' strategies. There is a history of uncertainty surrounding grants such as the LCBP as grant levels and available money change over time (Bergman and Jardine, 2009a; REA, 2008). Currently the market is undergoing another shift as feed-in tariffs and the renewable heat incentive are being introduced, which are planned to reduce payback time to 8–12 years for all renewable electricity technologies; if the payments do indeed live up to this estimate, and the scheme is not changed drastically in the coming years, this will increase certainty in the market. These uncertainties affect the micro-generation market, and most likely play a part in the observed aggressive sales tactics, including PV installers focusing on maximizing sales, not maximizing energy savings (Keirstead, 2007), inflated prices in the solar thermal industry (Bibbings, 2006), or advertising unrealistically short payback times (Staffell et al, 2010). This risks damage to the young market, raising concerns among stakeholders (Genus, 2008) and potentially risking the reputation of micro-generation among the public (Bibbings, 2006).

The problems of the small micro-generation market are also observed in the lack of the expected drop in prices over time. For example, during the first two years of the LCBP (2006–2008), there was no significant drop in prices for solar thermal, ground source heat pump or micro-wind installations and only a possible drop in price for the tiny wood-fuelled boiler market. Only PV showed a drop of 10 per cent in price with reasonable confidence. This might be due to a phenomenon called a 'price umbrella', first described by the Boston Consulting Group (BCG, 1968): with only a few players in a new market, there is little incentive to lower prices even as costs fall over time, because there is little competition and therefore little chance of being undercut, leaving prices fairly constant with high profit margins. When the market grows and more actors enter it, the situation eventually becomes unstable leading to a 'shakeout phase' with a rapid drop in prices and profit margins. It could be that the micro-generation market in the UK, with too few players to be truly competitive, is in such an umbrella phase, which suggests prices might drop more readily as the market grows. However, the relatively large solar thermal market should have already had a shakeout phase, suggesting further investigation is warranted into the nature of the market and the actors in it (Bergman and Jardine, 2009).

Finally, installers play an important part in providing information to householders. For solar thermal installations, research shows that most people accepted the recommendation of the installer, even if they did not understand the technology (Caird et al, 2008). This could well mean that if one's plumber recommended against installing solar thermal panels, people would accept this, like the 43 per cent of the people for whom the reputation of unreliability was a barrier to purchasing a condensing boiler; this reputation persisted until condensing boilers became virtually mandatory under UK building regulations in 2005 (Caird et al, 2008). Installers who specialize in one technology only, and perhaps only know one brand, cannot give householders good whole-house information on the best micro-generation installation for their home. Moreover, installers do not always have, let alone impart, information on best use of the system to consumers. A stark example is that many consumers who install solar thermal systems are told there is no behaviour change required ('solar systems work for you whatever you do'), when in fact adjusting boiler operation timing can significantly increase savings – and many users are not aware of this (Hill, 2009).

Overall, this paints a complex picture of the micro-generation industry and the actors therein. While there are undoubtedly some 'cowboy salesmen' and poor or inappropriate installations, the overall picture is of installers, salespeople and manufacturers competing in an uncertain market with changing policies and new technologies, without sufficient regulation for broadening the skills base. In the current market and regulatory system, it is unrealistic to expect installers to supply consumers with the best information on purchase and use of micro-generation systems, and this is clearly an information gap which needs to be addressed. It may be that the recent introduction of FITs will increase consumer demand for high quality installations and more information, but this remains to be seen. The state of the industry, including the obligations and incentives of installers and manufacturers, must be considered in future policy.

Who purchases micro-generation and why?

We next turn our attention to the householders as consumers, looking at who purchases micro-generation, and what their main motivations and barriers are.

Diffusion of innovation theory describes the spread of new technologies. It classifies the adopters by five categories: innovators, early adopters, early majority, late majority and laggards, as shown in Figure 8.6. Current purchasers of micro-generation in the UK are best described as innovators and early adopters. These groups choose to purchase largely through personal motivations, which could include (environmentalist) ideology, love of technology,

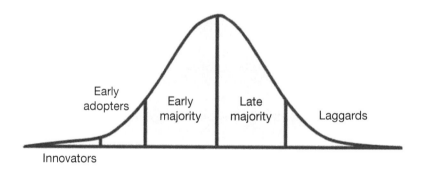

Source: Rogers, 1995

Figure 8.6 *Categories of adoption according to diffusion of innovation theory*

self-sufficiency and saving money. They are willing to take more risks and pay more when buying new technologies. The majority of consumers put more weight on social motivations, such as social status of the technology, or friends and relatives or even celebrities who have purchased it. They are not willing to pay a very high price, nor accept a long payback time, and demand that the technology have a good, safe reputation (Rogers, 1995). Uptake must move on from early adopters to the early majority if micro-generation is to become a mainstream part of the UK energy system. Making this leap is non-trivial for any technology, due to these different motivations, and has been described as 'crossing the chasm' (Moore, 1991).

Models of the energy system often use economic 'rational actors' to describe consumers' energy-related behaviour. This means consumers maximize their 'economic utility' in terms of money savings or payback (although utility could also be comfort, happiness, etc.). Perfect information is also assumed, that is, a full understanding of cost, payback and any other factors affecting utility. This can lead to simplistic tools that predict emissions (and energy) savings through micro-generation, as displayed in Figure 8.7.

However, whilst economic performance is clearly important, the idea that consumers act as rational economic agents has been questioned.

Research into energy-related behaviour shows that people tend to heavily discount or devalue delayed incomes, and generally do not consider future savings or revenues from energy, in contrast to a strictly economic calculation that a business might take (Jager, 2006; Oxera, 2006). In the dilemma between short-term negative outcome (high upfront cost, disruption of installation) and long-term positive outcome (saving energy, saving money, helping the environment), people tend to disregard the long-term benefits and accentuate short-term

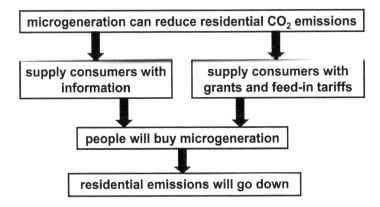

Figure 8.7 *A simplistic model of micro-generation diffusion in the residential sector*

problems such as high cost, excessive bureaucracy, negligible environmental contribution, and so on (Jager, 2006). Thus, it is not a foregone conclusion that households *will* invest in micro-generation if the payback time is a few years, and *will not* invest if the payback time is longer (Sauter and Watson, 2007). Counter examples show that most UK households do not invest in energy-saving measures which have a relatively short payback, such as cavity wall insulation, while many invest in double glazing, which has a considerably longer payback time. This most likely has to do with the different images of the two technologies, with double glazing being familiar and visible, and having the immediate benefit of noise reduction, while cavity wall insulation is thought of as disruptive and expensive, and is not visible. Even if consumers do wish to maximize their utility, it is not straightforward to predict whether this will lead to purchase of micro-generation. For example, if a household is debating between a PV installation and building a new kitchen, maximizing the utility cannot be calculated simply by the payback time of the PV system. It is therefore important to consider what the main motivations and barriers to micro-generation purchase are. The 'rational' motivations are usually considered to be cost, environmental concern, technological interest and self-sufficiency (Allegra Strategies, 2006; Green Alliance, 2006), but other motivations include familiarity of the different technologies and their social status. Cost is undoubtedly a major consideration, both in terms of upfront payment and payback time. The former can be reduced by grants and loans, the latter depends on tariffs for energy generation or export. As discussed above, upfront costs and short payback times are more important than long-term gains to mainstream consumers. In the UK, upfront costs are still high enough to be a real barrier, excluding large parts of the population from purchase (Green Alliance, 2006; Jager, 2006; Oxera, 2006). Payback times can also be very long; for example for PV systems estimates range from 20 to 60 years, possibly longer than the lifetime of the system (Staffell et al, 2010). For consumers to buy micro-generation based solely on economic considerations, lower upfront prices and better tariffs are needed, and are now being introduced in the UK.

Environmental values are often a motivation for purchase of micro-generation, but the increased willingness to save energy is considered alongside perceived costs and benefits (Whitmarsh et al, 2006). The perceived costs are not always accurate, for example one study found that surveyed Londoners overestimated the price of solar thermal, but underestimated the price of PV (Ellison, 2004).

Personal reasons for purchasing micro-generation include love of technology, home comfort, and self-sufficiency. It is worth noting that while for many innovators and early adopters technological novelty acts as a motivation, for the mainstream majority it is more likely to be a barrier, as micro-generation technologies are perceived to be risky, immature or unreliable (Owens and Driffill, 2006). Personal reasons are often intertwined with social motivations, such as familiarity and interest in a micro-generation technology owned by family, friends or celebrities, or more broadly the social status of a specific technology or 'green' behaviour.

We draw one main conclusion from this section: *maximizing domestic uptake of micro-generation is non-trivial, and is not merely a question of cost.* We next turn to the question of whether maximizing uptake is enough to maximize energy and emission savings.

Maximizing uptake versus maximizing emission and energy savings

In the complex domestic energy sector, analysis of energy and emission savings from micro-generation does not end at purchase. It is important to see how the householders use the installed system and whether there are other changes to their energy-related behaviour. Possible behaviours include disappointment or disillusionment with the installed system, which could lead to disuse; a 'fit-and-forget' approach, in which there is no behaviour change and the system is not necessarily used to its full potential; increased energy awareness, which could lead to other energy-related behaviour changes; or a 'rebound effect', where perceived energy or money savings from micro-generation lead to higher energy use (Sorrell, 2009).

Studies of household behaviour after installation paint a mixed picture. One UK survey of 110 households who had installed micro-generation found no significant behaviour change (Element Energy, 2008). Another study, surveying 118 UK households with PV, found that 34 per cent showed small energy savings and 8 per cent large energy savings after installation (Keirstead, 2007). An Austrian study found that households with high electricity consumption tended to reduce their usage after PV installation, while those with low consumption tended to increase their usage after PV installation (Haas et al, 1999). A smaller UK study with in-depth interviews found there were changes in energy awareness, attitudes and behaviour when people installed micro-generation or moved into a house with micro-generation (Dobbyn and Thomas, 2005). Interestingly, this last study shows how living with micro-generation led people to have a greater understanding of energy issues, and more control over their energy use. Furthermore, the changes they describe, such as switching appliances off at the mains or using washing machines at peak (micro-generation) production periods suggest a change to routine behaviour. This contrasts with mainstream *routine* behaviour which shows little change in consumption patterns and low awareness of the connection between everyday behaviour and energy consumption.

Dobbyn and Thomas (2005) also found that households whose purchase of micro-generation was motivated by love of technology or self-sufficiency exhibited a greater behavioural shift than those motivated by environmental beliefs. This prompts the question of whether it matters *why* people buy micro-generation. Socially motivated purchasing of 'green' products promotes visible action over environmentally effective action (Crompton, 2008), which in the case of micro-generation could lead to installations with higher status or visibility taking precedence over those which would save the most energy. If micro-generation became mainstream, we might expect less efficient installations as the majority act more on a social basis. Moreover, it is not clear whether the greater energy literacy and awareness would hold true for the mainstream majority who purchase largely through social motivations; they might be more prone to 'fit-and-forget' or there could be more of a rebound effect. We conclude that *better understanding of people's motivations and behaviour is vital to predicting and maximizing energy savings from micro-generation.*

A powerful example of the difference between potential (technical) energy savings and real savings was found by Hill (2009), who surveyed households with solar thermal installations. Hill estimated current energy savings from solar thermal installations at 16–17 per cent, only *one-quarter* of the industry's predicted 60–70 per cent. There were many reasons for this, including: some installations were not fully compatible with existing boiler systems, due to incompatible products or sub-optimal installations; savings depend heavily on boiler timing, but most households were unaware of this, while others didn't have separate control for space heating and water heating, so couldn't optimize boiler timing; and the maximum grant per installation did not allow larger households to install larger solar thermal systems which would have saved more energy. The low savings were thus due to a complex mixture of technical, institutional, economic and behavioural reasons. The industry does not take a holistic view of household energy, but rather focuses on maximizing sales rather than savings, and does not provide households with the best information. Meanwhile, householders lack access to information, and many are not inclined to investigate best purchase or optimal usage of their system. This highlights our next conclusion: *maximizing uptake does not necessarily realize the 'technical potential' of energy and emission reductions from micro-generation.*

Recalling the 'rational actor' model, we see that consumers lack the presumed 'perfect information', and this can lead to poor purchase and use, reducing energy savings and lengthening the payback time of an installation. Our last conclusion for this section is therefore that *there is a shortage of independent, expert advice for householders as to the best investments in terms of energy and monetary savings.* This forces householders to research, evaluate and choose between the different technologies, which have different attributes, costs and carbon saving potential. Such trade-offs are not trivial calculations to make.

Potential for greater change

Finally, we consider the role of micro-generation at a time when the UK has to meet tough CO_2 emission reduction targets, which must include a significant reduction in domestic sector emissions. In recent decades, domestic energy use has grown fast enough to cancel out gains from efficiency and lower-carbon technologies, suggesting that a more radical change is needed – a transition to modes of behaviour and business which are compatible with low carbon emissions, and the infrastructure and institutions to support them. We explore the idea that the full potential of micro-generation is not as a basket of technologies providing a small percentage of the UK's energy, but as a catalyst for change or part of a new domestic energy system. In other words, 'It would be naive to think of micro-generation simply as another generation technology' (Watson et al, 2006).

The difficulty in accomplishing radical change in a 'socio-technical system' – consisting of infrastructure, institutions, manifold actors, regulations and culture – is known as *lock-in* (Geels, 2005; Smith et al, 2005). Lock-in results from mainstream actors focusing on system optimization through incremental change, rather than system innovation. This is due to vested interests and short-term advantages; existing infrastructures and technical skills; regulations and institutions; and habits, norms and culture. All of these 'lock the system' into limited trajectories of change, which tend to support existing patterns.

There is much evidence that the domestic energy system in the UK is in a state of lock-in (Bergman et al, 2009). The current energy system benefits incumbents, making it hard for new companies to enter the market, and deters large investment in distributed energy systems such as micro-generation, due to cost, risk, and institutional and infrastructural barriers (Genus, 2008; Mitchell and Connor, 2004; Smith et al, 2005). Policy focuses on incremental change, missing opportunities to invest in micro-generation as part of a broader shift to demand reduction and behaviour change (Watson et al, 2006). Market uncertainty causes micro-generation companies to focus on maximizing sales, often unaware of their ability to influence consumers' behaviour (Keirstead, 2007). Patterns of energy consumption are reinforced, with household energy usage growing faster than efficiency savings. People feel disempowered in relation to climate change, expecting the government to act, and often do not connect their own behaviour with energy generation and climate change. All of these describe a system into which radical change, including major reduction in energy consumption or a large micro-generation sector, does not easily fit. Meanwhile, there is a danger that 'if behavioural responses to micro-generation technologies are not considered now, when consumer technologies and protocols are still being developed, then the industry could find that households become locked into behaviours that may be undesirable in the longer term' (Keirstead, 2007, p4137). In other words, if a successful micro-generation market does develop without major changes to the system, it could strengthen the lock-in, and not make a significant contribution to energy and emission savings, as demonstrated earlier in the solar thermal example.

While the exact nature of a more sustainable energy system is unknown, we can provide some ideas. A reoriented domestic energy system, with a prominent role for micro-generation, would be based on increased knowledge, participation and awareness of householders who would share responsibility for the grid, and possibly an economic interest in it (Bergman, 2009; Watson et al, 2006). This could shift the focus of the market from energy production to energy services. Micro-generation households play a non-traditional role in the energy system: they are not just consumers, but also producers and investors, in energy and even infrastructure. This new system could include both household micro-generation, and larger-scale community energy production. Individual energy-producing households would be seen as 'energy citizens', with new rights and responsibilities, or might form energy cooperatives, representing groups of energy-producing households. Larger-scale production could include community micro-grids, which could both import or export energy to the grid. Household and community responsibility for energy could lead to various (as yet unknown) models of energy services.

Studies (Bergman, 2009; Watson et al, 2006) suggest several areas of policy which could help in the envisioned reorientation of the domestic energy system. *Economic* policies would ensure householders investing in any micro-generation technology have access to capital allowances just as energy companies do, and are rewarded for the energy they generate (e.g. feed-in tariffs). Several models exist for this 'levelling the playing field' (Watson et al, 2006). *Institutional* policies would increase the skills base, both in terms of skilled installers for individual technologies, and in terms of overall household energy expertise. These policies would engage the industry to ensure appropriate, professional installations, maximizing consumers' understanding and control of the systems, rather than maximizing uptake as soon as possible. *Regulatory* policies would ensure flexibility and financial protection for emerging technologies and other low carbon innovations, such as installing micro-generation-based micro-grids in new developments. *Educational* and *informational* policies would ensure consumers had access to independent, 'whole house energy' information, and would ideally create a culture of energy literacy, linking household routine behaviour to energy use and climate change; smart meters could play a part in this. While the proposed policies and envisioned reoriented energy system are not an exact science, they do address some of the problems of the simplistic micro-generation diffusion model presented earlier (Figure 8.7). Figure 8.8 is an attempt at a more complex, but more realistic model of micro-generation diffusion and energy and emission savings in the residential sector.

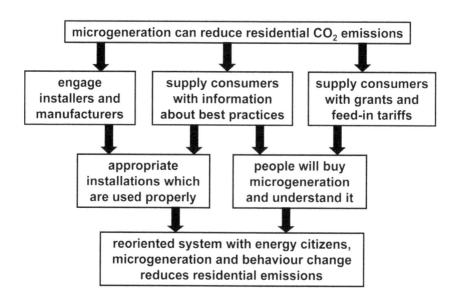

Figure 8.8 *A model of micro-generation diffusion in a reoriented residential energy sector*

Policy challenges for distributed energy resources

In the context of the technical opportunities and constraints offered by each of the technologies and the human dimensions of their installation and operation as discussed above, this section examines policies relevant to micro-generation and the demand-side in the UK. A new policy perspective on the importance and potential of the demand-side could better aid its involvement in the future energy system and simultaneously go some way towards handling the unique opportunities and barriers faced in this sector.

Energy system change and the role of distributed energy resources (DER)

Over the past decade decentralized generation and other smaller-scale energy resources have benefited from increased attention due to rapid technical development and the potential for economic and environmental benefits. These benefits predominantly (but not solely) revolve around co-location of generation with demand and associated efficiencies regarding infrastructure investment and operation, access to otherwise-stranded primary energy resources, and the ability to utilize thermal energy output in the case of combined heat and power. Attention also refocused on the built environment as a sector where DER could make an impact, and broadened to consider demand-side flexibility as a complementary resource.

In the few years up to 2010 there was a rapid development of energy policy and strategy regarding low carbon futures. In contrast to developments regarding DER, these have largely concentrated on a vision of major changes to the large-scale centralized supply-side of the system, and by and large assume that demand is an inflexible quantity to be met, albeit with substantial passive efficiency improvements in some studies. As expounded in this chapter, these mainstream changes to thinking regarding the future energy system could benefit from considering a much more active and integrated demand-side, where investment in and operation of DER could lead to a more rapid and cost-effective low carbon energy future.

In response to the establishment of national targets for greenhouse gas emissions reduction (HMG, 2008a, 2009), a number of recent modelling and policy development efforts have focused on the nature of energy system transition required to meet them. A consistent result of these studies, including the research reported in other chapters here, is that 'successful' scenarios often involve rapid decarbonization of electricity supply over the next few decades (CCC, 2008; DECC, 2009d; UKERC, 2009). What these scenarios have difficulty capturing is their impact on the characteristics of demand-side technology that are more likely to be successful. This is because some demand-side interventions align better with opportunities for emission reduction provided on the supply-side (for example, electric versus primary-fuel-based space heating in the context of supply-side decarbonization of electricity). As such, it is clear that there is a dynamic interaction between the demand-side and supply-side in terms of successful low carbon futures.

The fundamental nature, performance characteristics, and controllability of DER including micro-generation could therefore be a valuable part of meeting the mitigation challenge in coming decades. Rapid development of these small-scale generation and storage technologies, along with associated developments in smart metering and ICT, combine to provide an appealing opportunity for a more cost-effective response to energy system CO_2 mitigation than the supply-side actions alone.

Developments in policy mechanisms for the residential sector

Possibly the most influential policy instrument applicable to the residential sector is the building regulations, providing minimum energy-related standards for a variety of aspects of design and construction, predominantly related to the building shell and lighting. The building regulations are an effective tool for raising the standard of new-build dwellings via implementing what are for the most part 'no-regrets' measures. However, it is clear that they will be unable to deliver the radical change required by 2050 alone. This is mainly because roughly two-thirds of existing homes will still be in use in 2050, but the building regulations are most effective in their application to new build, and only apply to the existing housing stock for specific types and scale of changes. Therefore, other mechanisms must be developed to address existing homes. A variety of instruments are designed to tackle this area including:

- white certificates (or similar) for energy demand or CO_2 reduction;
- tax incentives;
- capital grants to aid purchase and installation;
- financial performance-based or utilization-based incentives;
- smart metering.

As noted in the previous sections, both demand and generation at the residential level could be a valuable resource that could be actively managed in the future energy system. Therefore the structure of all of these policy instruments is of great relevance for broader energy system transformation, and they should be formulated in a way that enables the least expected-cost and least-risk low carbon future to emerge, whilst also taking account of the unique nature of actors in the sectors as described above. This requires a raft of alterations to policy that could challenge the conventional role of the demand-side.

The following sub-sections discuss a selected set of policy measures in this context, highlighting where they may be ineffective, and exploring their possible role in the future.

Net zero carbon new buildings

Perhaps the most important policy instrument in relation to the future of micro-generation in the UK is the upcoming requirement for all new dwellings to be built to 'net zero carbon' standards from 2016 (CLG, 2006). This standard will be implemented via the building regulations, although it is not yet clear exactly what they will require in that the definition of zero carbon has been the subject of consultation, and the final policy instruments do not yet exist. However, assuming that net zero carbon refers to net annual zero CO_2 emissions from heating, cooling and lighting, one way to achieve it is via installation of multiple micro-generation measures (e.g. heat pump and solar PV) in a single dwelling. When measured over an entire year, the net carbon footprint of the dwelling may be reduced to zero because the quantity of electricity used by the heat pump (and other sources of consumption except for appliances) could be smaller than the quantity produced by the solar PV, resulting in *net* zero carbon status. Although a strong boost would be experienced by the micro-generation industry if such a method to achieve the net zero carbon standard became pervasive, the issues raised regarding installers in this chapter may hinder such a development, particularly where they focus on one technology alone. Whilst the building regulations offer an avenue to mandate change and thus circumvent some issues relating to the installers, they still only focus on uptake rather than appropriate application, which as discussed earlier in this chapter can lead to poor outcomes.

Furthermore, whilst the building regulations are clearly a mainstay of residential sector energy policy, they do not play such an important role in capitalizing on the potential of demand-side management (DSM). This is because these regulations apply only at the time of build or installation, and do not relate directly to the *use* of the dwellings and the operation of devices and appliances within it. Whilst it is conceivable that future incarnations of the regulations could provide some credit to DSM-enabled installations, particularly where that DSM is of a passive nature, it is probably wiser to use the regulations to achieve maximum energy efficiency standards, and leave active DSM to those parties most exposed to the costs associated with uncontrolled demand (e.g. suppliers, network operators, and their supply chains).

White certificates

The UK employs what is in essence a white certificate scheme where the retail elements of the liberalized market (i.e. suppliers) are regulated to attempt to reduce the energy consumption of their residential customers. Initially this occurred over two 'energy efficiency commitment' (EEC) periods between April 2002 and March 2008. From 2008 this policy mechanism was extended, the focus changed to carbon emissions rather than energy consumption, and called the Carbon Emissions Reduction Target (CERT). It obliges suppliers to incentivize their customers to install measures from a set of 'qualifying actions', each of which has an associated deemed carbon emissions reduction. The aggregate (deemed) emissions reductions from these interventions must equal or exceed that supplier's target over the commitment period. The target levels of EEC and CERT have approximately doubled each three-year period, and in 2009 it was confirmed that the CERT will be increased by 20 per cent and extended by one year (with prorated target) to deliver $185MtCO_2$ lifetime emissions reductions by 2012. Alongside this, a new scheme called the Community Energy Saving Programme (CESP) was introduced, with the intention of encouraging a whole-house and street-by-street approach in specific priority areas. Whilst this scheme is currently small, it is a welcome addition to the policy mix in that it should go some way towards addressing the institutional and installation-related issues discussed above.

These white certificate schemes emulate typical cap-and-trade systems (with the notable difference of a regulated deemed reduction in emissions as opposed to a cap on total emissions) in that:

- the cumulative amount of a supplier's qualifying actions must equal or exceed their allocated portion of the total target;
- suppliers are allowed trade actions and commitments;
- savings in excess of the target can carry over to the next commitment period.

Therefore it can be argued that CERT is a relatively market-friendly mechanism for the promotion of emissions reduction in the sector. It gives suppliers the flexibility to find the most cost-effective set of actions and/or commitment trading to meet their targets. The primary issues with CERT and many white certificate schemes in general is that the means by which actors can meet emissions reductions leaves the 'additionality' of savings questionable.

With regard to the questionable additionality of savings, suppliers are allowed to meet their targets in any way they choose, subject to the approval of the regulator. Whilst this is laudable in terms of promoting innovative approaches, it also introduces potential for difficulty in verifying the actual impact of actions. Examples of approaches used by suppliers to meet targets include providing energy-efficient light bulbs to customers, subsidizing installation of insulation, and working with manufacturers and retailers of energy-consuming equipment to subsidize purchase of more efficient models. To a greater or lesser extent depending on the nature of the intervention, it is debateable whether some of the CO_2 saving actions would have occurred anyway if funding via the CERT programme were not present (a typical additionality argument). Perhaps the most obvious example of this is that of subsidizing purchase of energy-efficient appliances; there is little scope to argue that this action is additional because the buyer may have already decided to purchase the more efficient appliance. Furthermore, it is important to note that the performance of measures in CERT is deemed; there is no easy way to check whether or not a measure was actually installed in a dwelling, or if it is being used in a way that achieves the goal of reducing aggregate greenhouse gas emissions. An example of this is where energy-efficient light bulbs were mailed to customers who did not need or want them, and many of them may not have replaced less efficient lighting modes, as was intended.

Overall, similarly to the building regulations, white certificate schemes are not well suited to providing reward for in-situ performance. It is an instrument that is directed at installation alone, with little or no regard to ex-post assessment.

Feed-in tariffs (FITs) and the renewable heat incentive (RHI)
Both of the policy instruments discussed above can provide significant economic support for micro-generation, but focus on uptake of technologies rather than their appropriate application. New methods of support – a feed-in tariff for micro-generation (also known as the cash-back scheme) and a renewable heat incentive – were put forward in the Energy Act 2008 (HMG, 2008b). Both of these instruments (in principle) focus on appropriate application of technologies.

In reality, the FIT, which has been in operation in the UK since April 2010, is the only policy instrument for the residential sector that partially rewards actual performance of a device (although the major component of the FIT payment is based on deemed or predicted performance, additional payments are made for measured exports of electricity to the grid). Such instruments and associated technologies such as smart metering form the building blocks of a system that can provide reward to micro-generators commensurate with their actual impact on the system, and are a step towards demand-side transparency in energy markets. The FIT is structured such that an investor in small-scale generation (from household scale up to 5MW$_e$) should receive a certain return on their investment. Therefore different technologies benefit from different levels of reward (e.g. more than 40p/kWh for residential solar PV, to 10p/kWh for gas-fuelled micro-CHP). This structure of reward is based purely on financial issues, and therefore the total cost of carbon savings can be high. But this can be justified on the basis that it supports a new industry in the UK.

The structure of the RHI has been proposed and was under consultation at the time of writing (DECC, 2009b). Due to the flexible nature of residential heating and propensity for comfort-taking (where thermostat set points are raised in response to a lower cost of heating), the proposed structure of the RHI does not reward actual use, but rather deems usage based on the thermal properties of the dwelling. Whilst this is not ideal, it is arguably necessary in the case of heating, and at least there is a direct link between level of support and level of reward.

Despite the advantages of the FIT and RHI, it is important to note that unconsidered application of instruments such as feed-in tariffs can also have negative impacts, and can even create perverse incentives as in the case of gas-fuelled micro-CHP: one study (Hawkes and Leach, 2008) showed that where micro-CHP is incentivized via a feed-in tariff for electricity, gratuitous operation of the device to benefit from the feed-in tariff can result in higher CO_2 emissions than the case where no support is offered. Examples like this show that care must be taken with the structure of such instruments to ensure they help to meet policy aims.

Smart metering
A key enabler in any move towards performance-based reward of demand-side interventions is smart metering. When supported by appropriate information and communications technology, the actual performance of a raft of interventions and actions can be measured. On one level such information could be used to reward relatively passive demand-side interventions (e.g. insulation or low

energy lighting), but it is not difficult to imagine extension of functionality to enable two-way communication and thus real-time demand-side management.

The initial roll out of smart metering in the UK is focused on providing information to consumers such that they may make more informed choices regarding investment and operation of energy-consuming devices, allow faster and more accurate billing and time-of-use metering, and a number of other prospective advantages (DECC, 2009c). The importance of smart metering as an enabling technology towards smart grids should not be underestimated, and their functionality should be left flexible enough to enable such developments.

Performance assessment and policy support

A common theme in all of the above discussions, and indeed a fundamental characteristic of a system with a more integrated demand-side, is the possibilities for measurement of the actual performance of interventions as opposed to assumption of some deemed performance. As such, the basis and intricacies of performance assessment are also an important part of enabling and rewarding beneficial demand-side interventions.

Specifically relevant to this is the method by which the CO_2 performance credentials of a low carbon intervention are examined. The key element of such assessment is definition of the 'reference' or 'baseline' system that is displaced by the intervention. This is often based on a 'business as usual' scenario. This reference system is essentially a portrayal of what would happen if the intervention were not to go ahead, such as electricity generation from other plant or investment in a more polluting technology, and so on. Therefore, by definition, this displaced reference system can only be estimated and is consequently challenging to quantify defensibly.

Furthermore, changes to the baseline system are 'marginal' by nature in that low carbon interventions do not bear upon all aspects of the energy system proportionally. It is the incremental change in the system that defines their actual impact. For that reason, use of system-average statistics in lieu of the estimated performance of the marginal reference system, as is commonly done in carbon abatement calculations, can provide misleading conclusions regarding the reduction afforded by the intervention. In the worst case this could lead to promising low carbon technology development being abandoned, or the introduction of poorly structured policy support or none at all. Given these issues, it is important to develop and maintain a high-quality understanding of the marginal reference system. This will ensure that predicted and actual abatement remain strongly correlated, and more effective support can be afforded to the most promising interventions.

An example of such a situation arises for the case of residential heating as detailed in Figure 8.9.

Figure 8.9 presents the case of the low heat-to-power ratio solid oxide fuel cell micro-combined heat and power (micro-CHP) versus high-temperature air source heat pumps, investigating the sensitivity of performance to the assumed composition of the reference system. The horizontal axis is the CO_2 intensity of the marginal grid electricity reference system, and the plotted lines map the

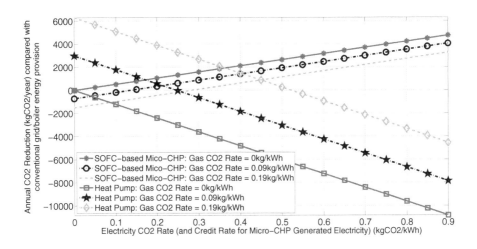

Figure 8.9 *Modelled CO_2 emissions reduction provided by a $1kW_e$ solid oxide fuel cell (SOFC) micro-CHP system versus a high temperature air source heat pump operating in a typical UK terrace house, with sensitivity to gas and electricity emissions rates*

sensitivity to varying CO_2 intensity in gas piped to the dwelling. If one were to assume that the marginal reference system were the current UK grid-average electricity system (approximately $0.52kgCO_2/kWh$) and natural gas were the only fuel option ($0.19kgCO_2/kWh$), then predicted emissions reduction is just above 1 tonne CO_2 per year for micro-CHP and approximately zero for the heat pump. However, if it could be guaranteed that coal-fired generation (approx $0.9kgCO_2/kWh$) were on the margin, emission reduction would be above 3 tonnes CO_2 per year for micro-CHP and negative 4 tonnes for the heat pump. At the other end of the spectrum, if the marginal CO_2 rate was as low as $0.06kgCO_2/kWh$ (i.e. the *Energy 2050 Low Carbon* scenario for 2035, in grid-average terms), emissions reduction would be negative (i.e. an increase) by more than a tonne CO_2 for micro-CHP, and a reduction of over 5 tonnes for the heat pump. Clearly an extremely wide range of outcomes is possible depending on the CO_2 intensity of marginal grid electricity, and high-quality information on this quantity is required to determine the actual impact of demand-side interventions.

This analysis also highlights the opportunities and importance of co-ordination between supply- and demand-side decarbonization strategies. In terms of 'fit-and-forget' demand-side intervention it seems clear that micro-CHP could be challenged to help meet low carbon policy goals in scenarios of rapid (marginal) electricity grid decarbonization. However, on the other hand it is also apparent that if (for example) micro-CHP could be shown to, on average, displace high carbon centralized generation, it could continue to provide substantial emissions savings. Equivalently, and at the other end of the spectrum, heat pumps have strong synergies with low carbon marginal electricity. In the extreme of this concept, it is possible to visualize an actively managed system (both on demand and supply sides) where different demand-side interventions interact specifically with the supply-side technologies with which they have synergies.

Whilst this is a specific example, it serves to demonstrate the issue and also brings to light important complementarities with other policy challenges for the residential sector. It is apparent that there are manifold benefits to better performance assessment and feedback to residential energy consumers, because not only does it allow for appropriately directed reward, but also it could help catalyse appropriate uptake and operation. Delivering these benefits revolves around collection and appropriate use of information regarding demand and onsite generation, alongside tackling further institutional and social issues. Policy instruments would do well to take these matters into account, and whilst developments such as smart metering, feed-in tariffs and the CESP appear to be constructive, they still only address part of the issue.

Further development of policy supporting micro-generation and associated technologies could be a stepping-stone towards much more active involvement of the demand-side in the future energy system, and this could ultimately lead to a more energy-efficient, cost-effective and smarter low carbon future.

Conclusions

This chapter has explored the technologies, policies and human dimension of micro-generation and associated technology developments. These small-scale energy resources and related information and communication systems could play an important role in the future energy system, providing an additional source of low carbon supply as well as the potential to leverage behavioural change and give rise to more awareness of people's day-to-day interaction with the energy system. Taken to the extreme, there is a possibility that the demand-side flexibility potential of micro-generation could facilitate a lower-cost and lower-risk transition to a low carbon energy system.

Despite these potential opportunities, it has also been demonstrated that small-scale energy resources in the residential sector face a unique set of circumstances, and certainly cannot be considered to be interchangeable with or equivalent to other aspects of energy system infrastructure. Technically, performance is highly specific to the technology type, siting, quality of installation, and usage pattern as dictated by the behaviour of dwelling occupiers. As such, the existing institutions and market structures that support the introduction of micro-generation need attention to ensure consistent high-quality advice is given and proper installation and operation are achieved. Without this there is a risk of mediocre or even counterproductive outcomes. Attention should focus on tackling the whole house as a system rather than focusing on any single aspect of efficiency, comfort or micro-generation.

The policy framework to support micro-generation has recently benefited from increased attention in the UK, but significantly more effort is required to ensure a level playing field between decentralized and centralized energy resources, and to enable much more active inclusion of micro-generation and other demand-side interventions in system transformation. Specifically, much of existing policy focuses on the uptake of measures, rather than their performance after adoption or the nature of their interaction with broader energy system change. Instruments are also not well joined up in the way that they address the mechanisms by which beneficial interventions might be adopted and used (e.g. the links between institutions, installers and markets), and largely ignore the potential for unlocking further benefits of consumer engagement. Indeed links between behaviour, technology, and the structure of policy and regulation could create opportunities for inspiring the lifestyle change as described in Chapter 9. Developments such as whole-house approaches in the Community Energy Savings Programme, smart metering, and performance-based support via feed-in tariffs are steps in the right direction in this regard – though their precise impact is still unclear at this stage.

Micro-generation and associated technologies should not be considered in competition with large-scale change to the supply-side, but rather as a distinct and complementary intervention that can leverage both the impact of supply-side change and possibly motivate cultural shift. It is evident that the issues associated with achieving their full potential are by no means trivial, but the possible benefits warrant serious consideration and action. Given likely developments with the building regulations, whole-house approaches in white certificate programmes and introduction of feed-in tariffs, it seems that micro-generation and associated systems are likely to materialize in some form in the UK. The only question is whether their full potential can be leveraged to realize this aim.

References

Aki, H. (2007) 'The Penetration of Micro CHP in Residential Dwellings in Japan', *IEEE Power Engineering Society General Meeting*, pp1–4

Allegra Strategies (2006) *Project Renew: UK Consumer Perspectives on Renewable Energy – Strategic Analysis*, Allegra Strategies, London

Baxi (2008) *Baxi micro-CHP: Clean and efficient energy for the home*, Baxi, Warwick

BCG (1968) *Perspectives on Experience*, Boston Consulting Group Inc., Boston, MA

Bergman, N. (2009) 'Can micro-generation catalyse behaviour change in the domestic energy sector in the UK?' in: Broussous, C. and Jover, C., (eds) eceee 2009 Summer Study, 2009 La Colle sur Loup, France, pp21–32, www.eceee.org/conference_proceedings/eceee/2009/Panel_1/1.029/

Bergman, N., Hawkes, A., Brett, D.J.L., Baker, P., Barton, J., Blanchard, R., Brandon, N.P., Infield, D., Jardine, C., Kelly, N., Leach, M., Matian, M., Peacock, A. D., Staffel, I., Sudtharalingam, S. and Woodman, B. (2009) 'UK micro-generation. Part I: policy and behavioural aspects', *Proceedings of the Institution of Civil Engineerg: Energy*, vol 162, pp23–36

Bergman, N. and Jardine, C. (2009) *Power from the People: Domestic Micro-generation and the Low Carbon Building Programme*, Environmental Change Institute, University of Oxford, Oxford

Bibbings, J. (2006) *Powerhouses? Widening Micro-generation in Wales,* Cardiff, Welsh Consumer Council, Cardiff

BRE Ltd, Environmental Change Institute, EnergiE sas, Innovation Energie Développement, Irish Energy Centre, Omvarden Konsult AB / Four Elements Consulting, Wuppertal Institute, Van Holsteijn and Kemna, Danish Gas Technology Centre, Consult GB Ltd, George Henderson Associates and Advantica Technologies Ltd (2002) *DG for Energy and Transport: Final Technical Report*, European Commission, Brussels

BWEA (2009) *UK Wind Speed Database* [Online], available www.bwea.com/noabl/ [Accessed 2009 August]

Caird, S., Roy, R. and Herring, H. (2008) 'Improving the energy performance of UK households: Results from surveys of consumer adoption and use of low- and zero-carbon technologies', *Energy Efficiency*, vol 1, pp149–166

Carbon Trust (2007) *Micro-CHP Accelerator: Interim Report*, Carbon Trust

CCC (2008) *Building a low-carbon economy – the UK's contribution to tackling climate change*, Committee on Climate Change, London, UK

Clark, P., Gallani, M., Hollis, D., Thomson, D. and Wilson, C. (2008). *Small-scale Wind Energy – Technical Report,* Urban Wind Energy Research Project Part 2 – Estimating the Wind Energy Resource, Carbon Trust, London

CLG (2006) *Code for Sustainable Homes: A step-change in sustainable home building practice*, In: Department for Communities and Local Government (ed.). Communities and Local Government Publications, Wetherby, UK

Crompton, T. (2008) *Weathercocks and Signposts: The environment movement at a crossroads,* WWF-UK, Godalming, Surrey

DECC (2009a) *Digest of United Kingdom Energy Statistics (DUKES)*, Department of Energy and Climate Change, London, UK

DECC (2009b) *Renewable Heat Incentive: Consultation on the proposed RHI financial support scheme*. Department of Energy and Climate Change, London, UK

DECC (2009c) *Towards a Smarter Future: Government Response to the Consultation on Electricity and Gas Smart Metering*. Department of Energy and Climate Change, London, UK

DECC (2009d) *The UK Low Carbon Transition Plan: National Strategy for Climate and Energy*, Department of Energy and Climate Change, London, UK

DEFRA (2008) *Boiler Efficiency Database* [Online], available www.sedbuk.com/ [Accessed June 2008]

Dobbyn, J. and Thomas, G. (2005) *Seeing the light: the impact of micro-generation on our use of energy*, Sustainable Consumption Roundtable, London

Ekins-Daukes, N. (2009) *Solar energy for heat and electricity: the potential for mitigating climate change*, Grantham Institute for Climate Change, Briefing paper No 1, London

Element Energy (2008) *The growth potential for micro-generation in England, Wales and Scotland*, Commissioned by consortium led by BERR, London, UK

Ellison, G. (2004) *Renewable Energy Survey 2004: Draft summary report of findings*, ORC International, London

Enatec Micro-cogen B. V. (2007) *Enatec micro-cogen signs agreement with Bosch Thermotechnik*, MTS and Rinnai (press release dated 12th September 2007)

Encraft (2009) *Warwick Wind Trials – Final Report*, Encraft

Energy Saving Trust (2000) *Heat Pumps in the UK – a monitoring report* (GIR72), Energy Saving Trust, London http://www.energysavingtrust.org.uk/business/Global-Data/Publications/Heat-Pumps-in-the-UK-a-monitoring-report-GIR72 [Accessed 29 June 2010]

Energy Saving Trust (2009) *Location, location, location: Domestic small-scale wind field trial report*, Energy Saving Trust, London

Fisher, J., Jessop, C., McGuire, K. and Waddelove, A. (2008) *A review of microgeneration and renewable energy technologies*, NHBC Foundation and BRE Press, Milton Keynes

Forward, D., Perry, A., Swainson, M., Davies, N. and Elmes, S. (2008) *Viridian Solar – Clearline Solar Thermal Field Trial*, BRE Report number 225314, BRE, Watford

Geels, F.W. (2005) 'Processes and patterns in transitions and system innovations: Refining the co-evolutionary multi-level perspective', *Technological Forecasting and Social Change,* vol 72, pp681–696

Genus, A. (2008) 'Changing the Rules? Regimes, Niches and the Transition to Micro-generation', DIME International Conference: 'Innovation, sustainability and policy', Bordeaux, France

Green Alliance (2006) *Achieving a step-change in environmental behaviours: A report from three Green Alliance workshops held with civil society organisations in October and November 2006*, Green Alliance, London

Haas, R., Ornetzeder, M., Hametner, K., Wroblewski, A. and Hübner, M. (1999) 'Socio-economic aspects of the Autsrian 200kWp-photovoltaic-rooftop programme', *Solar Energy,* vol 66, pp183–191

Hawkes, A., Staffell, I., Brett, D. and Brandon, N. (2009a) 'Fuel Cells for Micro-Combined Heat and Power Generation', *Energy and Environmental Science,* vol 2, pp729–744

Hawkes, A., Brett, D. and Brandon, N. (2009b) 'Fuel Cell Micro-CHP Techno-Economics: Part 2 – Model Application to Consider the Economic and Environmental Impact of Stack Degradation', *International Journal of Hydrogen Energy,* vol 34, pp9558–9569

Hawkes, A.D. and Leach, M.A. (2008) 'Utilisation-Based Policy Instruments to Support Residential Micro-CHP', 7th British Institute of Energy Economists (BIEE) Academic Conference. Oxford, UK

Hill, F. (2009) Consumer Impacts on Dividends from Solar Water Heating, unpublished MSc dissertation, University of East London, London

HMG (HM Government) (2008a) *Climate Change Act 2008*, The Stationery Office (TSO), London, UK

HMG (HM Government) (2008b) *Energy Act 2008*. The Stationery Office (TSO), London, UK

HMG (HM Government) (2009) *The Climate Change Act 2008 (2020 Target, Credit Limit and Definitions),* Order 2009, United Kingdom

Horio, M., Suri, A., Asahara, J., Sagawa, S. and Aida, C. (2009) 'Development of Biomass Charcoal Combustion Heater for Household Utilization', *Industrial and Engineering Chemistry Research,* vol 48, pp361–372

IPCC (2007) *IPCC 4th Assessment Report: Synthesis Report*, www.ipcc.ch/, Intergovernmental Panel on Climate Change, Geneva

Jager, W. (2006) 'Stimulating the diffusion of photovoltaic systems: A behavioural perspective', *Energy Policy,* vol 34, pp1935–1943

Kayahara, Y. (2002) 'The Development of 1kW Residential Gas Engine Cogeneration System', *Energy and Resources,* vol 23, pp173–176

Keirstead, J. (2007) 'Behavioural responses to photovoltaic systems in the UK domestic sector', *Energy Policy,* vol 35, pp4128–4141

Martin, C. and Watson, M. (2001) *Side by side testing of eight solar water heating systems*, DTI, URN 01/1292

Met Office (2008) *Climate Indicators* [Online], available: http://tinyurl.com/yfk4ybe [Accessed 2009 May], UK Met Office, Exeter

Mitchell, C. and Connor, P. (2004) 'Renewable energy policy in the UK 1990–2003', *Energy Policy,* vol 32, pp1935–1947

Moore, G.A. (1991) *Crossing the Chasm*, Harper Business Essentials, New York

Munzinger, M., Crick, F., Dayan, E.J., Pearsall, N. and Martin, C. (2006) *Domestic Photovoltaic Field Trials: Final Technical Report*, BRE and DTI, URN: 06/2218, BRE, Watford

Omata, T., Kimura, T., Yamamoto, Y. and Nishikawa, S. (2007) *Current Status of the Large-Scale Stationary Fuel Cell Demonstration Project in Japan*, 10th Grove Fuel Cell Symposium, 2007, London, UK

Onovwiona, H.I. and Ugursal, V.I. (2006) 'Residential cogeneration systems: review of the current technology', *Renewable and Sustainable Energy Reviews,* vol 10, pp389–431

Orr, G., Lelyveld, T. and Burton, S. (2009) *In situ monitoring of efficiencies of condensing boilers and use of secondary heating*, Department of Energy and Climate Change, London

Owens, S. and Driffill, L. (2006) *How to Change Attitudes and Behaviours in the Context of Energy*, UK Government's Foresight Programme, London, UK

Oxera (2006) *Policies for energy efficiency in the UK household sector*, Oxera Consulting Ltd, Oxford

Pehnt, M. (2003) 'Life-cycle analysis of fuel cell system components', in Vielstich, W., Lamm, A. and Gasteiger, H.A. (eds) *Handbook of Fuel Cells – Fundamentals, Technology and Applications,* John Wiley and Sons, Chichester

Phillips, R., Blackmore, P., Anderson, J., Clift, M., Aguilo-Rullan, A. and Pester, S. (2007) *Micro-wind turbines in urban environments – an assessment*, BRE

REA (2008) *Renewable Energy Association* [Online], available www.r-e-a.net/ [Accessed 13 October 2008], REA, London

Rogers, E.M. (1995) *Diffusion of Innovations*, Free Press, New York

Sauter, R. and Watson, J. (2007) 'Strategies for the deployment of micro-generation: Implications for social acceptance', *Energy Policy,* vol 35, pp2770–2779

Smith, A., Stirling, A. and Berkhout, F. (2005) 'The governance of sustainable socio-technical transitions', *Research Policy,* vol 34, pp1491–1510

SolarFocus GmbH (2008) *Log wood + pellets: Therminator II, the combi boiler*, SolarFocus GmbH, St Ulrich/Steyr, Austria

Sorrell, S. (2007) *The Rebound Effect: an assessment of the evidence for economy-wide energy savings from improved energy efficiency*, UK Energy Research Centre, London

South West of England Regional Development Agency (2008) *Biomass Case Study: The Science Museum and Library, at the Old Wroughton Airfield, near Swindon*, South West of England Regional Development Agency, Exeter, available www.knowledgewest.org.uk/cases/?id=60

Staffell, I. (2009a) *Fuel cells for domestic heat and power: Are they worth it?*, PhD Thesis, Chemical Engineering, University of Birmingham

Staffell, I. (2009b) *A Review of Domestic Heat Pump Coefficient of Performance* [Online], available: http://wogone.com/iq/review_of_domestic_heat_pump_cop.pdf [Accessed April 2009]

Staffell, I., Baker, P., Barton, J., Bergman, N., Blanchard, R., Brandon, N.P., Brett, D.J.L., Hawkes, A., Infield, D., Jardine, C., Kelly, N., Leach, M., Matian, M., Peacock, A. D., Sudtharalingam, S. and Woodman, B. (2010) 'UK Micro-generation. Part II: Technology Overviews', *Proceedings of the Institution of Civil Engineers: Energy*, vol 163(4), pp143–165

Staffell, I. and Green, R.J. (2009) 'Estimating future prices for stationary fuel cells with empirically derived learning curves', *International Journal of Hydrogen Energy*, vol 34, pp5617–5628

Teekaram, A. (2005) *Installation and Monitoring of a DACHS Mini CHP unit at BSRIA*, Building Services Research and Information Association, West Bracknell, Berkshire

The Energy Monitoring Company Ltd (2001) *Analysis of Performance Data from Four Active Solar Water Heating Installations*, DTI, UK, URN 01/781, Department of Trade and Industry, London

Thomas, B. (2008) 'Benchmark testing of Micro-CHP units', *Applied Thermal Engineering*, vol 28, pp2049–2054

UKERC (2009) *Making the Transition to a Secure and Low-Carbon Energy System*, Synthesis Report, UK Energy Research Centre, London, UK

Utley, J.I. and Shorrock, L.D. (2008) *Domestic Energy Fact File*, BRE, London, UK

Watson, J., Sauter, R., Bahaj, B., James, P.A., Myers, L. and Wing, R. (2006) *Unlocking the Power House: Policy and System Change for Domestic Micro-generation in the UK*, SPRU, Brighton, UK

Watson, S.J., Infield, D.G. and Harding, M.R. (2007) *Predicting the yield of micro-wind turbines in the roof-top urban environment*, EWEC2007, Milan, Italy, www.ewec2007.info/

Whisper Tech (2003) *AC WhisperGen System – Product Specification*, Whisper Tech, Christchurch, New Zealand

Whitmarsh, L., Haxeltine, A., Köhler, J., Bergman, N. and Valkering, P. (2006) *Simulating consumer decision-making in transitions to sustainable transport*, MATISSE WP9 18-month report (October 2006)

Wohlgemuth, J.H., Cunningham, D.W., Nguyen, A.M. and Miller, J. (2005) *Long Term Reliability of PV Modules*, BP Solar International, BP, London

9
The Way We Live From Now On: Lifestyle and Energy Consumption

Nick Eyre, Jillian Anable, Christian Brand,
Russell Layberry and Neil Strachan

Introduction

What are lifestyles?

The notion that people's 'lifestyle' may need to move in more sustainable directions has rapidly become a focus of environmental policy and popular commentary on environmental issues. There is considerable speculation around the possibility of a 'cultural shift' affecting the scale and patterns of consumption and behaviour in ways that will lead to a lower impact, less energy intensive and potentially more community oriented society (Defra, 2008; Thogersen, 2005). This transition in the discourse from sustainable 'consumption' to sustainable 'lifestyles' implies a shift in the salient source of meaning away from consumption towards specific values, rules and social practices which are shared by groups of persons and constitute their 'way of life' (Evans and Jackson, 2007).

Yet, despite a widely agreed consensus that societal energy consumption and related emissions are not only influenced by technical efficiency but also by lifestyles and socio-cultural factors, there is a methodological gap between the perceived importance of these factors for energy demand and practice in many quantitative modelling exercises. Modelling studies such as the Japan–UK Low Carbon Societies project (Strachan et al, 2008) have identified improved treatment of behaviour in quantitative analysis as a key priority. Indeed, there is much less consensus as to the character and extent of these influences, particularly when broadened out to include psycho-social factors such as wellbeing, cultural norms and values, and few attempts have been made to operationalize these insights into models of future energy demand.

This chapter addresses that gap by contrasting the techno-economically driven core scenarios with one in which social change is strongly influenced by concerns about energy use, the environment and wellbeing. It takes as its starting point the notion that lifestyles, whilst closely intertwined with consumption, encompass more than economically justifiable preferences and the accumulation of material goods (Reusswig et al, 2003). Not only are non-price determinants of behaviour recognized, such as values, norms, fashion, identity, trust and knowledge, but non-consumptive elements of behaviour such as patterns of time use, mobility, social networking, expectations and policy acceptance are considered in our characterization of future patterns of energy service demand.

In addition to consumers, people are also seen as ethical and political actors who are responsible for reflexive and political preferences as well as market choices (Reusswig et al, 2003). Consequently, lifestyles are viewed as more than transient fashions or trends. They encapsulate ethical commitment so that they straddle both notions of individuality and identity on the one hand and community or sociality on the other (Evans and Jackson, 2007). This allows our scenario approach to pay attention to the *interaction* between society and technology (Elzen et al, 2002) and underlines the role that policy can play in working with attitudes, opportunities and impacts to exert a positive influence on the type of society that develops and the nature of the technical system that co-evolves with it.

The drivers of lifestyle change

History demonstrates that major shifts in societal attitudes and behaviour around perceived collective goods like the environment, national security and multiculturalism do take place (Rajan, 2006). Firstly, however, we need to understand the ways in which positive behaviours and ethical commitments are adopted in the first place as well as how they can be maintained and reinforced over time. Moreover, lifestyle change as a driver of social change is more than a shift in attitudes or behaviours and cannot be measured in single dimensions. In particular, our understanding needs to be informed by a sophisticated appreciation of the ways in which modern lifestyles operate not just at the material level but also at the psychological, social and cultural levels (Jackson, 2005).

What is clear from social psychology is that multiple barriers and drivers all impact on behaviour in combination, but that at a societal level, behaviour change could take place at a sufficiently large scale given the right circumstances (Rajan, 2006). Without repeating that complex discussion here, suffice to say that the most widely-adopted models show many factors, both internal and external, collective and individual, impacting from different directions on an eventual behaviour or collective choice (Anable et al, 2006; Darnton et al, 2006). Many of the factors are non-rational, for instance relating to opportunities or infrastructure, rather than intentional motive. In this way, simple linear ('information deficit') models showing how increased awareness of an issue leads to a reasoned decision and appropriate action, are overturned.

The notion of 'lifestyle' as defined above suggests people can choose to reduce their environmental impacts as part of their 'life project'. Despite this emphasis on consumption and matters of personal or collective choice, consumers are often 'locked-in' to unsustainable patterns of living 'by a combination of perverse incentives, institutional structures, social norms and sheer habit' (Jackson, 2005). This highlights the interrelationship between consumer behaviour, the structure of the market, technologies and physical infrastructure, institutions and public policy. The implication for policy-makers is that policy needs both to help empower consumers to change lifestyles and to loosen some of the external constraints that make changes towards a more sustainable lifestyle difficult (Thogersen, 2005).

However, as Darby (2007) points out, the major challenge is that the balance of what is judged to be acceptable or optional (needs or wants) changes with culture, time and in response to technological developments. Individuals will tend to have strong and differing views on what levels of energy services are sufficient, based on their life experiences and on cultural norms. Darby illustrates this with the example of the way in which an abundant, highly predictable supply of electricity has become a 'need' in Europe over the last hundred years. Yet, normative judgements on sufficiency, identifying 'how much is enough' and intervening in lifestyle choices is anathema to modern neo-liberal politics of governance (Jackson, 2005; Darby 2007). Identifying the likely impact of lifestyle as a driver of social change, source of social pressure and influence on policy in this context is not easy. Nevertheless, we set out to test the concept of 'lifestyle' change to understand the extent to which such shifts could contribute to meet carbon reduction targets and avoid the need to employ new supply-side technologies.

Why lifestyle matters for energy use

Energy use in the UK, as in other developed countries, is central to our current way of life. It fuels our manufacturing processes and high levels of mobility, keeps warm (and sometimes cool) our buildings and powers a huge array of electrical appliances from lighting and refrigeration through to the proliferation of modern consumer electronics.

Ultimately, all energy use results from consumption decisions, as it is demand that drives production. So it is theoretically possible to account for all energy use at the point of consumption, as either used directly by consumers or as energy used in production and therefore 'embedded' in purchased goods and services. But it is helpful to distinguish between these two categories. In the early stages of industrialization, it was industry that dominated our energy use. But the long-term trend is towards an increased share being used in the sectors of the economy where energy is primarily associated with consumption rather than production – in households and transport. These sectors now use 29 per cent and 38 per cent of UK final energy respectively (DECC, 2009a). Increasingly it is direct consumption that drives UK national energy use. That is the focus of this chapter.

Personal energy consumption is determined by two factors – the energy services that we demand (for example comfort, mobility and entertainment) and the energy efficiency of the energy conversion devices that provide these services. Lifestyle therefore drives direct energy consumption in different ways.

Most importantly, lifestyle determines the type and quantity of energy services we use. Size and scale matter: for example, keeping a large home at a comfortable temperature tends to use more energy than delivering a similar quality of service to a smaller home. And some energy services are intrinsically more energy intensive than others – in particular services that involve heating, cooling or accelerating large amounts of matter have minimum energy requirements determined by the laws of thermodynamics.

Lifestyle is also related to the propensity to use more energy more efficiently. As the environmental impacts of energy use become better known and the ecological limits better understood, different conversion technologies are increasingly associated with lifestyle choices. The 'social statement' made by the purchase of a hybrid vehicle is different from that of a 4×4.

Potentially, lifestyle may also affect carbon emissions through the carbon intensity of the energy sources we use. Traditionally, the role of individual choice in this factor has been somewhat limited: mains gas, grid electricity and petrol at the pump have been dominant energy carriers, with carbon intensities determined upstream in the energy system. But the commercialization of different transport fuels and small-scale renewables may be expected to alter this and add an extra degree of freedom to the relationship between lifestyle and carbon.

In principle, lifestyle change can affect energy use and carbon emissions in either direction. Greater wealth, higher levels of consumption and new energy services all tend to increase energy use. These trends have more than offset technological improvements in energy efficiency in most countries at most stages of economic development, so that energy consumption has generally risen with increasing affluence (IEA, 2007) and it is quite possible that future energy use in the UK could easily follow such a trend. But it is not inevitable; energy use in the UK has declined modestly in recent years (DECC, 2009a), and this trend is continued in the baseline energy scenarios considered in this book. This chapter considers the potential for lifestyle change to amplify and drive stronger reductions in energy use.

Quantifying lifestyle

The challenge of quantifying lifestyle

Lifestyle, as defined above, is a qualitative concept. In contrast, the role of energy use in the energy system is only meaningful if quantified, yet, for the reasons set out above, it is a function of lifestyle. To quantify energy use we need to use metrics of the key energy-using behaviours (service demands and technologies) that result from a given lifestyle, such as room temperatures, kilometres travelled and conversion technologies used. Both the energy service demands and conversion efficiencies are, by definition, quantifiable.

In practice, we cannot define scenarios for every energy service demand and technology. We choose to focus on those that are quantitatively important for energy use, either now or potentially in the period to 2050. Both the simplifying assumptions made and the plausible range in parameters inevitably introduce uncertainties. These are particularly significant for the end of the period to 2050, by which time significant social, cultural and technical change is expected. However these uncertainties apply to all scenarios, not just those that consider pro-environment behaviour change. All the scenarios set out in this book and elsewhere implicitly use assumptions about future behaviour change, even where these are hidden or modelled very simplistically, for example as an extrapolation of existing trends. All that is conceptually distinct about the pro-environmental scenario set out in this chapter is that the implications for consumer energy use are explicit.

Modelling the energy implications of lifestyle change involves not only detailed assumptions about both the demand for energy services and the choice of energy-using technology, but also how these interact with other decisions relating to the energy system. For example, if car buyers purchase more electric vehicles, this will increase demand for electricity and, other things being equal, lead to more power station construction. However, the choice of new power station is not made by the car buyer, and so, as well as describing future consumer behaviour in some detail, we need to model the implications for the whole energy system. We have addressed this challenge by using three models with different capabilities:

- the UK Domestic Carbon Model (UKDCM) (Palmer, 2006);
- the UK Transport Carbon Model (UKTCM) (Brand, 2010; Brand et al, 2010);
- the MARKAL energy system model (Strachan and Kannan, 2008; Strachan et al, 2008; Kannan et al, 2009).

The use of these models is described in more detail below.

Using these tools, we have modelled a 'lifestyle scenario', but with two variants: one (denoted LS-REF) in which carbon emissions are not constrained, and another (denoted LS-LC) in which carbon emissions from the whole energy system are constrained to fall by 80 per cent relative to 1990 levels by 2050. These two lifestyle variants have been contrasted with the corresponding core UKERC Energy 2050 scenarios (REF and LC), as described in Chapters 4 and 5. The relationship between the four scenarios is shown in Table 9.1.

Table 9.1 *The lifestyle scenarios related to the core scenarios*

| | | System-wide carbon constraint in 2050 | |
		None	−80%
Social/lifestyle assumption	Business as usual	REF	LC
	'Lifestyle'	LS-REF	LS-LC

The use of the different models is set out in Figure 9.1. In essence, we have used the detailed sectoral models (UKTCM and UKDCM) to model the energy service demands and details of some end-use technologies, and the UK energy system model (MARKAL) to model wider interactions with the energy system including the impact of a national carbon emissions constraint. More details on the assumptions in the UKDCM and UKTCM models and how energy service demands and technology link into an energy system framework are provided below and in a UKERC Working Paper (Anable et al, 2010), with the full range of key energy system assumptions for these runs given in Anandarajah et al (2009).

In judging what rate and scale of change seem plausible in the Lifestyle scenario we have given most weight to the existing variation in lifestyle observed in societies like our own, that is, technologically advanced, liberal democracies. Subject to some obvious constraints imposed by age, wealth and location, for example, it seems reasonable to suppose that if a significant fraction of the population (say 5–10 per cent) somewhere in the OECD already behave in a particular way, then it is plausible for this to become a majority behaviour in the UK within the timeframe to 2050.

Quantifying residential energy demand

UK Domestic Carbon Model

This is a model of energy use in the UK housing stock. It contains many categories of residential buildings, each category representing a number of real-world dwellings in the country in the model base year of 1996. For each building category, there is data about the building form and fabric and other properties related to energy use (windows, walls, lofts, air change rates and internal temperatures). The majority of the information was taken from the English House Condition Survey (EHCS) 1996 which contains structural information

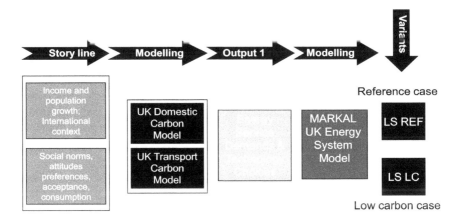

Figure 9.1 *Outline of the modelling process*

for almost 30,000 representative dwellings (DETR, 2000). The 1996 EHCS contained an energy sub-module which later surveys have lacked.

For each building category, the model calculates the energy demand placed on the space heating system in each month to keep the required mean internal temperature, assuming the mean monthly UK external temperature for the period 1970–2000. This is done by calculation of energy flows out of the building, from the information set described above on building fabric areas, U-values and air change rates, using the same approach as the BREDEM-8 model (Henderson and Shorrock, 1986). The calculations take account of incidental gains from cooking, human metabolism, solar gain (through windows) and waste heat from hot water, lights and appliances. In common with most similar models for the UK, there is no explicit modelling of space cooling, as this is (at least currently) a very small component of demand.

UKDCM can be used to calculate the final demand for space heating and water heating over the stock, using inputs on the type and efficiency of heating systems, including gas boilers, electric heating, solar thermal and solid fuels, and in future years such technologies as CHP (Stirling engine, fuel cell, district heating), heat pumps and biomass heating.

The model allows building electricity demands to be offset by on-site generation (from CHP, micro-wind and solar PV). The calculations result in monthly fuel demands (gas, electricity, coal/oil, biomass) from the UK housing stock. For each year, the housing stock is updated (new build, demolition and retrofits) and the heating system type, demand for hot water, demand for lights and appliances and internal temperatures may also be varied annually. The result, when aggregated, is the end-use demand (for space heating, cooking, lights and appliances and hot water) and the fuel demand (gas, electricity, coal, oil and biomass) of the UK housing stock on a yearly basis from 1996 to 2050. From these data, carbon emissions due to UK household energy use in each year are calculated. The model is described in more detail in Palmer et al, 2006.

Scenarios can be constructed for alternative rates of house-building, demolition, fabric improvements, micro-generation installation and efficiencies, improved efficiency in lights and appliances, as well as changes in internal temperature, hot water use and other energy-using behaviours. The scenarios can explore the potential for energy use and carbon emissions in the period to 2050, including the impact of changes in external mean temperatures. For this research, UKDCM was used to specify the energy service demands for space and water heating, with MARKAL (see below) then used for technology choice. For other energy end uses, UKDCM was used to specify the final energy demand. In all cases, MARKAL readjusts demand levels to allow for responses to price changes.

MARKAL model

MARKAL is a widely applied technology-rich, multi-time period optimization model (Loulou et al, 2004). As already described in Chapter 4, it portrays the entire energy system from imports and national production of fuel resources, through fuel processing and supply, explicit representation of infrastructures, conversion of fuels to secondary energy carriers (including electricity, heat and hydrogen), to end-use technologies and energy service demands of the entire economy. As a perfect foresight partial equilibrium optimization model, MARKAL in its elastic demand mode minimizes the sum of producer and consumer surplus – as a metric of social welfare – by considering the investment and operation levels of all the interconnected system elements as well as resultant demand changes. The inclusion of a range of policies and physical constraints, the implementation of all taxes and subsidies, and calibration of the model to base-year capital stocks and flows of energy, enables the evolution of the energy system under different scenarios to be plausibly represented.

Model choice

A comparison of UK MARKAL with UK building stock models is given in Kannan and Strachan (2009). UK MARKAL and UKDCM have very different strengths. UKDCM provides a very detailed simulation of the types and fabrics of the housing stock. It contains information on costs of individual technologies that allows calculation of the costs of different investment scenarios and, if fuel prices are added exogenously, their cost effectiveness. It is a simulation model and therefore technology choices are not determined by economic criteria (such as least-cost optimization); these are exogenous inputs relying on modeller judgement. MARKAL has a much less detailed description of the housing stock and its technologies. Fuel prices are calculated in other modules of the model and, with technology costs (and where included shadow prices, e.g. for carbon), determine the choice of technology by economic optimization, subject to any external constraints imposed by the modeller at either the household sector or whole system level.

Investments in the housing stock are difficult to describe using rational actor economic models (Sanstad and Howarth, 1994; OXERA, 2006). If MARKAL is used, it needs to be heavily constrained which reduces the benefits of its economic insights. In this project we have therefore chosen to use explicit modeller judgement and UKDCM for most housing related variables. The exception is that we have used MARKAL to model the choice of heating system to 2050. This is because we expect this choice to depend not only on the characteristics of the technologies and the people and buildings they serve, but also upon the development of the wider energy system, in particular the use of electricity and biomass in other sectors, relative fuel prices and the carbon content of electricity. Use of electricity or gas for heating in carbon-constrained scenarios is particularly sensitive to these factors and extremely important in relation to the wider energy system. However, none of these factors will affect so strongly investments in building fabric, energy efficiency measures or appliance choice, which we therefore have modelled with the higher resolution available in UKDCM.

Quantifying mobility energy demand

The quantification of mobility energy demands for these scenarios involves:

- storyline development and bottom-up spreadsheet modelling to develop an alternative set of transport energy service demands;

- sectoral modelling using the UK Transport Carbon Model (UKTCM) for vehicle ownership, vehicle technology choice (size, performance, preference, market potentials) and vehicle use (in-use fuel consumption);
- the translation of UKTCM outputs of fuel consumption and vehicle fleet evolution by technology into MARKAL inputs, through specification of technical energy efficiency, and technology deployment constraints and bounds.

Storyline modelling

Transport energy demand is a function of transport mode, technology and fuel choice, total distance travelled, driving style and vehicle occupancy. Distance travelled is itself a function of land use patterns, destination, route choice and trip frequency. Most travel behaviour modelling and forecasting is based on principles of utility maximization of discrete choices and on the principle that travel-time budgets are fixed (Metz, 2002). However, based on evidence relating to actual travel choices, the lifestyle variant scenarios modelled here explored a world in which social change is strongly influenced by concerns relating to health, quality of life, energy use and environmental implications. As such, non-price driven behaviour, which has already been found to play a significant role in transport choices (Anable, 2005; Steg, 2004; Turrentine and Kurani, 2007), becomes a dominant driver of energy service demands from transport.

The 'lifestyle' consumer is more aware of the whole cost of travel and the energy and emissions implications of travel choices and is sensitive to the rapid normative shifts which alter the bounds of socially acceptable behaviour. Consequently, the 'Lifestyle' variant scenarios assumed the focus would shift away from mobility towards accessibility. In other words, the quality of the journey experience rather than the quantity and speed of travel would become more important. Social norms elevate active modes and low carbon vehicles in status and demote large cars, single-occupancy car travel, speeding and air travel.

The consequences for travel patterns of these shifts were first analysed using a spreadsheet model which took as its starting point the figures for current individual travel patterns based on the UK National Travel Survey (DfT, 2008a). Figures for each journey purpose (commuting, travel in the course of work, shopping, education, local leisure, distance leisure and other) in terms of average number of trips, average distance (together producing average journey length), mode share and average occupancy were altered based on an evidence review relating to the impact of transport policies and current variation in travel patterns within and outside the UK (see Anable et al, 2010a, for a detailed overview of the calculations).

UK Transport Carbon Model

The UKTCM is a strategic transport-energy-environment simulation model which provides annual projections of transport supply and demand, and calculates the corresponding energy use, life cycle emissions and environmental impacts year-by-year up to 2050. It simulates passenger and freight transport across all transport modes, built around exogenous scenarios of socio-economic and political developments. It integrates simulation and forecasting models of elastic demand, vehicle ownership, technology choice (using a discrete choice modelling framework), stock turnover, energy use and emissions, life cycle inventory and impacts, and valuation of external costs. The UKTCM is described in detail in Brand (2010) and Brand et al (2010).

A number of UKTCM runs produced detailed fuel consumption and vehicle stock projections for the Core and Lifestyle scenarios, based on the shifts in travel patterns and energy service demands. These then act as inputs to the MARKAL model.

Lifestyle change at home

Influences on residential energy demand

Residential energy demand includes all of the energy services that households require within the premises of their home. The most important historically has been thermal comfort provided by space heating, which now uses 58 per cent of household energy (Utley and Shorrock, 2008). This is followed by two other heating services: water heating (largely for personal hygiene) and for cooking food. These remain the main uses for fuel (as opposed to electricity). The key drivers are the level of service required and the efficiency with which it is provided.

For space heating, energy service demand is driven by heated floor area and internal temperature. Energy efficiency is determined by both heating device efficiency, but also very importantly by the thermal properties of the home (external surface area, insulation and air-tightness), which vary by large factors across the building stock. In most cases, water heating is provided by the same device – a gas boiler in 80 per cent of UK homes (Utley and Shorrock, 2008), and therefore water heating efficiency depends on the energy efficiency of the boiler and the water use efficiency of the device.

The development of electricity grids and their extension to give nearly universal coverage by the mid-20th century provided the stimulus for electrically provided services. Although electricity still provides only a small share of final household energy (22 per cent), its share in both costs and emissions is much larger. Only 10 per cent of homes rely on electricity as the main heating fuel, partly because use of direct resistance heating is inefficient (due to power station losses) and expensive. There has been a larger shift in cooking.

The predominant uses of electricity are for other services provided by lighting and electrical appliances. Traditionally appliances have been segmented into 'cold appliances' (refrigerators and freezers), 'wet appliances' (washing machines, dryers and dishwashers) and 'brown appliances' (radios, televisions and other entertainment) with a more diverse group of other 'minor appliances' e.g. irons, vacuum cleaners. In all cases, the electricity use is essentially determined by the product of appliance numbers, running hours and specific energy consumption.

In recent years, the use of electronic appliances has grown and diversified. As with other appliances, key drivers of electricity use include efficiency and the number of appliances. For many electronic devices there is the option of 'standby'. Although standby power demand is relatively small, very long running hours in this mode can result in standby energy consumption forming a significant fraction of energy use in such devices.

Recent trends

The trend in household energy use from 1970 to 2007 is shown in Figure 9.2. For most of the period, there was a trend of rising energy use at ~1 per cent annually. There is significant inter-annual variability due primarily to weather, but this is adjusted in Figure 9.2 to show the long-term trend. This rate is broadly similar to the rate of increase in household numbers; that is, annual energy use per household has been broadly constant over the period. This is due to the combination of different counteracting drivers. In general, the level of energy services (internal temperature, hot water volume, lighting levels etc.) has increased, but the efficiency with which these are provided has risen.

Figure 9.2 shows that space heating remains the dominant component in household energy use, even though its relative importance in the most recent years has fallen. The rise in space heating demand is driven by increasing internal temperatures that masks the strong opposing effect of major improvements in

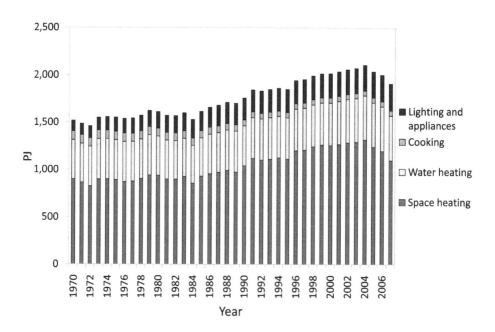

Source: Based on DECC, 2009b

Figure 9.2 *UK household energy use 1970–2007*

home insulation and heating system efficiency, without which space heating energy use would have doubled since 1970 (Utley and Shorrock, 2008). Space heating energy use has declined substantially since 2004, over a period in which energy prices have risen sharply and household energy efficiency programmes have increased in scale substantially.

Demand in other end uses has risen faster over recent decades, with the exception of cooking, where it has changed very little. Hot water use has increased faster than heating system efficiency, leading to modest growth in energy use.

Energy use for lighting and appliances (almost exclusively electricity) has risen fastest of all (see Anable et al, 2010). In 1970, electricity use was dominated by lighting and cooking. In these end uses demand for electricity has only risen slowly. In cooking this has been due entirely to electricity substituting for gas (DECC, 2009b). In lighting, there has been significant growth in the number of fittings, but this has been largely offset by improved efficiency, especially with use of compact fluorescent lamp technology which has reduced lighting electricity demand since 2002 – a trend which is expected to continue and intensify with the phase out of incandescent bulbs, except in specialist applications.

Increased demand for electricity in households has been driven much more strongly by the wider use of appliances delivering new energy services. For each new energy service demand, there is a tendency for energy demand to rise quickly as the market develops, then stabilize as it saturates and even begin to fall as appliance efficiency improvements overtake increased use.

There was growth in energy use in cold appliances (predominantly through the introduction of freezers) in the 1970s and 1980s, although this has now stabilized and fallen slightly in the last decade. Energy use in wet appliances is still rising slowly, primarily due to rising use of dishwashers. Since 1990, demand growth has been driven primarily by consumer electronics, joined in the last decade by home information and communications technologies, ICT (primarily home computing). Together they have grown from using only 11 per cent of household electricity to 32 per cent in just two decades (DECC, 2009b).

Future trends in a low energy lifestyle

The trends set out above and illustrated in Figure 9.2 show that household energy demand has fallen rapidly in recent years. It is clear that most of the main energy service demands are rising only slowly (if at all) and that efficiency improvement is more than offsetting these.

However, it would be unwise to project this trend uncritically. Rising environmental awareness and expanded energy efficiency programmes have been major contributors to the change, and these may well continue. But there has also been an impact from a very sharp rise in energy prices of between 50 per cent and 100 per cent for the main household fuels, which is a trend that is unlikely to continue, certainly at this rate.

The key changes that affect residential energy use in the Lifestyle scenario are as follows. The scenario assumes increasing use of low energy technologies and that this continues to be accompanied by stabilizing and/or declining levels of energy service demands through to 2050. This is driven by a combination of increasing energy awareness, modestly higher prices and improved real-time information. These increasingly allow more pro-environmental attitudes to be reflected in behaviour. Social norms increasingly emphasize environmental performance and this is reflected in market values, as building and product labelling make reliable information available. The same factors make conspicuous consumption of energy socially unacceptable.

Insulation of the building stock to a high standard becomes a social expectation. New homes rapidly conform to best European practice (Passivhaus standard). However, housing demolition rates remain at the low level of recent years, reflecting a desire to maintain existing neighbourhoods and reuse existing capital assets, and therefore refurbishment of the existing stock is crucial.

Government and energy company programmes allow the basic insulation measures to reach close to saturation levels by 2020. Adoption of more expensive and difficult measures is slower, but 'whole house' and 'whole street' mechanisms increasingly support very high efficiency retrofits. Refurbishment is initially through green 'able to pay' markets and low-income programmes in areas of multiple deprivation in cities and then to rural solid-walled properties. After 2020, high quality retrofits are required as properties are sold. The building trades who undertake repair and maintenance of residential properties develop the requisite new knowledge and skills. There is increased use of external wall insulation both for solid wall insulation and cladding cavity walls. Windows are replaced at the end of their design lifetimes, with new windows rapidly improving to a standard broadly equivalent to current best available.

Smart meters are deployed rapidly after 2010, with rapid improvements in technology to provide information on energy use by individual devices, in real time. These provide comparisons with historical data and similar households, via a choice of dedicated displays, TV, SMS and internet.

Over-heating of buildings becomes socially unacceptable. The long-term trend in rising internal temperatures ends. Internal demand temperature peaks at 20°C, then falls back at 0.2°C/year to 17°C in 2025 and stabilizes there, i.e. at the levels experienced in well-heated homes in the 1990s. Hot-water use also declines as water-conserving and lower temperature washing technologies are introduced. Its use falls by 1.25 per cent annually from 2010 to 2050 to levels found currently in many other European countries.

The dominance of gas and oil boilers continues for some years. However, new low carbon heating systems enter the market and take a large market share around 2020. In urban centres, district heating schemes using CHP (some of it waste and biomass fired) become common, initially driven by building regulations for new developments, but then also in mixed use retrofit schemes. District CHP supplies about 10 per cent of homes by 2050.

In the remaining stock there is initially strong competition between gas-fired micro-CHP systems and electrically powered heat pumps. As electricity production is decarbonized, the economic and environmental balance swings in favour of heat pumps, for which take-up reaches 60 per cent by 2050. Biomass boilers also play a role, particularly in larger properties and those off the gas grid.

Incandescent lighting is successfully phased out, with a wide range of solid-state lighting systems rapidly entering mass consumer markets as efficiencies rise and costs fall. These initially encourage the existing trends to lighting proliferation, but as conspicuous consumption becomes less socially acceptable, the emphasis tends towards design quality for good performance.

Appliance labels and standards are improved at an EU level with a focus on consumption, which prevents markets for US-size appliances. Appliance efficiencies continue to improve up to 2050, especially through the use of advanced insulation in refrigeration. With market saturation, appliance energy demand falls.

Continued improvement in ICT is enabled by greater processing speed. Every home is internet connected before 2020, linked to changes in mobility, but growth in electricity use for home computing ends, as remote processing of data with low-power clients for home access is introduced. The 'One Watt Initiative' for standby (IEA, 2010) is widely implemented globally. With increased use of 'all off' switching and automatic low-power modes, standby electricity use decreases.

The period 2010 to 2020 sees the continued development of a niche market for air conditioning as the frequency of warm summers increases in the UK. However, building standards prevent its use for new homes and stimulate wider use of passive and low-tech approaches, such as shading, shutters and ceiling fans. These practices are then reflected in retrofit designs. The use of air conditioning in housing is limited to some reversible air source heat pumps in peak summer conditions.

Social attitudes prevent significant growth in markets for new high energy devices such as patio heaters, hot tubs and large plasma screens, initially through peer pressure. This is reflected in the CSR (corporate social responsibility) policies of major retailers and then in regulation.

Early use of micro-renewables is highly dependent on green energy innovators. Trends towards zero-carbon new homes (and then retrofits) accompanied by more generous financial incentives then make micro-generation increasingly popular. First solar water heating, then photovoltaics and micro-wind in specific locations become common as technology cost and performance improve. Solar thermal is used on 50 per cent of dwellings providing 25 per cent of domestic hot water by 2050. Solar PV panels are installed on 15 per cent of dwellings and micro-wind turbines on 5 per cent of dwellings by 2050.

The precise assumptions used, more detailed justifications of them and the approach to modelling are set out in a UKERC Working Paper (Anable et al, 2010a).

Lifestyle change in mobility and transport

Influences on mobility and transport

Energy demand, whether for private or commercial transport purposes, is essentially the product of four factors:

1   the demand for movement (distance), itself derived from the need to access facilities, services and goods and determined by land use patterns, trip frequency and route choice;
2   the mode of transport used to meet that demand;
3   the technical efficiency of vehicles used to power the vehicles; and
4   the operational efficiency with which vehicles are used (e.g. how they are driven and how much of their carrying capacity is used).

Each of these areas in turn is influenced by a wide range of factors that help explain transport emission trends to date. For instance, the UK Department for Transport makes use of its National Transport Model (NTM) to forecast future levels of traffic. It notes that 'key drivers of traffic growth in the NTM are changes in income, population, employment, and travel costs' (DfT, 2008a). Current mid-range forecasts are that traffic will be 31 per cent higher in 2025 and car ownership 33 per cent higher per capita than in 2003.

Traditionally, transport activity, economic activity and transport energy demand have been strongly correlated (Banister and Stead, 2002). As incomes grow, the demand for goods and services increases, as does the demand for travel. These trends can be influenced by individual preferences as well as social and cultural norms that have an impact on journey purposes (e.g. more travel for leisure), journey lengths and modes used – we travel further and faster, choosing to purchase vehicles with greater power and additional features, thus increasing vehicle weight and offsetting efficiency gains (Sorrell, 2007). The type of land use that accompanies economic growth is also important. The trend towards centralization of service, distribution and retail provision often at edge-of-town developments, together with less dense housing provision, have all contributed towards increasing demand for transport.

The last 50 years have also seen some dramatic changes to the socio-demographic structure of Great Britain with associated impacts on travel patterns. Whilst the number of households has increased by almost 8 million since 1961, average household size has declined from 3.1 to 2.4 over the same period (Jeffries, 2005). The UK's demographic structure is expected to change significantly in future, in particular through an increasingly ageing population. This could lead to an increase in future transport energy demand as, unlike past generations, these older cohorts may have higher incomes, will have grown up being dependent on the car and may have a higher propensity to travel by air. For example, of those aged over 70, over half hold a driving licence (51 per cent) compared to only 15 per cent in 1975–1976 (DfT, 2008b).

There has also been a marked increase in the number of women in the workplace, and 63 per cent of women now hold a full driving licence, up from 29 per cent in 1975–1976 (DfT, 2008b). Overall, however, driving licence holding has stabilized at around 70 per cent of adults since 2000, in part due to a slowdown in driving licence uptake by younger people.

However, while only three out of ten households in Great Britain in 1961 had a car, by 2004, one in four households did *not* have a car, whilst almost one in three had two or more (DfT, 2008b). This, in turn, has been driven by a reduction in real terms in the overall costs of motoring in the last 20 years (Green Fiscal Commission, 2009). In addition, over this period increases in public transport fares above the rate of inflation have made travel by car relatively cheaper.

Similar drivers support the future forecast growth in aviation (Pearce, 2008). Since deregulation of the airline industry in 1996, the development of the low cost aviation sector has introduced low and unrestricted fares and has opened up the range of destinations and airports available. There is some debate as to whether the growing affordability of air travel has led to an increase in the overall passenger growth rate or whether it may well have happened anyway, particularly due to income growth. Since 1996, annual growth rates have averaged around 5–6 per cent, which represent strong growth but are similar to the rates experienced prior to deregulation (Dargay et al, 2006). What is clear, is that most of the current air passenger demand is for leisure purposes and the availability of low cost flights has not in fact significantly altered the type of people who are flying (Dargay et al, 2006; CAA, 2006). The growth is composed of existing passengers flying more than in the past, particularly those from middle and higher income bands travelling short-haul.

Recent trends

It is questionable how certain we can be that historical relationships between travel, income, demographic composition and employment will hold true in future decades. For instance, Bayliss et al (2008) identify that actual traffic levels (up to 2006) have been well below the mid-range forecasts provided by both the 1989 and 1997 National Road Traffic Forecasts. It would appear that traffic growth is already decoupling from economic growth. Since the mid-1990s, the rate of growth in both passenger and goods transport has halved even though economic growth has been 50 per cent higher (DfT, 2008b). Between 1996 and 2006 car travel increased by only 11 per cent compared with a 34 per cent increase during the previous decade. By contrast, travel by all other passenger modes increased by 34 per cent (mostly rail) compared with a decline of 4.5 per cent previously. The long-term decline in walking and bus use has been stemmed, but not reversed, and there has been no overall increase in cycling levels. As a result the share of private trips by car has *fallen* over the last decade (from 94 per cent to 92 per cent) – an unprecedented occurrence (Headicar, 2009).

Data from the National Travel Survey shows that the total distance travelled per person (by all modes) has levelled off since 1999. In the two decades prior to this, the large increase in overall distance travelled was primarily a function of increased trip length rather than additional trips. Over the last decade, average trip lengths by car have continued to increase but the number of trips made has fallen, resulting in only a slight increase in distance travelled overall. Yet, car ownership has continued to increase quite steeply over the last decade which makes the levelling off in car use all the more notable. As a result, the use made of individual cars has fallen – the annual mileage per car fell by 10 per cent during the last decade after increases over several decades previously.

Some believe it may not be a coincidence that decoupling began to coincide with introduction of the internet in the early 1990s (Lyons et al, 2008). The information age is unfolding around us far more rapidly than the motor age did before it. In 1998 only 9 per cent of households had access to the internet. By 2007 this had increased to 61 per cent, with 52 per cent having broadband access (Marsden et al, 2010). We have passed the point where there are more mobile phones than people in the UK. In the early 1990s, commentators had said 'there is no natural way for grocery teleshopping to evolve alongside superstore retailing' (Hepworth and Ducatel, 1992) and yet today online grocery shopping is very much making its presence felt with over 20 million people shopping online in 2005 and internet sales representing 10 per cent of the value of all sales of UK non-financial sector businesses in 2008 (OFT, 2007; ONS, 2009).

Together, these trends have led to speculation that ICT will continue to weaken the temporal and spatial fixity of participation in activities and, since much if not all travel is derived from such participation, it follows that ICTs will impact on the demand for mobility. ICTs can impact upon travel by substituting for trips, stimulating more trips and enriching the experience of travel itself through travel time use (Lyons et al, 2008). The more radical changes are likely to take place through changes in work patterns. The impacts of teleworking are known to be complex, but potentially important. Currently, 3 per cent of workers say they always work at home but an additional 15 per cent occasionally work from home or say it would be possible for them to do so (DfT, 2007). This latter group are working at home more often and, in the future, the composition of the labour market may change to facilitate more home-working. ICTs allow us to do things differently. What is uncertain is how such opportunity permeates into society and everyday social practices to redefine norms of behaviour. The question for policy-makers is whether they should be inactive, reactive or proactive in policy response.

Despite the relative decoupling of car travel to income, improvements to vehicle efficiency over the period have only resulted in a stabilization in energy demand from personal car travel. Improvements in engine efficiency during this period have been essentially negated by the increased traffic levels and uptake of more powerful vehicles. There is some evidence, however, that car-buying habits in the UK may be changing as in 2009, for the first time, more small cars were sold than larger models and the annual rate of improvement of average new car CO_2 emissions was the best on record at 4.2 per cent in 2008 (SMMT, 2009). At the same time, the membership of car clubs, albeit still small at 100,000 people in the UK, is doubling year on year (Carplus, 2009).

Future trends in a low energy lifestyle

Given the apparent breakdown in traditional relationships between income growth and travel demand, how far might this trend go? For instance, what impact might continuing volatility in the oil market have on lifestyle choices? Will future generations cease to see congestion increases and carbon reduction as the major economic drain that they are conceptualized as today (Goodwin and Lyons, 2009)?

In spite of the constraints imposed by the organizational and spatial structures in society, most individuals and households do have some room to alter their behaviour in ways that may reduce energy used from travel through altered mode choice and trip lengths, by better coordination of their daily activities, or by adjusting their housing location (Rajan, 2006; Goodwin 2008). They also have choices about which car to purchase and how to drive them. Thus, there is large scope for travel behaviour change even without radical lifestyle change or reliance on fundamental changes in land use patterns (Gross et al, 2009). These could result from a curbing of some of the many inefficient patterns of current demand which are ripe for change within relatively short timescales.

A longer-term shift requires the reconceptualization of mobility as a means and not an end. This is the distinction between mobility (physical movement) and accessibility (the ability to reach desired goods, services, activities and destinations). Focusing on enhancing accessibility represents a fundamental shift from the traditional focus on enhancing mobility through infrastructure development and instead requires attention to the spatial distribution of land use, the quality of the journey experience and satisfying demand by making use of alternative solutions and mobility services.

Consistent with this emphasis on efficiency and accessibility, the Lifestyle scenario assumes that low-energy and zero-energy (non-motorized) transport systems will gradually replace current petrol and diesel car-based systems. The increased uptake of slower, active modes reduces average distances travelled as distance horizons change. Localism means people work, shop and relax closer to home and long-distance travel will move from fast modes (primarily air and the car) to slow-speed modes covering shorter distances overall (local rail and walking and cycling).

The novelty of air travel wanes as not only does it become socially unacceptable to fly short distances, airport capacity constraints also mean it becomes less convenient. Weekends abroad are replaced by more domestic leisure travel but this is increasingly carried out by low carbon hired vehicles, rail and luxury coach and walking and cycling trips closer to home. It also becomes socially unacceptable to drive children to school.

However, capacity constraints limit the pace of change so that mode shift to buses and rail will be moderated. New models of car ownership are embraced. This includes car clubs ('pay as you go' car hire schemes known as 'car sharing' in many other European countries) and the tendency to own smaller vehicles for everyday family use and to hire vehicles for longer distance travel. These are niche markets in which new technology is fostered. Lower car ownership is correlated with lower car use.

The new modes, in turn, will result in a new spatial order towards compact cities, mixed land uses and self-contained cities and regions. Some services return to rural areas, but it becomes more common to carry out personal business by internet. Small-scale technology and ICT facilitate relatively rapid behavioural change. Telematics, in-car instrumentation, video conferencing, smartcards and e-commerce make cost and energy use transparent to users and change everything from destination choice, car choice and driving style to paying for travel, including in the freight sector. A more radical change takes place through changes in work patterns and business travel. Teleworking particularly affects the longer commute trips and thus has a disproportionately large impact on average trip lengths. Increased internet shopping and restrictions on heavy goods vehicles, particularly in town centres, increases the use of vans. There is some shift towards rail freight.

Combined with the shifts towards active modes and different models of car ownership, this amounts to significant lifestyle shift. Modelling of these changes in lifestyle led to a 74 per cent reduction in distance travelled by car by 2050. The use of all other surface transport modes increases, apart from a 12 per cent fall in distance travelled by trucks. The reduction in car travel comes about as a result of significant mode shifts, particularly to bus travel towards the latter half of the period (184 per cent increase in vehicle kilometres) and cycling and walking. The take-up of cycling as a mode of transport reaches the same level in terms of mode split by 2050 as is the norm in the Netherlands today (40 per cent of all trips). However, mode shift is combined with destination shifting as trips are either totally abstracted from the system through virtual travel or shorter as a result of localization.

The LS-REF scenario implies that 10 per cent of the UK car fleet will be able to connect to the grid by 2020 and 26 per cent of road transport energy demand is met by plug-in electric hybrids (PHEV) by 2050. There is no change compared to the REF scenario in the short term, as the numbers remain constrained by the lack of vehicle and infrastructure availability. Car owners downsize and drivers respond to the on-road fuel efficiency programme and speed limit enforcement as the car fleet alone uses 5–6 per cent (2020) and 11–12 per cent (2050) less energy per kilometre driven. For road freight, all of the scenarios imply that nearly half of the UK van and HGV fleets will be able to connect to the grid by 2030. Overall, the LS-REF scenario results in a 26 per cent and 58 per cent reduction in transport CO_2 emissions by 2020 and 2050 from baseline (REF) levels.

The key outputs are summarized in Table 9.2.

The precise assumptions used, more detailed justifications of them and the approach to modelling are set out in a UKERC Working Paper (Anable et al, 2010).

Lifestyle change for a low carbon world

Implications of lifestyle for the energy system

The scenario changes outlined earlier in the chapter, under 'Future trends in a low energy lifestyle' for home and transport settings, were incorporated into the UK MARKAL elastic demand model to assess broader energy system impacts (see Anable et al, 2010a, for a complete discussion). As all energy production and transformation is directly derived from energy demands, the most significant impacts of lifestyle change on the wider energy system result from reductions in the overall demand for final energy. Lifestyle change alone has a similar

Table 9.2 *Summary results of the Lifestyle scenario (LS-REF)*

	2007	2020	2050
Average distance travelled (km pppa)	11,484	10,756	9035
Avg. car occupancy	1.58	1.75	1.94
Mode split (% distance)			
• cars and motorcycles	83%	71%	40%
• slow modes	3%	9%	28%
• bus and rail	14%	20%	32%
Share of new large cars	15%	0%	0%
'On-road fuel efficiency':	*km affected*	*km affected*	*km affected*
• cars, 8% better per km	4%	46%	62%
• vans, 8% better per km	4%	50%	70%
• trucks, 4% better per km	4%	50%	70%
Air demand growth (pa)	2.5%	1.0%	0.0%
Technology choice, e.g. share of *new* cars by propulsion	99% ICE	16% EV 28% HEV 8% PHEV	9% EV 18% HEV 46% PHEV
Transport CO_2 at source, reduction over baseline (REF)	n/a	−26%	−58%

effect on total final energy demand in 2050 as an 80 per cent carbon constraint with no lifestyle change. Final energy demand falls by 30 per cent by 2050 in the Lifestyle scenario, compared to a modest rise in the Reference scenario (see Figure 9.3). Because of our assumption that these changes are lifestyle driven, this demand reduction occurs in the household and transport sectors.

Within these sectors, the most important reductions are in the use of natural gas (households) and oil-derived fuels (transport). Gas demand under lifestyle change falls from year 2000 levels by 20 per cent in 2020 and 60 per cent in 2050. Oil demand reductions are even larger – 40 per cent by 2020 and 70 per cent by 2050 (rising to 90 per cent in the Lifestyle carbon-constrained scenario). In both cases, however, the changes are earlier than in low carbon scenarios without lifestyle change. However by 2050, scenarios with a carbon constraint, but without lifestyle change, require similar or even greater reductions in oil and gas use. In other words, lifestyle change facilitates an earlier start in supply-side changes, but with less dramatic changes in the second quarter of the century.

Demand for electricity is affected far less than other fuels by the lifestyle scenarios, remaining broadly constant through 2020 and then rising by 5–10 per cent from 2000 levels by 2050. Although lifestyle change involves reductions in demand and improvements in the efficiency of electricity use, it also includes significant fuel switching to electric technologies, notably plug-in hybrid vehicles and electric heat pumps. However, this electrification effect is much less pronounced than in carbon-constrained scenarios without pro-environmental

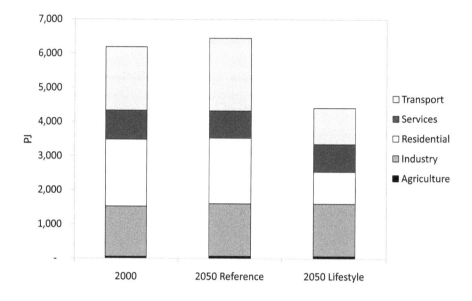

Figure 9.3 *Final energy use in the Reference and Lifestyle scenarios*

lifestyle change (e.g. the carbon-constrained scenarios set out in Chapter 5), where electricity demand rises by approximately 55 per cent by 2050.

This moderation in demand for electricity in the Lifestyle scenario implies a significant easing of required investment in zero-carbon electricity by 2050. However, the power sector generation mix in 2050 still depends on detailed assumptions about the relative costs of different power generation options, which are discussed more in Chapter 5, as well as the social acceptability of different large-scale technologies, which was not explicitly considered in the Lifestyle scenario, but the environmental dimensions of which are discussed in Chapter 10. However, it is clear that the effect of demand reduction is to decrease pressures for investment across a mix of carbon capture and sequestration (CCS), nuclear and large-scale renewables. In the carbon-constrained scenarios, although there is still a high penetration of electricity in the transport sector, electricity demand is reduced by 32 per cent in the residential sector in the Lifestyle scenario (compared to a 90 per cent increase in the LC scenario). This will have a very significant impact on required investments in the low-voltage distribution system.

These changes in final energy and fuel mix feed through to major changes in primary energy in the Lifestyle scenarios. From 2000 levels, primary energy demand falls more than 20 per cent by 2020 and more than 30 per cent by 2050. In the LS-LC scenario this demand reduction reaches 44 per cent in 2050. Furthermore, specific resource requirements are moderated, notably a halving in the demand for biomass, primarily through demand reduction and modal switch in the transport sector.

The implications of this analysis are very important for delivery of carbon emission reduction targets. Lifestyle changes alone result in emissions of $407MtCO_2$, a 31 per cent decline from 1990 levels by 2050. Clearly this is not sufficient alone to deliver UK carbon emissions reduction goals, but it is a very significant contribution. It has two important implications for the energy supply system in delivering ambitious carbon targets. First, it allows progress towards delivering carbon targets to begin earlier, resulting in lower cumulative emissions. Secondly, it requires less change in the supply side by 2050. The combination means that there is less reliance on some of the very dramatic changes implied by the carbon-constrained scenarios set out in Chapter 5. Increases in use of CCS, nuclear power, renewable electricity, bio-fuels, heat pumps and electric vehicles may all still be needed, but not with the speed implied in other carbon-constrained scenarios. This has important implications for climate mitigation policy. A scenario that involves voluntary lifestyle change will place much less pressure on policy to require rapid (and potentially disruptive) technical change. Encouraging lifestyle change presents a different set of challenges for policy-makers: it requires more emphasis on early behaviour change, but, as a result, requires less radical change to the energy supply system by 2050 than only relying on a 'top down' technical solution.

Implications for energy security

Definitions of resilience and energy security are discussed in Chapter 6 and incorporate improvements in energy intensity, diversification and reliability. The largest implications for energy security in the UK of the Lifestyle scenario flow directly from the earlier and larger reductions in primary energy demand set out above. The effects are also strongest for the fuels where import dependence is potentially most problematic – oil and gas. In both cases, the UK has become a net importer in recent years, and there are expectations, within the period of time addressed in this work, of reliance on sources from areas with significant geopolitical risks – Russia and West Asia for natural gas, and the Middle East for oil. The nature of potential energy security risks for the UK is set out in more detail in Chapter 6. For oil, there are also geological and economic concerns about resource availability after 2020 (Sorrell et al, 2009).

The similarity between these conclusions and those set out in Chapter 6 for scenarios driven by energy security concerns is not a coincidence. The common theme is reductions in energy demand. In 'resilience' scenarios this is driven explicitly by the goal of reducing reliance on imported oil and gas via reductions in energy intensity; in Lifestyle scenarios the reduction is achieved primarily through pro-environmental behaviour, but achieves the same end of reducing use of oil and gas.

Energy security concerns do not relate solely to availability of imported oil and gas. Indeed, historically (as noted in Chapter 6), most disruptions to continuity of energy supplies in the UK have other causes. Lifestyle scenarios should be advantageous with respect to availability of other fuel sources, whether imported or not. These include nuclear fuel and bio-fuels. They are also less sensitive to failures in new supply-side technology, including nuclear accidents, breaches of carbon dioxide transportation and storage systems, and systemic failures in offshore renewables and transmission. Finally, an easing in the requirement for new energy infrastructure, especially low voltage distribution, will assist in ensuring reliability.

However, Lifestyle scenarios do not remove energy security concerns. Like all low carbon scenarios they imply some increased electrification, with heating and transport systems highly dependent on electricity supply continuity. The greater decentralization of electricity supplies in a Lifestyle scenario has complex implications here. It potentially allows operation of local grids in an 'islanded mode' in the event of wider system problems. On the other hand, it also implies increased dependence on highly decentralized generation, requiring a new 'active management' approach to distribution grids with new risks as well as new benefits (Woodman and Baker, 2008).

Economic implications

The standard economic technique, used for example elsewhere in this book, to make economic comparisons between scenarios is to measure the 'social welfare', defined to be the aggregate of producer and consumer surpluses. Such comparisons assume a given set of consumer preferences, that is, the same willingness to pay for energy services in the scenarios under comparison. This is not the case when comparing the Lifestyle and Reference scenarios – consumer preferences differ very explicitly because of different lifestyles and not solely price signals – and therefore a welfare comparison cannot be made.

What can be meaningfully compared are the costs of the energy systems implied by different scenarios. Total energy system costs are £16 billion lower in 2020 and £89 billion lower in 2050 in the Lifestyle scenario compared to the Reference scenario. (These numbers rise to £17 billion and £94 billion respectively in the carbon-constrained scenarios.) This is primarily driven by the much smaller size of the energy system. Another relevant metric is the cost of delivering a low carbon system in 2050 – this falls from £17 billion for a Reference scenario to £12 billion in a Lifestyle scenario. In other words, due to a much smaller energy supply system in the Lifestyle scenario, the cost of a low carbon energy system is lower.

These figures need to be subject to the caveat that our cost modelling approach excludes the costs of some (but not all) demand-side investments (see Anable et al, 2010a, for details). However, off-model calculations indicate that these are substantially smaller than the modelled cost increases. For example the additional annual cost of household insulation measures in the Lifestyle scenarios might be £0.35 billion in 2050. The difference in unmodelled transport sector costs would consist of a combination of increased investment in public transport systems and reduced expenditure on other transport infrastructure. We would expect the latter to be larger. The estimate that the Lifestyle scenario energy system costs in 2050 are at least £90 billion lower than the reference comparators is therefore reasonably robust.

Although these are not GDP costs (as this depends on the impact of changes in investment, consumption and the balance of trade), as a context,

UK GDP was £1.46 trillion in 2008 and, with an assumed average 2 per cent annual growth rate, is projected to be £3.3 trillion in 2050. The energy system costs are estimated to be 8 per cent of GDP in the Reference scenarios, but only about 5 per cent in the Lifestyle scenarios, similar to the current level. In other words, the energy system takes a rising share of GDP in the Reference scenarios, but can be held at current levels by lifestyle changes. In either case, the incremental cost of achieving an 80 per cent carbon emissions reduction target is modest – and certainly consistent with a lower bound of around 1 per cent as in the Stern Review (Stern, 2006) – with the lifestyle scenario costing around 50 per cent less than the reference CO_2 constrained case.

The marginal cost of carbon abatement in 2050 in the carbon-constrained scenarios falls from £169/tCO$_2$ in the Reference scenario to £163/tCO$_2$ in the Lifestyle scenario. This rather modest change shows that although fewer overall reductions are required, at the margin some rather expensive decarbonization options are still required to achieve 80 per cent carbon emissions reductions.

The reduced costs of the energy system in a Lifestyle scenario would be reflected in increased economic activity elsewhere in the economy, if overall economic activity (GDP) is to be the same (which is our assumption). Our models for this work are confined to analysis of the energy system, and therefore do not provide insights into the wider macro-economic implications of this reduction in size of the energy sector.

Public policy implications

The critical role of behaviour change in carbon emissions reduction policy is already established. The Stern Review identified a trio of types of intervention required to deliver a low carbon economy – pricing of carbon, low carbon technical innovation and behavioural change (Stern, 2006). Economic and technological analysis are insufficient to inform this third pillar, as neither seeks to explain changes in human behaviour due to factors other than technological change, prices or incomes. Analysis of policies to change behaviour needs to draw on a wider range of disciplinary traditions, including psychology and social sciences (e.g. Stern and Aronson, 1984; Lutzenhiser, 1993).

In the Lifestyle scenarios we assume an interaction between public policy and lifestyle change. This is not straightforward or uni-directional: public policy helps create the conditions in which different lifestyles are more or less acceptable, and pressure for particular lifestyles sets the parameters for public policy. Previous transitions in socio-technical systems show the inter-connectedness of social, technological and policy change (e.g. Geels, 2005).

In the Lifestyle scenario, environmental protection, and carbon emissions reduction in particular, continue to rise in prominence as policy objectives. Carbon reduction becomes a social norm, in a similar way to health and safety, with an expectation of Government leadership and regulation. The existing legislation for carbon budgets (HMG, 2009) can be seen as an initial step in that direction. In this context, the relevant questions for policy are which pro-environmental behaviours government can effectively promote, and how. The following sections set out the likely implications first for the broad framework of policy and then for the household and transport sectors.

Policy for social and behaviour change

There has been significant research in recent years on policy to promote pro-environmental behaviour change (e.g. Darnton et al, 2006; Halpern and Bates, 2004). A useful heuristic (Jackson, 2005) indicates that behaviour change is most likely when citizens face a set of influences that encourage, enable, exemplify and engage them in such change.

Many policy instruments focus on *encouragement*, in particular through economic instruments designed either to affect the prices of energy and carbon or to provide incentives for development and deployment of new low carbon technologies. In the Lifestyle scenario, the social acceptability of carbon pricing will be relatively high. The emphasis on individual and community action points to a downstream focus to target individual energy use, through end user taxes or downstream permits. The challenge of the regressive nature of carbon pricing will be addressed through explicit revenue recycling and increased support for behaviour change. Market-based instruments will continue to have a strong role to support innovation, but with more focus on citizen and community involvement in innovation. This will be through fixed-price or long-term capital subsidies, designed to reduce the risks of previous approaches (Bergman and Jardine, 2009).

Non-financial interventions are also required to *enable* desirable change, as behaviour is not solely driven by economic factors. This will involve providing the relevant practical information to enable positive intentions to be put into practice, as well as the education, skills and industrial capacity to deliver low carbon products and services.

Enabling also includes 'choice editing' through energy standards to ensure products meet high environmental standards; with the cultural attitudes prevailing in the Lifestyle scenario, this type of intervention will be expected and generally welcomed.

Action to reduce Government's own energy use is already part of UK carbon mitigation plans (HMG, 2009), but in the Lifestyle scenario this objective has additional importance to *exemplify* Government commitment. Government has the potential to lead both in terms of energy use investment, e.g. in social housing and via vehicle and ICT procurement, and in behaviour change, via energy management and travel substitution through telecommuting and teleconferencing.

Engagement of the general public in energy issues has traditionally been problematic for Government. Even with the growth of web-based information systems, significant behavioural change results primarily from more trusted role models, e.g. friends, family and community leaders. The implication is that similar principles to technological innovation need to be applied, with Government supporting rather than undertaking social innovation, e.g. by financing social entrepreneurs, in community projects and adopting a portfolio approach to recognize the inevitability of some innovation failure.

The implications of the Lifestyle scenario for policy extend far beyond simply adding in a set of new policies to 'deliver' behaviour change in the general population. The scenario envisions a society with significantly different attitudes, lifestyles and politics, which will have more far-reaching effects. Relationships between the state, market, communities and citizens will change, with implications for a range of public policies. We focus here on three: technology, taxation and governance; but there are potentially many more including health, education and foreign policy.

In the Lifestyle scenario, reduced energy use leads to less pressure for rapid innovation in energy supply than in low carbon scenarios without pro-environmental behaviour change (e.g. Chapter 5 and Anandarajah et al, 2009) or with an explicit focus on energy supply technology (e.g. Chapter 7 and Winskel et al, 2009). This results in a difference in emphasis in *technology policy*, with a much greater focus on technologies that facilitate lifestyle change. Increased investment is required in both 'low carbon community infrastructure', e.g. public transport and biomass community CHP, and for 'citizen-scale, low- and zero-carbon technologies', e.g. electric vehicles and micro-generation. The innovation challenge is recognized as socio-technical, i.e. not only to develop technologies but to deploy them in projects consistent with pro-environmental lifestyle choices.

There are major implications for *taxation* and public finances from reduced energy demand, particularly for highly taxed transport fuels. However, falling fuel use in this scenario may well be offset by greater acceptance of higher rates of fuel taxation and environmental taxes more generally, including energy and carbon taxes, and road user and parking charges. Increased social acceptance of the need for change and earlier implementation of energy demand reduction make a higher carbon price politically sustainable earlier. If current transport fuel tax rates were unchanged, demand changes would reduce tax revenues by £12 billion annually in 2050 (Anable et al, 2010). However, even with 80 per cent emissions reduction, this could be offset by an economy-wide carbon price of £100/tCO_2, which is less than the shadow price for carbon in the scenario (see section on 'Future trends in a low energy lifestyle' for mobility and transport, above). We conclude that the ability to levy new taxes and higher carbon charges earlier can more than offset the impact of declining fuel use on public finances. A broader discussion of energy demand changes and the taxation implications is given in the Green Fiscal Commission Report (2009).

Energy *governance* becomes more distributed in this scenario. Assuming similar trends in other countries, there will be an effective international framework for carbon emissions reduction and a framework for trade that discriminates positively in favour of the environment with, for example, a strong international product policy, including high standards for vehicles and electrical goods. However, the main implications are for policy more locally. Energy regulation and fiscal policy are likely to remain primarily national, but the greater role of communities in energy decision-making implies that governance (in its broad sense) is more distributed. It seems likely that this will be reflected in formal structures, with more locally-based decisions on infrastructure development, investment, incentives and advice. For the finance sector, the shift in investment to the demand side implies a greater emphasis on financing decentralized technology, and therefore more locally-based lending with reference to the sustainability of investments. The need for new infrastructure (including for public transport, cycling and walking, heat networks and smart grids) implies a greater active role for publically regulated economic actors, building on the roles currently played by electricity District Network Operators (e.g. to deliver advanced metering, infrastructure for vehicle recharging and real time demand response services) and Passenger Transport Executives (to increase mass transit capacity in major conurbations and improved low carbon bus services). The importance of public engagement implies much policy will need to be delivered primarily at a local level, e.g. through local government, third sector groups or community-based businesses. This represents a major change for energy policy that has traditionally been highly centralized. It also implies a broadening of the focus of energy policy to recognize the role of sectoral policy, notably in housing and transport, to facilitate change.

Housing policy

Measures to improve housing energy efficiency will be very important. Existing incentives for relatively low cost measures (notably loft and cavity wall insulation) should continue until their market penetration is close to complete. The Carbon Emissions Reduction Target (CERT) has proved to be a successful approach to delivering these, which has been replicated in other European countries (Eyre et al, 2009). However, given the scale of investment required to deliver the more expensive fabric and low- and zero-carbon technologies required to achieve a low carbon housing stock, funding solely from energy supply revenues is not plausible, and new instruments will be needed. New approaches based around 'whole house retrofit' and 'pay as you save' are already being developed (DECC, 2009c). These seek to lever investment against the very large fixed assets of the housing stock and energy infrastructure.

However, delivery of a housing stock that is low carbon will require more than just large investments. Refurbishment activities are complex, diverse and distributed. Making all refurbishment low carbon is a major challenge that needs to engage all of the building sector trades and professions in technologies and practices that are currently only very small niche markets (Killip, 2008). The re-skilling and attitudinal change for the sector is huge and requires an initiative of the type used to retrain gas heating engineers in condensing boiler technology in 2003–2005, but covering the whole sector.

The social housing stock can provide an early large niche market, with large contracts, predictable clients and consistent standards to reduce transaction costs. This is an example of where public procurement (local housing authorities) and the third sector (housing associations) can play an important role in exemplifying change.

The goal of much stronger building regulations is already well-established for new buildings, with very ambitious goals for 2016 that will challenge the technical capacity of the housebuilding sector and building control enforcement. Standards can also be applied to refurbishment, and are already used effectively for replacement boilers and glazing.

Whilst a whole house retrofit is a practical approach for major refurbishment, most home improvement is piecemeal retrofit with owner-occupiers in residence. 'Rational' project management (area-based, whole-house based and cost effectiveness driven) alone will not deal with this complexity. Policy also needs to engage with improvement that is cyclical, building-specific and owner-determined through support for a range of measures at appropriate 'trigger points' in house life cycle (e.g. sale or extension, as well as major refurbishment). Building regulation could achieve this through use of a performance standard (rather than a fixed set of measures) at such points. Such a policy has historically been seen as controversial and interventionist, but would be consistent with social expectations in the Lifestyle scenario. The first steps to such a policy are already in place with the implementation of Energy Performance Certificates at the point of occupier change, although some technical and procedural improvements are needed (Banks, 2008) before these will be able to be used as a regulatory framework.

In this scenario, people will expect product suppliers of all types to label the energy efficiency of products. However, there will also be demand for greater market intervention and product standards. The EU (or more global) level is likely to remain the main forum for product regulation, even with the greater emphasis on 'localization'. With similar lifestyle trends across Europe, EU policy-makers will be able to adopt more stringent product regulations and to base them on energy use rather than energy efficiency, for example to prevent the super-sizing of refrigerators and TV screens.

Carbon pricing can form part of the policy package. Currently, the EUETS provides no incentive for households except through electricity prices, and even here it is too small and insufficiently transparent to have any significant impact. More effective demand reduction incentives could be provided through transparent taxation of energy or carbon, through reformed (and therefore re-regulated) energy tariffs, or more radically through extending carbon trading to final users. Any of these options is likely to be more acceptable in this scenario than historically.

In the Lifestyle scenario, energy demand policy will not rely solely on investment, it will engage with energy users as well. With the rapid deployment of smart meters, there will be vastly improved energy billing and feedback. It is important that the short-term smart metering agenda is not captured by supply interests focused on load management through switching, thereby missing the potential for influencing behaviour. Future generations of electricity meters will allow identification of consumption profiles for individual appliances, and regulation of metering should require this as soon as practicable.

Energy efficiency advice has already proved highly cost effective (Defra, 2006). However, the next generation of technologies may not be amenable to the same low-cost telephone and web-based advice services. Face-to-face, in-home services are much more expensive, but will be needed at much larger scale. It is not yet possible to identify an 'optimum solution' for energy advice in a world of more complex home energy technologies. However, history indicates that it would be unwise to rely on a spontaneous energy services market; social innovation is required. In the first instance, a range of pilot approaches is needed – covering different potential providers, different funding mechanisms and even different energy control philosophies ('smart home' or 'smart person') – to encourage the diversity from which viable models might emerge.

Transport policy

Current western societies are based on high levels of mobility, facilitated by high-quality infrastructures and low transport costs. One obvious approach is to push for further improvements in vehicle and fuel technologies that will reduce the environmental impacts of motorized transport without limiting distances travelled. However, that leaves the problem that travel demand is growing faster than capacity possibly can. It also ignores the problem that efficiency gains can be offset by the uptake of vehicles with greater power and additional features and neglects the social issue that a significant share of the population cannot drive or does not have access to a car, for reasons of income, age, or ability (Handy, 2002).

Behaviour change is a strong natural force running through society and individuals as they move through their life course (e.g. changing locations of employment and residence). Yet, traditional forecasting models on which much current transport policy is based assumes business-as-usual behavioural choices and levels of mobility and rather implies that societal developments of significance to transport are 'external' to policy. By contrast, our Lifestyle scenario assumes that with appropriate and sufficiently robust policy levers, this behaviour change could be positively influenced for some immediately and substantially so that, over the course of the next 40 years, travel patterns are radically altered and the vehicle market transformed.

This requires taking the lessons from previous decades about the importance of price, quality and income so that policy can exert a positive influence on the type of society that is developing and the transport system required to support it.

Generally, the policy environment assumed in the Lifestyle scenario is one of 'push and pull' as fiscal and regulatory sticks are combined with the carrot of infrastructure investment (e.g. in car clubs, public transport, cycle infrastructure, railway capacity). In this context of more choice for local travel as the alternatives are improved, increasing acceptance of restrictive policies is assumed. These restrictions include the general phasing out of petrol and diesel vehicles in town and city centres through low-emission zones, increased parking charges and strict speed enforcement. This is balanced by the reallocation of road space towards public transport, walking and cycling as well as the recognition of telecommunications as a transport mode worthy of investment. To meet these demands requires transport policy-makers to focus on those policies which bring the most benefits at least cost (such as smarter travel choices, parking charges and investment in car clubs), to withdraw environmentally ineffective, and sometimes inequitable, subsidies (such as concessionary fares schemes and scrappage incentives), plans for energy-intensive modes (such as High Speed Rail) and to look to remove the many inefficiencies in the way we travel and move goods (Marsden et al, 2010).

To achieve the level of production and sales of low carbon vehicles demanded by the scenarios, market conditions and necessary infrastructure to support the rollout of grid-connected vehicles, particularly PHEV, beyond urban areas will need to be in place. The period after 2020 will need to see an increase in the range of vehicles available to consumers and freight operators in order to sustain the momentum. Market-based instruments and clear incentives for new technologies will continue to have a strong role for vehicles.

Delivering a radical change agenda will require a much better understanding than we currently have of how to engage the public with the various behaviour change initiatives which may be required. Non-price driven behaviour plays a significant role in transport choice. Although the relationship between attitudes and behaviour is by no means linear, it is likely that shifting attitudes in support of sustainable modes and practices will have a positive impact on actual behaviour, and allow more favourable responses to top-down measures. However, change requires both individual subjective responses in the form of self identity, moral norms and affective and instrumental attitudes, as well as collective emotional response in the form of social norms. These in turn need to be complemented by a change in the physical and social context to make such change possible, that is, investment in attractive alternatives and restraint of car use (Anable et al, 2006). Altering the existing patterns of car dependence therefore depends critically on a shift in the physical and social context at the local level, the policy and cultural context at the national level and changes in individual attitudes and habits.

The changes to transport systems lead to significant other benefits, such as better health (much more regular walking and cycling), reduced congestion, noise and accidents, and better local air quality. These are often key drivers of the case for change at a local level.

Conclusions

This chapter has investigated the role of pro-environmental lifestyle change for the UK energy system to 2050. We make two assumptions, both of which seem obvious when stated, but are frequently forgotten or ignored in energy futures work. The first is that the behaviour of energy users is not fixed, but rather the outcome of developments in society, and that these are uncertain with the level of uncertainty increasing over time. The second is that any policy framework that seeks to deliver major changes in the energy system, such as an 80 per cent reduction in CO_2 emissions, will be the outcome of a political process in which civil society, that is, energy users in other roles, will play a key role.

Analysis of lifestyle change needs to consider the interaction between personal decisions and the social context in which they are made. Our assessment is that they are intimately linked: energy-using behaviours are affected by, and contribute to, changes in their social and economic context, the available technologies, physical infrastructures and public policy. Our analysis is therefore socio-technical. In particular, this analysis implies that the role of policy is not restricted to influencing pricing and technological change. For good or ill, it also plays a role in shaping lifestyles and energy-using behaviours.

Quantifying the energy implications of a pro-environmental lifestyle scenario involves assumptions about a large number of energy-using decisions across the whole population. The key reason for using a scenario approach rather than modelling each energy-using behaviour separately as a sensitivity to business as usual, is that these behaviours are likely to be correlated. 'Lifestyle' is a property of the social system, not just a random collection of behaviours.

We have used an innovative methodology to combine the strengths of detailed end-use models (UKDCM and UKTCM) and an optimization model of the whole UK energy system (MARKAL). However, the models are individually well-established and have been tested extensively. We therefore have a high level of confidence that, given our assumptions, the energy system effects are broadly as modelled.

We have assumed changes to behaviour that we judge reasonable in an advanced economy, based on observation of energy-using activities across the developed world today. And we have assumed rates of change that seem feasible taking into account the need for both technologies and energy-using practices to diffuse and the external constraints to this, e.g. the need to change existing infrastructure.

Our results indicate that energy use in this sort of scenario might be expected to fall in both the household and transport sectors, by approximately 50 per cent in each by 2050. This implies rates of change (energy demand decreases) of just below 2 per cent annually. In the household sector, this is consistent with trends since 2004 (starting well before the recent recession) – demand has fallen approximately at this rate under the combined influence of rising prices and some stronger public policies. In transport, rising energy use trends have moderated for similar reasons, as well as some travel substitution by ICT.

The implications for energy demand in the economy are significant. Total final energy demand is projected to fall in the Lifestyle scenario by 30 per cent by 2050, even without any allowance for an externally imposed explicit carbon constraint. Impacts are strongest for natural gas and oil, which are the fuels for which there are the highest energy security concerns.

Implications for electricity are initially less significant, as lifestyle change includes earlier switching to electric technologies, notably heat pumps and plug-in hybrid cars, which partly offsets lower service demands and improved efficiency in other uses of electricity. However, in the longer term (towards 2050) energy demand reduction means that the electrification of the economy set out in Chapter 5, although it still occurs, does so to a lesser extent and more slowly.

The impacts on primary energy demand and carbon emissions are similar to those on final energy: a 30 per cent reduction without supply-side action, and with more early progress. We conclude that lifestyle change can make a significant contribution to delivering UK carbon emission goals, and assist early action, but that alone it is insufficient to deliver an 80 per cent reduction goal, as this requires a wider transformation of the energy system.

One of the major findings of our analysis is that the cost of decarbonization to the level of UK targets is much less in the Lifestyle scenario than other scenarios. Essentially this is because the energy system that needs to be decarbonized is smaller if energy service demands can be reduced and end-use efficiency improved through changing lifestyles. The direction of the effect is obvious, but the scale is more significant than identified in analyses that assume 'business as usual lifestyle change'.

The analysis in other chapters shows it is conceptually feasible to decarbonize the UK energy system within the context of a society which continues to be wastefully inefficient in energy use and primarily oriented towards consumerism. However, given that energy is a socio-technical system, such an outcome seems unlikely; in a democratic society some compatibility between the realms of public policy and social behaviour seems more probable. Energy consumers are also citizens capable of making intelligent choices about the future. Neglecting this in public policy risks forgoing the substantial opportunities for socio-economic benefits that are associated with decarbonizing through pro-environmental lifestyle change.

The policy agenda for lifestyle change is less well developed than the equivalents for pricing and technological change. But the broad principles of what works are increasingly well-understood. The traditional discourse of 'command and control' versus 'economic instruments' is not particularly helpful, as it neglects the diversity of drivers, agents and scales of influence on human behaviour. Broadening the energy debate to include 'energy citizens' will necessitate a similar broadening of the policy agenda.

References

Anable, J. (2005) '"Complacent Car Addicts" or "Aspiring Environmentalists"? Identifying travel behaviour segments using attitude theory', *Transport Policy*, vol 12, pp65–78

Anable, J., Lane, B. and Kelay, T. (2006) *An Evidence Base Review of Attitudes to Climate Change and Transport,* Report for the UK Department for Transport, London

Anable, J., Brand, C., Eyre, N., Layberry, R., Schmelev, S., Strachan, N., Bergman, N., Fawcett, T. and Tran, M. (2010) *Energy 2050 – The Lifestyle Scenarios,* UKERC Working Paper, forthcoming

Anandarajah, G., Strachan, N., Ekins, P., Kannan, R., and Hughes, N. (2009) *Pathways to a Low Carbon Economy: Energy Systems Modelling*, UKERC Energy 2050 Research Report 1, UKERC/RR/ESM/2009/001

Banister, D. and Stead, D. (2002) 'Reducing transport intensity', *European Journal of Transport Infrastructure Research*, vol 2 (2/3), pp161–178

Banks, N. (2008) *Implementation of Energy Performance Certificates in the Domestic Sector,* UKERC Working Paper, WP/DR/2008/001

Bayliss, D., Banks, N. and Glaister, S. (2008) 'A Pricing and Investment Strategy for National Roads', *Proceedings of the Institution of Civil Engineers: Transport*, vol 61 (3), pp103–109

Brand, C. (2010) *UK Transport Carbon Model: Reference Guide v1.0*. Oxford, UK Energy Research Centre, Energy Demand Theme

Brand, C., Tran, M. and Anable, J. (2010) 'The UK transport carbon model: An integrated lifecycle approach to explore low carbon futures', *Energy Policy*, doi:10.1016/j.enpol.2010.08.019

Bergman, N. and Jardine, C. (2009) *Power from the People,* Domestic Microgeneration and the Low Carbon Buildings Programme, ECI, Oxford

Carplus (2009) 'Lord Adonis Crowns 100 000th Car Club Member', Carplus Press Release, 26 November 2009

Civil Aviation Authority (2006) *No-frills carriers. A revolution or evolution? A study by the Civil Aviation Authority,* CAA, London

Darby, S. (2007) 'Enough is as good as a feast – sufficiency as policy', Paper presented to European Council for an Energy Efficient Economy (ECEEE), Summer Study, France, June 2007, www.eci.ox.ac.uk/research/energy/downloads/eceee07/darby.pdf

Dargay, J., Menaz, B. and Cairns, S. (2006) *Public attitudes towards aviation and climate change, Stage 1: Desk research,* Report to the Climate Change Working Group of the Commission for Integrated Transport, CfIT, London

Darnton, A., Elster-Jones, J., Lucas, K. and Brooks, M. (2006) *Promoting Pro-Environmental Behaviour: Existing Evidence to Inform Better Policy Making*, Report to Defra, Department of Environment and Rural Affairs, London

DECC (2009a) *Digest of UK Energy Statistics*, Department of Energy and Climate Change, The Stationery Office, London

DECC (2009b) *Energy Consumption in the UK,* Department of Energy and Climate Change, London, www.decc.gov.uk/en/content/cms/statistics/publications/ecuk/ecuk.aspx

DECC (2009c) *Heat and Energy Saving Strategy: a consultation*, Department of Energy and Climate Change, London

Defra (2006) *Synthesis of Climate Change Policy Evaluations,* Department of Environment and Rural Affairs, London

Defra (2008) *A framework for pro-environmental behaviours*, Department of Environment and Rural Affairs, London

DETR (2000) *English House Condition Survey 1996: Energy Report,* Department of the Environment, Transport and the Regions, London

DfT (2007) *Personal Travel Factsheet*, Department for Transport, July 2007, London

DfT (2008a) *Road Transport Forecasts 2008 – Results from the Department for Transport's National Transport Model,* Department for Transport, London

DfT (2008b) *Transport Statistics Great Britain 2008 Edition*, Department for Transport, November 2008, TSO, London

Elzen, B., Geels, F., Hofman, P. and Green, K. (2002) 'Socio-technical scenarios as a tool for transition policy: an example from the traffic and transport domain', Paper for 10th International Conference of the Greening of Industry Network, Gothenburg, Sweden, June 2002

Evans, D. and Jackson, T. (2007) *Towards a sociology of sustainable lifestyles*, RESOLVE Working Paper 03-07, University of Surrey, Guildford

Eyre, N., Pavan, M. and Bodineau, L. (2009) 'Energy Company Obligations to Save Energy in Italy, the UK and France: What have we learnt?', European Council for an Energy Efficiency Economy Summer Study, 1–6 June 2009, La Colle sur Loup, France

Geels, F.W. (2005) *Technological Transitions and System Innovation: A Coevolutionary and Socio-Technical Analysis*, Edward Elgar, Cheltenham

Goodwin, P. (2008) 'Policy Incentives to Change Behaviour in Passenger Transport', in *Transport and Energy: The Challenge of Climate Change*, OECD International Transport Forum, OECD, Paris

Goodwin, P. and Lyons, G. (2009) 'Public attitudes to transport: scrutinizing the evidence'. Proceedings of the 41st Universities Transport Study Group Conference, London

Green Fiscal Commission (2009) *The Case for Green Fiscal Reform – The Final Report of the Green Fiscal Commission*, London, October 2009

Gross, R., Heptonstall, P., Anable, J., Greenacre, P. and E4Tech (2009) *What policies are effective at reducing carbon emissions from surface passenger transport? A review of interventions to encourage behavioural and technological change*, UKERC Report, ISBN 1 903144 0 7 8, UKERC, London

Handy, S. (2002) *Accessibility vs. Mobility: Enhancing strategies for addressing automobile dependence in the U.S*, Prepared for the European Conference of Ministers of Transport

Halpern, D. and Bates, C. (2004) *Personal Responsibility and Changing Behaviour: the state of knowledge and its implications for public policy*, Prime Minister's Strategy Unit, Cabinet Office, London

Headicar, P. (2009) *Transport Policy and Planning in Great Britain*, Routledge, Abingdon

Henderson, G. and Shorrock, L.D. (1986) 'BREDEM – the BRE Domestic Energy Model: Testing the predictions of a two-zone version', *Building Services Engineering Research and Technology*, vol 7, no 2, pp87–91

Hepworth, M. and Ducatel, K. (1992) *Transport in the information age. Wheels and wires*, Belhaven Press, London

HMG (HM Government) (2009) *The UK Low Carbon Transition Plan – National Strategy for Energy and Climate*, The Stationery Office, London

IEA (2007) *Energy for the New Millenium – Trends in Energy Use in IEA Countries*, International Energy Agency, Paris

IEA (2010) *Standby Power Use and the IEA One Watt Plan*, International Energy Agency, Paris, www.iea.org/subjectqueries/standby.asp

Jackson, T. (2005) *Motivating Sustainable Consumption: A Review Of Evidence on Consumer Behaviour and Behavioural Change*, Policy Studies Institute, London

Jeffries, J. (2005) 'The UK population, past present and future', in *Focus on People and Migration*, Office for National Statistics, London

Kannan, R., Ekins, P. and Strachan, N. (2008) 'The Structure and Use of the UK MARKAL Model', in Hunt, L. and Evans, J. (eds) *International Handbook on the Economics of Energy*, Edward Elgar, Cheltenham

Kannan, R. and Strachan, N. (2009) 'Modelling the UK Residential Energy Sector under Long-term Decarbonisation Scenarios: Comparison between Energy Systems and Sectoral Modelling Approaches', *Applied Energy*, vol 86 (4), pp416–428

Killip, G. (2008) *Transforming the UK's existing housing stock*, client report for Federation of Master Builders, July 2008

Loulou, R., Goldstein, G. and Noble, K. (2004) *Documentation for the MARKAL family of models*, Energy Technology Systems Analysis Program, www.etsap.org/

Lutzenhiser, L. (1993) 'Social and Behavioural Aspects of Energy Use', *Annual Review of Energy and Environment*, vol 18, pp247–289

Lyons, G., Farag, S. and Haddad, H. (2008) *The substitution of communications for travel?* in Ison, S. and Rye, T. (eds) *The Implementation and Effectiveness of Transport Demand Management measures: An International Perspective*, Ashgate, Farnham, Surrey

Marsden, G., Lyons, G., Anable, J., Ison, S., Cherret, T. and Lucas, K. (2010) *Opportunities and Options for Transport Policy*, Universities Transport Study Group 42nd Annual Conference, Plymouth, 5–7 January 2010

Metz, D. (2002) 'Limitations of Transport Policy', *Transport Reviews*, vol 22, pp134–145

OFT (2007) *Internet Shopping, An OFT Market Study*, Office of Fair Trading, London

ONS (2009) *E-commerce and information and communication (ICT) activity*, 2008, Statistical Bulletin, Office for National Statistics, 27 November 2009

Oxera (2006) *Policies for energy efficiency in the UK household sector*, Report to Defra, Oxera Consulting Ltd, Oxford

Palmer, J., Boardman, B., Bottrill, C., Darby, S., Hinnells, M., Killip, G., Layberry, R. and Lovell, H. (2006) *Reducing the environmental impact of housing. Final report*, Consultancy study in support of the Royal Commission on Environmental Pollution's 26th Report on the Urban Environment, Environmental Change Institute, University of Oxford, Oxford

Pearce, B. (2008) 'What is driving travel demand? Managing Travel's Climate Impacts IATA', in *The Travel & Tourism Competitiveness Report*, World Economic Forum, Geneva

Rajan, S.C. (2006) 'Climate change dilemma: technology, social change or both? An examination of long-term transport policy choices in the United States', *Energy Policy*, vol 34 (6), pp664–679

Reusswig, R., Lotze-Campen, H. and Gerlinger, K. (2003) *Changing Global Lifestyles and Consumption Patterns: the case of energy and food*, PERN Workshop: Population, Consumption and Environment Dynamics: Theory and Method, Wyndham Hotel, Montréal, Canada, 19 October 2003

Sanstad, A.H. and Howarth, R.B (1994) '"Normal" Markets, Market Imperfections and Energy Efficiency', *Energy Policy*, vol 22 (10), pp811–818

SMMT (2009) *New Car CO$_2$ Report 2009*, The Society of Manufacturers and Traders, London

Sorrell, S. (2007) *The Rebound Effect: An Assessment of the Evidence for Economy-wide Energy Savings from Improved Energy Efficiency*, UKERC Report, ISBN 1903144035, UKERC, London

Sorrell, S., Speirs, J., Bentley, R., Brandt, A. and Miller, R. (2009) *An assessment of the evidence for a near-term peak in global oil production*, UK Energy Research Centre, London

Steg, L. (2004) 'Car use: lust and must. Instrumental, symbolic and affective motives for car use', *Transportation Research Part A*, vol 39, pp147–162

Stern, N. (2006) *The Economics of Climate Change*, HM Treasury, London

Stern, P.C. and Aronson, E. (eds) (1984) *Energy Use: the Human Dimension*, Freeman, New York

Strachan, N., Kannan, R. and Pye, S. (2008) *Scenarios and Sensitivities on Long-term UK Carbon Reductions using the UK MARKAL and MARKAL-Macro Energy System Models*, UKERC Research Report 2, UKERC, London

Strachan, N., Foxon, T. and Fujino, J. (2008) 'Policy Implications from Modelling Long-term Scenarios for Low Carbon Societies', *Climate Policy*, vol 8, S17–S29

Strachan, N. and Kannan, R. (2008) 'Hybrid Modelling of Long-Term Carbon Reduction Scenarios for the UK', *Energy Economics*, vol 30 (6), pp2947–2963

Strachan N., Balta-Ozkan, N., Joffe, D., McGeevor, K. and Hughes, N. (2009) 'Soft-Linking Energy Systems and GIS Models to Investigate Spatial Hydrogen Infrastructure Development in a Low Carbon UK Energy System', *International Journal of Hydrogen Energy*, vol 34 (2), pp642–657

Thogersen, J. (2005) 'How may consumer policy empower consumers for sustainable lifestyles?', *Journal of Consumer Policy*, vol 28 (2), pp143–177

Turrentine, T.S. and Kurani, K.S. (2007) 'Car buyers and fuel economy', *Energy Policy*, vol 25, pp1213–1223

Utley, J. and Shorrock, L. (2008) *Domestic Energy Fact File*, BRE Group, Watford

Winskel, M., Markusson, N., Moran, B., Jeffrey, H., Anandarajah, G., Hughes, N., Candelise, C., Clarke, D., Taylor, G., Chalmers, H., Dutton, G., Howarth, P., Jablonski, S., Kalyvas, C. and Ward, D. (2009) *Decarbonising the UK Energy System: Accelerated Development of Low Carbon Energy Supply Technologies*, UKERC Energy 2050 Research Report No. 2, UKERC, London

Woodman, B. and Baker, P. (2008) 'Regulatory frameworks for decentralised energy', *Energy Policy*, vol 12, pp4527–4531

10
Not Just Climate Change: Other Social and Environmental Perspectives

David Howard, Brighid Jay, Jeanette Whitaker, Joey Talbot,
Nick Hughes and Mark Winskel

Introduction

The environmental consequences of climate change are one of the major factors shaping energy policy around the world. As a driver of change, energy policy often unwittingly creates stresses on the environment that can be seen as environmental pressures such as atmospheric emissions. However, the stresses of capturing and using energy are more widespread than just the release of greenhouse gases and include emissions of other pollutants, the conversion and destruction of ecosystems and habitats through land take and the consumption of natural resources including water. Equally important are people's perception of potential risks and impacts of low carbon technologies; this should play a part, alongside scientific evidence, in shaping energy policy.

This chapter explores key environmental and socio-environmental issues arising from scenarios of UK energy system change. The two topics discussed are:

- *Environmental pressures*: the environmental pressures and their consequences, other than climate change, that arise from the deployment of different energy generation and use strategies identified by the core scenarios (REF, LC, R and LCR) developed in Chapter 4.
- *Socio-environmental sensitivities*: where socio-environmental perspectives are used to modify the Low Carbon (LC) scenario so as to constrain energy resources and technologies. The scenario variants are used to explore the impact of societal resistance to new technologies on the energy system, and possible pathways to decarbonization.

The final section offers a concluding discussion that draws together the wider importance of both social and environmental factors in shaping the future energy system.

Environmental pressures

Environmental consequences of changes in the energy system

There are profound environmental impacts arising from the energy system – from obtaining energy feedstock, generating energy, using energy, and also from all the supporting activities and infrastructure such as manufacturing devices and installing grids and pipelines. Energy systems have shaped the world we live in, helping to determine and control the distribution and activities of the human race and constraining our opportunities to exploit the natural environment. The tripartite interactions between the energy system, our activities and the environment are well documented and have resulted in both research and monitoring (e.g. Millennium Ecosystem Assessment, 2005). The consequences of the interactions vary depending upon where and when the interaction takes place and are determined not only by the form and magnitude of the energy system driver, but also the presence and condition of the ecosystem or organism being impacted.

The importance of forces such as our demand for and use of energy in driving change led to the development of a framework to structure evidence for policy development and assessment. In 1995, as part of the European Environment Agency's first assessment of the state of Europe's environment the DPSIR model was developed (EEA, 1995). The model is an extension of the traditional Pressure-State-Response (PSR) model that describes how an action impacts a system component to cause a change (or response). While PSR captures the kernel of the physical change well it does not comfortably map the socio-political components either in driving the action or in reacting to it.

The DPSIR model forms a loop that relates the *Drivers* of change (initially termed human activities) to the *Pressures* they generate on the environment. *Pressures* are forces and processes such as the release of damaging pollutants, the consumption of environmental components and the control or management of ecosystems. These interact with the environmental conditions (*State*) to generate an environmental *Impact* or problem. The *Response* to the impact is new or modified policy that actions change in *Drivers*, *Pressures* or *States*. A modification of the original framework diagram is presented as Figure 10.1; Pressure-State-Impact equates to Pressure-State-Response.

The scenarios provide a good description of the *Drivers* of change by detailing changes in activities relating to energy systems. To interpret the environmental consequences of change we first identified the changes in *Pressures* resulting from the *Drivers*. The European Environment Agency (EEA, 2008) identified the dominant environmental pressures from the production and use of energy to be primarily the release of greenhouse gases and their precursors, followed by emissions of pollutants that are acidifying, increasing eutrophication or toxic.

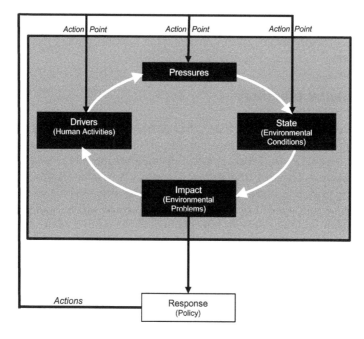

Figure 10.1 *The DPSIR (Driver-Pressure-State-Impact-Response) model as originally defined by EEA (1995)*

Estimates of emissions for the core scenarios are made by taking detailed technologies and activities from the results of the MARKAL-MED model and applying emission factors taken from authoritative sources, such as the National Atmospheric Emissions Inventory. The results can be used to compare the environmental pressures associated with different energy generation and use strategies, although it must be borne in mind that there are two important energy uses not included in the core scenarios, namely international aviation and shipping. The comparison described below reveals that there are some common trends across all scenarios, but there are also important differences. With care, these pressures can be interpreted as the implications of different energy decisions. Although changes in the magnitude of the pressures have been calculated, the impacts of the pollutants can only be described in general terms as the model has no spatial component and employs a coarse temporal representation (5-year time steps). Consequently, the analysis should be viewed as indicative rather than definitive.

The analysis of emissions is built on the current understanding of energy technologies. The most appropriate emissions factors have been used for each different source and their associated abatement technologies but the figures describe technologies as they operate today. There are one or two exceptions, for example where we know emissions reduction policies will have an effect such as the EC Large Combustion Plant Directive (EC, 2001). However, as technology improvements are likely to lead to lower levels of emissions, the results presented here may show a 'worst case' interpretation and should be read as relative rather than absolute values.

For each scenario we have, in detail, the magnitude and change in environmental pressures, that we then interpret through using our knowledge of the environmental state to suggest differences in impacts. However, even if we knew the location and function of each power plant and the timing of its activities (daily, seasonal, peaks and troughs, etc.) it would still not be possible to effectively describe what impact they will have or how the environment will respond, as in 40 years' time our natural and semi-natural environment will have changed, reflecting the impact not only of the myriad of drivers of change over that period but also the delayed effect of events that have happened over past decades. This means that whilst valuable descriptions of the environmental pressures can be made, expert judgement has to be employed to identify potential consequences or impacts. Using a pragmatic approach described below an initial assessment of the environmental pressures generated by the core scenarios has been made.

The analysis concentrates on environmental pressures arising during the operational phase of energy technologies and provides a consistent methodology to compare different scenarios. The main analysis summarized below assesses eight major pollutants arising from the UK energy system between 2000 and 2050 – namely carbon dioxide (CO_2), carbon monoxide (CO), oxides of nitrogen other than nitrous oxide (NO_x), sulphur dioxide (SO_2), methane (CH_4), nitrous oxide (N_2O), particulates (represented by PM_{10}) and radioactivity. The relative changes in pollutant emissions are described for each of the core scenarios presented in Chapter 4 (*Reference* (REF), *Low Carbon* (LC), *Resilient* (R) and *Low Carbon Resilient* (LCR)).

Although the energy system is a major source of the pollutants assessed in this chapter, a variety of other non-energy sources also contribute; the analysis covers only emissions from the energy system. In addition to the pollutants, a preliminary assessment of the altered water demand and land take for each scenario was conducted.

To illustrate the relative importance of emissions from the UK energy system, as modelled in the core scenarios, Figure 10.2 shows the relative proportion of energy-related pollutants released compared with those from non-fuel sources at the start of the 21st century (NAEI, 2006; BERR, 2008). For one group of pollutants (CO_2, CO, NO_x and SO_2) energy generation and use is currently responsible for a large percentage of total emissions and this situation is unlikely to change radically. For other pollutants, such as CH_4 and N_2O as described in Chapter 2, energy generation is responsible for only a small part of total emissions. The dominant sources for these pollutants are agriculture, through livestock for CH_4 (Lockyer and Jarvis, 1995), nitrate fertilizers for N_2O (McElroy et al, 1977; Thiemens and Trogler, 1991; Lockyer and Jarvis, 1995), along with industrial manufacturing (e.g. nylon for N_2O) (Thiemens and Trogler, 1991). Although these sources are unlikely to disappear, they have shown considerable fluctuations in the past and may show significant change over the next 40 years as reported in Chapter 2.

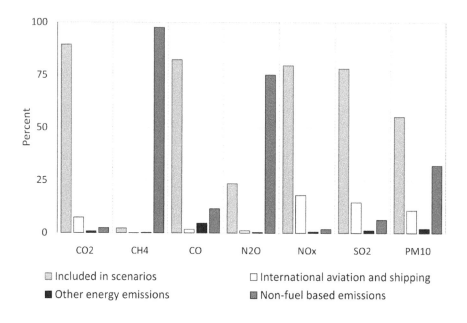

Figure 10.2 *The proportion of UK pollutant attributed to different sources*

Pollutant emissions under different energy core scenarios

Overall, emissions of the eight pollutants considered here are lower in 2020 than at the start of the modelling in 2000, with little variation between the four core scenarios. Irrespective of the strategy, policies that manage our consumption of fossil fuels to deliver low carbon energy or a secure supply also reduce emissions of other non-CO_2 pollutants. The overall interpretation is that society's use of energy does not have to be as environmentally damaging as it has been in the past; current policies are delivering improvements and it will take some time for the impact of different strategies to be apparent. After 2020, the Low Carbon Resilient (LCR) scenario has significantly lower total pollution emissions compared with the other core scenarios. This suggests that while both strategies (low carbon and resilience pathways) benefit the environment, the greatest improvement is seen when they operate in combination, suggesting they are not delivering the environmental benefits through the same processes. The Resilient (R) scenario has similar reductions in pollutant emissions as the LCR scenario for seven of the pollutants, but does not achieve the 80 per cent CO_2 reduction target; to achieve this target is a difficult challenge that is unlikely to be achieved unless specifically addressed.

When the pollutants are considered individually it is clear that there are key areas where the LCR scenario goes beyond the Low Carbon (LC) scenario in reducing emissions. Emissions of CO, N_2O, NO_x and PM_{10} are significantly lower in the LCR scenario post-2020, mainly due to changes in the transport and residential sectors. The main causes of differences between LC and LCR are that LCR has a greater reduction in energy consumption, particularly in the residential sector, and greater penetration of hybrid and electric cars for transport. In contrast, LC shows a smaller decline in the use of energy in all sectors; greater biomass use for heating in the residential and service sectors (increasing PM_{10}, CH_4 and CO); and greater use of transport bio-fuels, as opposed to hybrid or electric cars. Although these changes reduce CO_2, they do increase the emissions of other pollutants. There is a general rule that is clear from these analyses: the more energy you use, the more pollution you create. The analysis detailed below investigates these trends for each of the pollutants studied.

Sulphur dioxide (SO₂)

Coal has been a major fuel in Britain for over 200 years. When burned, it releases sulphur dioxide (SO_2), nitrogen dioxide (NO_2), methane (CH_4) and particulates as well as CO_2, with different mixtures depending upon the form and temperature of the combustion. Emissions of sulphur dioxide, which contribute to acid rain (Fowler et al, 1982; RGAR, 1997), have been reduced through regulation by over 90 per cent since the 1970s (see for example UNECE, 2009). Sulphur dioxide also contributes to eutrophication (impacting the growth of vegetation) and toxic air pollution (harming human health as well as that of other organisms).

Britain's terrestrial sources of atmospheric SO_2 emissions are dominated by conventional coal-fired power stations (over 60 per cent in 2009); other sources include coal used in industry, and fuel oil and petroleum coke use in oil refineries. The current major source of SO_2 deposited in Britain is from shipping, which still uses sulphur-rich diesel as a fuel (Fowler et al, 2005). Coal power stations with carbon capture and storage (CCS) release very little SO_2, as it has to be removed to prevent it impeding the capture process. Consequently, the model shows SO_2 emissions falling sharply in the LC and LCR scenarios (Figure 10.3) as CCS is introduced to coal-fired power stations and becomes a dominant technology between 2020 and 2035. Emissions of SO_2 also fall in the REF and R scenarios, but to a lesser extent, due to continued use of conventional coal-fired power stations. However, the requirement from the EU Large Combustion Plant Directive for flue gas desulphurization (FGD) in conventional power stations after 2015, does reduce emissions by about 85 per cent.

To set the scenario results in the context of historic trends, in 2005 Britain's emissions were 688kt SO_2, nearly a tenfold decrease from their peak in 1970 (6370kt SO_2). By 2020 emissions are expected to fall to half of the 2005 level at 344kt SO_2 (Matejko et al, 2009). In the core scenarios, the values for SO_2 emissions in 2020 are lower than these predictions (Figure 10.3). Yet, as seen in Figure 10.2, the energy use only accounts for 75 per cent of current emissions (with 25 per cent from industrial processes such as cement and brick production, combined with other sources including global imports). Taking this into consideration makes the published predictions and model figures very close (within about 25kt).

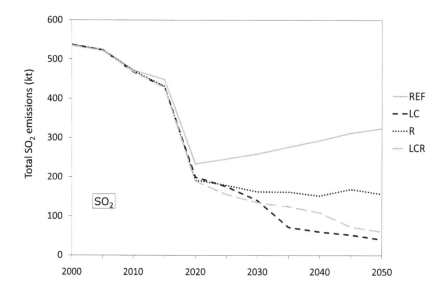

Figure 10.3 *Total emissions of sulphur dioxide (SO₂) over time in the core scenarios*

Methane (CH₄)

It is important to recognize that currently only 2 per cent of Britain's methane (CH_4) emissions are generated by the UK energy systems (NAEI, 2006), with 80 per cent of existing emissions from waste decomposition and livestock. The main concern about CH_4 as a pollutant is its role as a greenhouse gas; it has over 20 times the global warming potential of CO_2.

Its flammability makes methane an attractive fuel and it is a major component of natural gas used for domestic and commercial heating. It is also naturally released from carbon-rich rocks such as coal. Consequently it can be released through coal mining; this release is not included in our estimates as it is relatively small. Habitats such as bogs also generate CH_4, as do animals (especially ruminants) as they digest their food. These releases are supplemented by rotting landfill waste, where the methane is now captured and used to generate electricity (currently Britain's major source of bio-energy).

In the core scenarios, CH_4 emissions are dominated by the residential sector's use of coal and solid smokeless fuel for heating. In all four scenarios this is phased out by 2025–2030, resulting in a steady decline in emissions.

Particulates

Particulates are classified by their size; fine particles of less than 10 micrometres (μm) diameter damage the lungs if inhaled and are reported as PM_{10}. This category may be broken down further into smaller particles less than 5μm and 2.5μm which can be reported separately, as they have different toxicities.

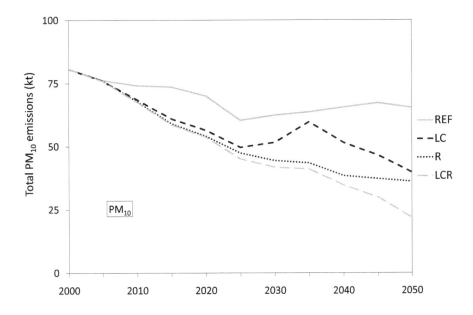

Figure 10.4 *Total emissions of particulates (PM$_{10}$) in the core scenarios*

Interactions between particles and other pollutants may enhance their toxicity by focusing their delivery. For this exercise we only examined PM$_{10}$ as an indicator of particulate emission.

Most PM$_{10}$ emissions are from the transport and residential sectors; within transport, diesel vehicles are the main source of emissions. In the LC, R and LCR scenarios, total PM$_{10}$ emissions halve by 2050, in part through reduced diesel consumption (Figure 10.4). However, in all scenarios future technology developments may reduce particulate emissions further.

In the residential sector PM$_{10}$ emissions fall in all scenarios by approximately 95 per cent between 2000 and 2030, due to the phasing out of coal, oil and wood for heating. However, in the LC scenario, increased use of biomass fuel in the residential sector causes total emissions to rise by around 15 per cent between 2025 and 2035 (Figure 10.4). The extent of this rise will depend on the specific technologies used to reduce PM$_{10}$ emissions, and the associated health impacts in modern biomass boilers or stoves.

Carbon monoxide (CO)
Energy use in transport gives rise to a number of different pollutants, and is the dominant anthropogenic source of carbon monoxide (CO) and the oxides of nitrogen (nitrous oxide (N$_2$O) and NO$_x$). However, each transport mode and fuel type has its own distinct footprint: for example, CO comes mostly from petrol cars whilst NO$_x$ splits more evenly between all liquid fuel

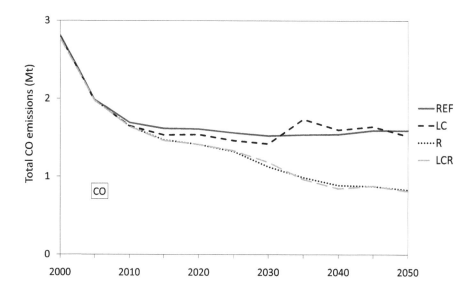

Figure 10.5 *Total emissions of carbon monoxide (CO) in the core scenarios*

vehicles. The increasing use of catalytic converters in petrol cars caused an initial decrease in CO emissions in 2000–2005 in all scenarios. The downward trend continues through the addition of bio-ethanol to the petrol fuel mix (Figure 10.5).

The impact of CO is best known for its toxicity to humans; in mammals it combines with haemoglobin to disrupt the delivery of oxygen to the tissues. However, concentrations are spatially very variable. Being only slightly lighter than air, it accumulates in enclosed spaces (such as buildings) or in urban streets especially under specific climatic conditions. CO emissions result from inefficient combustion and usually occur in small devices such as car engines and household fires. The residential sector accounts for about one-fifth of the CO emissions (approximately 20 per cent in 2000). Phasing out coal and solid smokeless fuel use between 2000 and 2030 reduces CO emissions in all the core scenarios (Figure 10.5). Only the LC scenario shows any reversal in the trend, due to the use of wood in the residential and service sectors in 2035–2050. The lowest CO emissions are found in the two resilient scenarios (R, LCR), due to the introduction of hybrid and plug-in cars and transport sector demand reductions.

Oxides of nitrogen (NO$_x$) and nitrous oxide (N$_2$O)
As with methane, energy is not the major contributor to UK emissions of nitrous oxide (N$_2$O), being responsible for only 20 per cent (NAEI, 2006); over half of UK emissions are currently derived from agricultural fertilizers.

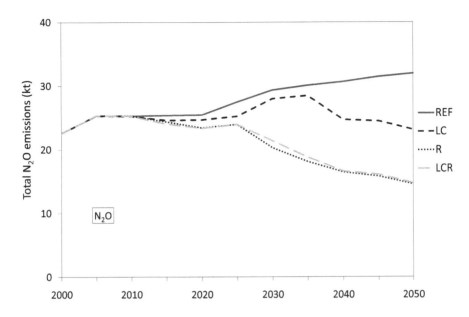

Figure 10.6 *Total emissions of nitrous oxide (N$_2$O) in the core scenarios*

Nitrous oxide is the most toxic of the oxides of nitrogen, at high concentrations affecting the functions of the human heart and brain (it is still used as an anaesthetic: laughing gas); it also is injurious to plants. Emissions of N$_2$O increase in all scenarios, initially by the uptake of catalytic converters in cars, the opposite of the effect seen with CO (Figures 10.5, 10.6); the REF scenario shows a continuing rise. Demand reduction and the use of hybrid and plug-in cars reduce N$_2$O emissions in the R and LCR scenarios by 2025. There is a later and smaller fall in emissions in the LC scenario for similar reasons.

Energy technologies and uses covered by the scenarios are responsible for 80 per cent of the UK's current emissions of other oxides of nitrogen (NO$_x$). Emissions are predominantly from the transport sector (roughly 50 per cent of emissions), particularly cars and HGVs. In all scenarios, emissions decline; emissions in the LCR scenario fall to approximately 65 per cent of 2000 emissions by 2050, while there is a smaller reduction of 20 per cent in the REF scenario. Emissions excluded are mostly from international aviation and shipping.

Radioactive releases

Radioactivity in the environment comes from several sources, including natural radiation, residues from the Chernobyl accident and atmospheric testing, plus radioactive discharges and emissions from nuclear and non-nuclear sites (authorized premises). Radioactive releases considered here are all planned releases from nuclear power stations, coal-fired power stations and other sources such as oil and gas platforms. Radioactive releases decline in the REF scenario, as nuclear power stations coming to the end of their life are not replaced (Figure 10.7). In the other scenarios, new nuclear power stations are built so emissions rise to varying extents after a time lag due to the long planning and construction time required. The highest estimated discharge occurs in the LC scenario resulting in a nearly three-fold increase in discharges by 2050. All new discharges would be regulated by various bodies and authorized release limits would be specified, the aim being to ensure that exposures to humans from the releases are below acceptable limits.

An extensive monitoring programme for radionuclides in food is carried out and results are published annually in the Radioactivity in Food and the Environment report (RIFE) which is used to estimate radiation doses. Radiation doses are not directly related to the amount of Becquerels released (as estimated for the scenarios) since the effective dose depends on the type of radiation and its biological effect. The monitoring programme provides independent UK-wide data of radioactivity in foods. In addition, the impact of facilities on Natura 2000[1] sites is explicitly considered by the Environment Agency under the EC Habitats and Birds Directive. The need for such assessments will depend upon where these increased discharges are occurring and the extent to which protected Natura 2000 sites are potentially impacted. The public acceptability of an increase in nuclear power plants is considered later in the socio-environmental scenarios and is influenced not only by regulated releases but also by concern about potential accidents and the environmental impacts of mining.

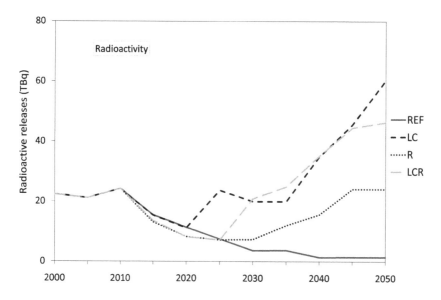

Figure 10.7 *Total radioactive releases in the core scenarios*

Other environmental pressures

Further environmental pressures resulting from energy generation and use relate to changing demand for water and land. Water is used as a power source (hydro and pumped storage), for cooling in power stations and for agricultural and forestry production of energy crops and bio-fuels. The demand for water includes use and consumption, the former being where it is discharged back into the environment in a similar state to when it was abstracted, the latter where its state has changed and it can be considered as lost to the system that provided it. Increased use of steam cycle generation of electricity using coal CCS and nuclear power along with the extensive production of bio-fuels and energy crops suggests that the LC scenario will result in the largest increase in water demand. Water demand also increases in the LCR scenario, for the agricultural production of energy crops; the REF and R scenarios show the smallest increases in water demand.

Land is another natural resource for which energy systems compete. Energy has helped shape Britain's landscape with regions dedicated to coal mining, flooded for hydro power and accommodating power stations. Whilst these systems are obvious and easily recognized, their spatial footprint is relatively small. As most renewable energy sources have low energy intensity, new technologies can be far more demanding of area in which to operate. For wind power, turbines are separated by four to five times the diameter of their blades and can operate over agricultural land in a multi-functional way; bio-energy crops replace the existing land use, usually as a monoculture. The land-take for bio-energy in the core scenarios is shown in Figure 10.8. The scenarios

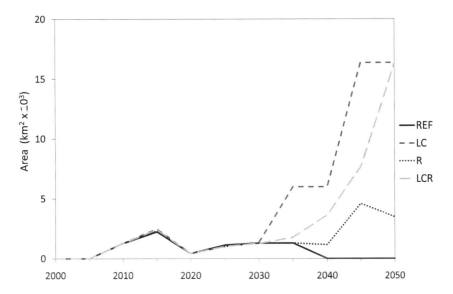

Figure 10.8 *Land-take for bio-energy in the core scenarios*

show similar trends through to 2030, when bio-energy starts to be significantly deployed in the LC and LCR scenarios, rising to about 8 per cent of the British land area by 2050 – around a third of the UK's current arable land, and more than 10 per cent of our total agricultural area. Clearly, the impact of this change will be dependent upon the location, condition and habitat history of the land replaced, but these location-specific impacts are not discussed here.

In conclusion, energy has gained a reputation as a polluting sector which is tolerated as it supports increasing incomes and desired lifestyles. The scenarios suggest that the prospects for the future need not be as damaging to the environment; the targets of low carbon emissions and reduced energy demand to secure delivery both offer opportunities for wider benefits. Of course, changes in the energy system will create new problems that will replace and in some instances exacerbate existing issues; the increased consumption of water and requirement for land seem to be the prime candidates.

Socio-environmental sensitivities

Introduction and methodology

As described in earlier chapters of this book, achieving the energy policy goals of decarbonization and resilience depends in large part on strategies to reduce energy demand and deploy low carbon technologies. So far, less attention has been given to how the public might respond to the introduction of low carbon technologies, and whether there would be public acceptance of the fundamental changes in energy supply that are envisaged in the *core* scenarios. Public acceptability is a key aspect of the possible pathways to decarbonization, and public resistance to certain technologies will alter the portfolio of options available. This section explores how public acceptance, reflecting particular 'socio-environmental sensitivities' could impact on the future energy system.

Taking the Low Carbon (LC) scenario as a baseline, three variant scenarios are presented here, representing distinct storylines about how people could respond to (and constrain) the deployment of particular energy supply technologies and fuels. These scenarios largely focus on the electricity sector. The scenarios were modelled in the MARKAL-MED energy system model described in Chapter 4. In this section, MARKAL-MED is used as a tool to explore socio-environmental constraints, and the modelling results are intended to illustrate relevant issues for further exploration; they are not offered as a prediction of the future. The MARKAL-MED model is informed by a techno-economic view of the energy system, including a high level of detail on costs and availability of technology and resource with less detail on social conditions. This section enhances the techno-economic approach by bringing socio-institutional enablers and barriers into consideration in a more explicit and detailed way.

The three socio-environmental sensitivities scenarios described in the next section – 'NIMBY', 'ECO', and 'DREAD'– are simplified illustrations of possible public responses to supply-side developments in the UK energy system. In reality, public responses could be a blend of the stylized responses characterized in these scenarios, and they would be expected to change over time. In addition, public opinion is not always a barrier; it can also be an important enabling factor for system change. Nevertheless, the scenarios provide a useful means to examine how public responses could affect the possible pathways to decarbonization of the energy system. Without considering social acceptance issues, the UK could face unexpected and possibly very significant barriers to the decarbonization pathways which appear attractive from a techno-economic view.

The socio-psychological motivations behind each of the three socio-environmental sensitivity scenarios are qualitatively characterized based on existing literature on public attitudes towards technologies, and case studies of particular developments and siting conflicts. As such, they are simplified characterizations of distinctive socio-environmental perspectives, to allow for the exploration of their possible impacts on energy technology change. Once the scenarios were qualitatively defined, consequences were quantified by constraining or blocking the deployment of specific technologies and translated into input parameters for the different scenarios. The impacts on the technologies described here are not meant to be comprehensive or 'scientifically objective' statements; instead, they reflect a hypothesized vision of how the general public might respond to low carbon energy resources and technologies.

Definition and quantification of the socio-environmental scenarios

NIMBY definition

The NIMBY scenario represents a storyline in which the public objects to certain energy developments when they perceive the development to have direct negative impacts on their lifestyle and community. The primary direct impact considered in this scenario is visual intrusion, which many studies have cited as a key reason for people's opposition (e.g. Wolsink, 2000; Wolsink, 2007). Accordingly, in the NIMBY scenario, people exhibit a strong resistance to major changes in their local landscape. Devine-Wright (2009, p426) has described this type of opposition 'as a form of place-protective action, which arises when new developments disrupt pre-existing emotional attachments and threaten place-related identity processes.' In a study of a hydropower development in Norway, Vorkinn and Riese (2001) found that place attachment explained a great deal of the variance in attitudes. Thus, in the NIMBY scenario, people's attachment to their visual landscape is the primary driver of opposition to the place-disruption that they believe would occur with a given energy development.

Based on this definition, in the NIMBY scenario there are strong constraints imposed on the deployment of technologies which have a history of (or for new technologies, an expectation of) public opposition based on negative visual impact. Accordingly, where the technology and infrastructure already exist, further developments will generally be accepted. However, developments will be limited in places where that type of development is unfamiliar, and raises objections based on disruptive visual landscape impact.

In the NIMBY scenario, the definition of 'landscape' is not exclusive to a person's immediate local area in terms of distance. Studies have shown that proximity to a development is not always directly correlated with support or opposition (Warren et al, 2005; Michaud et al, 2008). Rather, in this scenario, a person's landscape is more broadly defined to include places to which that person has some type of attachment. People may have attachments to their local home area as well as places such as holiday locations, National Parks or other landscapes of special interest.

It should be noted that this project's definition of NIMBY is very different from more established (and highly contested) descriptions of NIMBY, as 'not in my back yard'. This 'classic' definition envisions NIMBYism as opposition to a project based on self-interest. Many studies have suggested that this classic NIMBY definition is inaccurate and erroneous, because it fails to reflect people's actual motivations and masks a wide range of different motivational factors (Bell, 2005; Devine-Wright, 2005; van der Horst, 2007; Wolsink, 2007). Further, other studies have shown little or no evidence for the self-interest definition of NIMBYism (Wolsink, 2000; Braunholtz, 2003; Ek, 2005). It appears that NIMBY has also become a pejorative term, often used to dismiss people's valid concerns. Nevertheless, although NIMBY has become a contested and questionable term in the socio-psychological research community, it is used here as a recognizable term in popular usage. To avoid the pitfalls noted above, the NIMBY scenario has been carefully defined in terms of place-protective action, based on negative visual landscape impacts.

NIMBY quantification of impacts

In the NIMBY scenario, the deployable onshore wind resource is constrained by visual impact, which has been identified as a key reason behind objections to particular wind farms (Wolsink, 2000; Pasqualetti, 2001; Warren et al, 2005; Wolsink, 2007). The NIMBY scenario assumes such a strong opposition to wind developments on this basis that no new onshore wind applications receive planning approval. However, a number of the projects which are already under construction or have received planning consent are allowed to be built, based on 2009 data from the British Wind Energy Association (BWEA, 2009). This translates to a total of 8.9GW of onshore wind power available for use over the period to 2050.

Offshore wind developments in the NIMBY scenario are also subject to public objections based on visual impact. Therefore, in NIMBY, no developments are allowed within a 'highly visible' distance. However, developments are allowed beyond a buffer zone of 12 nautical miles from the coast, where there are likely to be fewer public objections (Bishop and Miller, 2007; DECC, 2009; Snyder and Kaiser, 2009). Using data from the UK Government's Offshore Energy Strategic Environmental Assessment Report (DECC, 2009), the available offshore wind power resource beyond a 12 nautical mile buffer zone is estimated at 80GW. This was used as the total constraint on deployable offshore wind power.

The Severn tidal barrage is not allowed in the NIMBY scenario, reflecting concerns over the landscape impact of the barrage such as the potential shifts in landscape type (such as mudflats and marshes), changes to local senses of place-identity and place-attachment as well as changes to historic ports (Sustainable Development Commission, 2007).

Nuclear power plants are not allowed at new locations in the NIMBY scenario. However, in areas around existing nuclear power plants, where facilities have already become a feature of the landscape and community, local communities are considered more likely to be willing to accept further development in the NIMBY scenario. There is also evidence of a more general 'reluctant acceptance' of nuclear power to help combat climate change (Bickerstaff et al, 2008). Therefore, in the NIMBY scenario, existing nuclear power plants can be replaced at the end of their lifetime. It is also assumed that the number of reactors can be expanded at certain sites that already host reactors. Assuming that new 1600MW power station units would be built on existing commercial nuclear power sites, this scenario allows for up to 30.4GW of total nuclear power capacity in the UK (Jay et al, 2010).

In the NIMBY scenario, fossil fuel powered plants fitted with carbon capture and storage (CCS) are only permitted at a few existing coastal power plant sites, or other major coastal industrial sites, due to possible public objections to the direct landscape impact of new power plants, capture plants and corresponding infrastructure such as pipelines and storage facilities. CCS technology appears to have low levels of public understanding (Curry et al, 2005) and remains unfamiliar to most people; only a few studies have examined attitudes to CCS, and they have found evidence of some negative attitudes towards CO_2 storage and pipelines (Huijts et al, 2007; Shackley et al, 2008). This suggests that within a NIMBY scenario, CCS would be constrained, especially at inland sites. Coastal sites could be less likely to be subject to objections and so, for the NIMBY scenario, seven coastal locations were considered to be consentable for fossil fuel plants fitted with CCS, providing 10.5GW of allowable capacity.

Bio-energy is also constrained in the NIMBY scenario, reflecting public concerns over large areas of land being switched to crops for bio-energy. This would noticeably alter the landscape. In addition, Heiskanen et al (2008) suggest that divergent interests over land use are one of the major conflicts found in case studies of public acceptance of bio-energy. Therefore, an analysis was undertaken using a spatial mapping approach to provide an estimation of the proportion of the UK productive land that might be allowed for bio-energy production under a NIMBY scenario. This analysis suggested that under strong NIMBY concerns, only 37 per cent of UK productive land would be allowed to be converted for energy feedstocks (Jay et al, 2010). Further, dedicated energy crops such as miscanthus and willow (which are visually unfamiliar in the landscape) were deemed to be unacceptable in the NIMBY scenario.

ECO definition

The ECO scenario seeks to represent public objections to certain technologies or resources based on perceptions of negative impacts on the natural environment and ecosystem services. In the ECO scenario, the deployment of low carbon technologies can be limited by environmental impacts other than carbon emissions. Some studies, such as Firestone and Kempton's (2007) study of the Cape Wind offshore wind project in the USA, have suggested that public objections are largely based on a perception of negative environmental impact.

This reflects a 'green on green' clash of environmental values, between the protection of the environment through increased use of low carbon energy technologies, and other environmental harm that could be caused by deployment of those technologies (Warren et al, 2005). This trade-off between different environmental concerns makes the deployment of certain technologies problematic (Wolsink, 2000).

The ECO scenario aims to represent resistance to key ecological impacts as perceived by the general public, as they evaluate scientific evidence of impact to produce their own assessment of risk. Public perceptions are mediated by, for example, media coverage, high profile scientists and NGOs. Different parts of the public can be seen to have differing levels of trust in key actors of an energy development, based on the perceived competence and motivation of those actors (Huijts et al, 2007). These varying levels of trust, understanding and media coverage all shape the public opinion of environmental impact.

ECO quantification of impacts

In the ECO scenario, the public objects to a number of proposed onshore wind farms due to concerns about the impact of the project on bird and bat mortality and damage to the land around the turbines during construction (especially for peat bogs). Studies have shown that concerns about birds can have a direct impact on objections to wind projects, even if not expressed in more general attitudes towards wind power (Wolsink, 2000). Yet, not all wind farms are perceived as an ecological threat. Therefore, in the ECO scenario, only a proportion of onshore developments are seen as contested on environmental grounds. Specifically, a limit of 15GW of onshore wind was permitted (based on 25 per cent of the UK onshore resource of 20GW being unavailable due to environmental concerns) (Jay et al, 2010).

Offshore wind power is also assumed to be constrained by public concerns about potential ecological impact. These concerns are likely to be lower than for onshore wind, as early evidence suggests reduced impact far offshore (DECC, 2009; Snyder and Kaiser, 2009). Thus, in the ECO scenario, offshore wind development is only allowed beyond a coastal buffer of 12 nautical miles, limiting potential ecological impacts on, for instance, bird life (DECC, 2009; Snyder and Kaiser, 2009). DECC (2009) has suggested that there are 80GW of offshore wind power available beyond 12 nautical miles; thus, this is used as the offshore wind installed capacity constraint.

The public perception of possible environmental impacts of wave and tidal power is not well understood. There is limited commercial development experience as devices are still being demonstrated and there has been little research on their environmental impact, and any associated public concerns. Thus there is currently little evidence to draw upon for how the public may respond to wave and tidal energy technologies. For the purposes of the ECO scenario, it is assumed that as more research emerges there may be some concerns over possible negative ecological impacts, such as damage to marine ecology from leaking fluids or other operational features (Boehlert et al, 2007). Accordingly, in the ECO scenario, the general public objects to certain technologies and developments in particularly sensitive marine areas. For this exercise, this is estimated as 25 per cent of the wave and tidal resource being unavailable for development due to ecological concerns.

The Severn tidal barrage would not be permitted in the ECO scenario because of public concerns about environmental impact. A coalition of well known environmental protection organizations, including the National Trust, RSPB and WWF oppose the barrage. The Sustainable Development Commission (2007) also reported possible serious impacts on the environment associated with the Severn Barrage (Cardiff-Weston scheme). The public is therefore considered to oppose the barrage in this scenario.

There are also strong public concerns about the sustainability merits of imported biomass in this scenario. Media attention has heightened these concerns which relate to the clearance of rainforests, the high carbon intensity of some crops and impacts on biodiversity. No imported biomass is permitted in the ECO scenario because of such public concerns. In order to safeguard the sustainability of crops domestically, only a certain percentage of the UK productive land is available for bio-energy; a spatial mapping analysis estimated that under the given ECO constraints 11 per cent of the UK productive land could be available for bio-energy (Jay et al, 2010). Public concerns over the sustainability of transport bio-fuels in the ECO scenario mean that no transport bio-fuels are used in the UK.

The recognized increasing damage caused by extracting fossil fuels, related to such issues as Arctic oil deposits and Canadian tar sands, is considered to create international agreement on controlling, mitigating and repairing the environmental harm. Consequently, the provision of global fossil fuel becomes more expensive and prices are taken to be much higher in the ECO scenario than in the Low Carbon (LC) core scenario, because, under ECO concerns, certain environmentally sensitive resources and methods of extraction are prohibited. Therefore, in the ECO scenario, the global prices of fossil fuels (coal, gas and oil) are increased substantially. Domestically, open cast coal mining is also prohibited as the public perceives it to be too damaging to the environment.

DREAD definition

The DREAD scenario is based on the concept of a 'dread' response, in which people perceive uncertain, involuntary and potentially catastrophic risks associated with a given technology (Slovic, 1987; Gregory and Mendelsohn, 1993; Singleton et al, 2009). Thus, in the DREAD scenario when the public perceives certain technologies to pose a serious risk to human health, those technologies are categorically rejected.

The potential risk to human health in this scenario is based on public perception of the risks, rather than statistics of fatalities or other quantitative measures of risk as experts might use. The public perception of risk is based on factors such as voluntariness, dread, knowledge, controllability and benefits (Slovic, 1987). For example, despite the fact that more people die in car accidents than in nuclear accidents, society seems more willing to accept the risks of cars than nuclear power stations. This is largely because the perception of risk of a nuclear accident is very different from that of a car accident; people feel more control over the risks of driving a car, the risk of a car accident is seen as less potentially catastrophic, and people receive a more obvious benefit from driving a car. For nuclear power, the benefits may be less obvious and the public is less willing to accept a high level of perceived risk for low levels of perceived benefits.

Thus, in this scenario, deployments of technologies which might evoke a DREAD response from the general public are prohibited. Rather than assessing whether the public's fears are well founded, or not, the scenario seeks to represent a highly risk-averse society.

DREAD quantification of impacts

New deployments of nuclear power are prohibited in the DREAD scenario due to fear of potential catastrophic consequences. Historic nuclear accidents such as Chernobyl and Three Mile Island have had a profound effect on society and people's perception of the nuclear risk (Slovic, 1987). Despite claims that new nuclear power would be safer, in the DREAD scenario, people still exhibit widespread concern over the potential for nuclear accidents and storage safety. According to psychometric studies of risk, nuclear reactor accidents, radioactive waste and other associated radioactive risks are seen as highly dreaded (Slovic, 1987). Hinman et al (1993) also show high levels of public concern and dread about the potential for catastrophic nuclear accidents and storage of radioactive waste products. Therefore, in DREAD, the public opposes all new nuclear developments. Existing nuclear power plants continue to operate until the end of their lifetime, but without lifetime extensions or any new or replacement nuclear power plants.

The public acceptability of coal CCS technologies is uncertain, but it may be a serious barrier to deployment. Some studies suggest that people may object to CCS technology based on the fear of the consequences of the technology on human health – i.e. a dread response (Wong-Parodi et al, 2008). Singleton et al (2009) use the psychometric theory of pubic risk perception to evaluate perceived risk from the geologic storage of carbon dioxide for CCS. They found that, as a new and unknown technology, there are public concerns associated with CCS storage; Singleton et al's rankings place geologic storage of carbon as having a higher dread risk than traditional fossil fuels, but lower than nuclear accidents (Singleton et al, 2009). Thus, in the DREAD scenario, the public is assumed to have a strong dread reaction to CCS, and no CCS power plants are allowed in the UK.

In the DREAD scenario the public is also assumed to have a powerful dread response to hydrogen and fuel cell technology, such that no deployment is permitted. There is a knowledge gap with regard to the public's understanding and perception of hydrogen and fuel cell technologies (Ricci et al, 2008). However, there are some persistent public notions about the dangers of hydrogen; for example, the vision of the Hindenburg disaster persists despite evidence that hydrogen was not the critical factor in that disaster (Stahl et al, 1988). This is an example of how media representation can affect public attitudes, and how historical precedent can shape public perceptions. Rhetoric about the dangers of hydrogen which suggests that a hydrogen fuel cell car could be made into a bomb by terrorists (such as in Shinnar, 2003) speaks explicitly to the dread factor. In the DREAD scenario, the risks of hydrogen fuel cell technologies are perceived as highly unknown and potentially catastrophic (i.e. highly dreaded). Therefore, no hydrogen fuel cell deployments are allowed in the DREAD scenario.

Scenario results: the impact of societal resistance to energy technologies

The energy systems deployed in the three socio-environmental scenarios illustrate different ways of meeting the UK's 80 per cent decarbonization target to 2050, while also addressing the additional constraints imposed by socio-environmental sensitivities. The three scenarios offer distinctive pathways to decarbonizing the energy system, in terms of the sequence, magnitude and rate of decarbonization of different sectors (transport, industry, residential, etc.). In all the scenarios (and in the LC scenario without socio-environmental constraints), there is an early emphasis on the decarbonization of the power sector, but to varying degrees and with different combinations of technologies. The degree of demand reduction also varies.

In terms of CO_2 emissions, there are significant differences in the sectoral breakdown of CO_2 emissions by 2050, with the ECO and DREAD scenarios showing the greatest divergence from the LC scenario, as seen in Figure 10.9. In the ECO scenario, the increased costs of fossil fuels and the prohibition of transport bio-fuels makes the transport sector more difficult and costly to decarbonize. As a result, the transport sector continues to use diesel and petrol, and greater emission reductions occur in other sectors, where there are lower cost decarbonization options (Figure 10.9). In the DREAD scenario, the power sector is almost completely decarbonized, with very high levels of wind power deployed, given constraints on other technologies. As a result, the residential sector accounts for a relatively high proportion of emissions in 2050 in the DREAD scenario.

The different socio-environmental scenarios are also associated with the selection of different supply technology mixes, particularly in electricity generation (see Figures 10.10 and 10.11). Socio-environmental constraints have impacts on overall demand, levels of installed capacity, the dominant technologies deployed and diversity of supply.

The DREAD scenario, for example, has very little supply diversity, given prohibitions on major supply options such as nuclear power and fossil fuels with CCS. Very high levels of wind power are deployed to meet the decarbonization

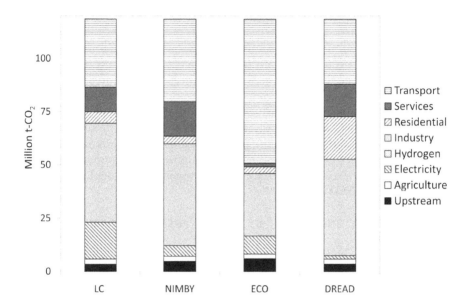

Figure 10.9 *Sectoral CO$_2$ emissions in 2050*

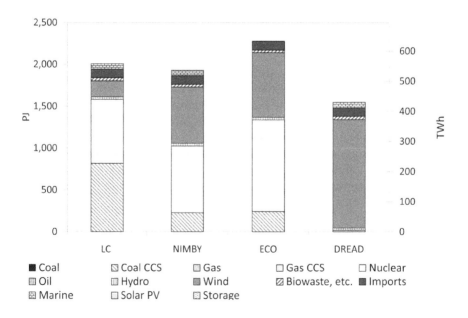

Figure 10.10 *Electricity generation in 2050*

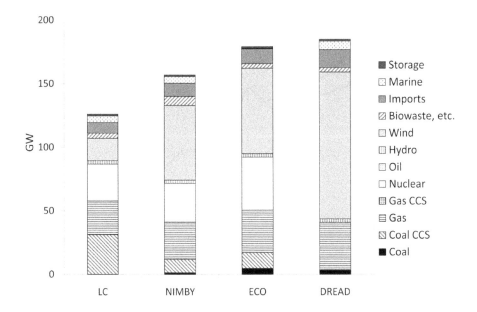

Figure 10.11 *Installed capacity in 2050*

target. Despite relatively low levels of overall generation, the installed capacity on the system in 2050 is much higher than the LC scenario and is the highest of all the socio-environmental sensitivity scenarios with high levels of gas-fired capacity needed to back up intermittent wind.

In all three socio-environmental scenarios, the constraints provoke additional demand reduction beyond that seen in the LC scenario across all sectors (industry, residential, services and transport). These changes can be illustrated with the examples of demand reduction in residential electricity and gas usage (Figures 10.12 and 10.13). While the NIMBY scenario shows a roughly similar demand reduction to LC, much higher levels of demand reduction occur in the ECO and DREAD scenarios, reflecting their more severe supply constraints.

A number of other interesting, and perhaps unexpected, results arise from the socio-environmental scenarios.

In both the NIMBY and ECO scenarios, bio-energy is severely constrained. Interestingly, in NIMBY, limited bio-energy resources are used in aviation; aviation jet fuel is completely replaced by bio-kerosene by 2050, whereas bio-kerosene is not deployed in the LC scenario or in any of the other core scenarios. In ECO, with imported crops and transport bio-fuels prohibited, bio-energy is more tightly constrained than in NIMBY. As a result, limited domestic bio-energy resources are used in the commercial service sector, in the form of wood and later pellets. This illustrates the flexibility of bio-energy resources despite limitations on their sourcing and usage.

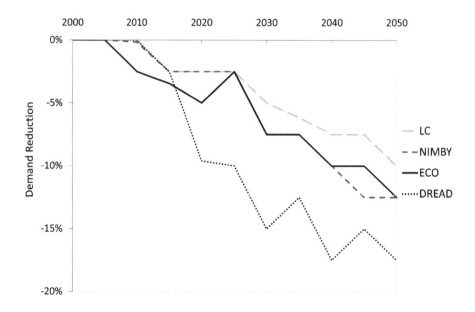

Figure 10.12 *Residential electricity demand reduction*

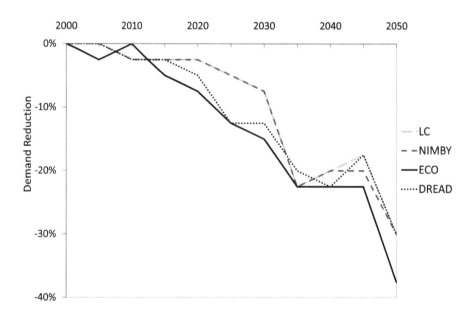

Figure 10.13 *Residential gas demand reduction*

Another interesting result can be seen in DREAD where 5GW of micro-wind generation is deployed; this is the only scenario in which wind micro-generation is deployed. The installed capacity of wind micro-generation is limited by very low capacity credits over 5GW (there is a 15GW total resource constraint on wind micro-generation based on figures from Energy Savings Trust (2005)). The deployment of micro-wind, despite low capacity credits, reflects the severe constraints on the electricity generation sector in the DREAD scenario.

A power supply system that is dominated by a single generating source, as in DREAD, is less likely to be resilient. In this case, the electricity supply system relies on very high levels of wind power. Although the operation is predominantly under UK control, it is exposed to the risk of prolonged periods of low wind generation due to high pressure areas. On the surface, storage appears to be an attractive solution for this type of power system, but less storage is employed in DREAD than in the LC scenario, indicating a lack of economically viable storage technologies. Emerging storage technologies may become affordable before 2050, but these have not been represented in the modelling of the scenarios.

All three socio-environmental scenarios explored (NIMBY, ECO and DREAD) impose higher costs on society than the LC scenario. However, the costs imposed by the ECO constraints are consistently the highest by 2050. This increased cost can be seen through three different measures: the marginal cost of CO_2, undiscounted total energy system cost and changed consumer and producer surplus (as a measure of social welfare). These measures are discussed in Chapter 4.

In the near term (2015–2030) the DREAD scenario has the highest marginal cost of CO_2. However, by 2050 the marginal cost of CO_2 is highest in the ECO scenario, as seen in Figure 10.14. The undiscounted energy system cost is similar in the LC, NIMBY and DREAD scenarios. However, in ECO, the energy system cost is consistently higher from 2015 onwards, ranging from roughly £2.5 billion more expensive in 2015 to £9 billion more expensive by 2050, suggesting that ECO constraints necessitate the adoption of more expensive measures to achieve decarbonization targets. The assumption about higher global energy prices also plays a role. Using consumer and producer surplus as a measure of social welfare, all three of the socio-environmental scenarios result in higher welfare losses than the LC scenario. Again, the ECO scenario shows a significantly greater decline in welfare from 2015 onwards than in the LC, DREAD or NIMBY scenarios (Figure 10.15).

This illustrates that public acceptance of energy technologies could have a substantial impact not only on the make-up of the energy system, but also on the financial and social cost of decarbonization. When the public rejects certain technologies for any of the reasons explored in these scenarios, decarbonization becomes more costly and more economically challenging.

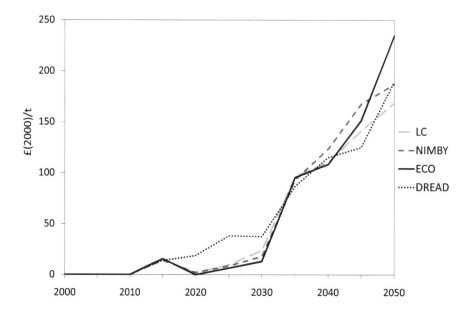

Figure 10.14 *Marginal cost of CO₂*

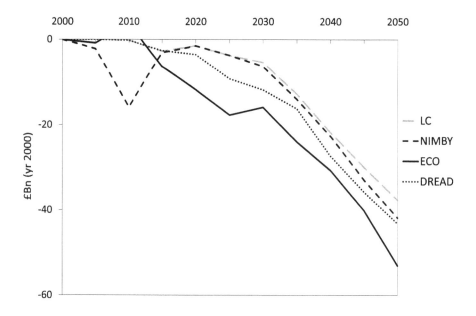

Figure 10.15 *Societal welfare expressed as consumer and producer surplus*

Key messages from the socio-environmental scenarios

The socio-environmental scenarios have illustrated how public attitudes and acceptance of technologies could impact on the possible pathways to decarbonization and energy system change. Despite the fact that these scenarios do not capture the full complexities of real world socio-technical relations, they do offer useful insights into the potential impacts on the energy system. In reality, attitudes and acceptance are multi-faceted, vary immensely amongst the members of the public and are not static over time; this makes it impossible to predict public acceptance issues. Nevertheless, by modelling a few illustrative scenarios of possible socio-environmental sensitivities, this work has highlighted possible types of impacts and changes.

Each of the scenarios illustrates different ways of meeting society's future energy needs and achieving decarbonization, while responding to a range of socio-environmental concerns. The differences in the scenarios vary predominantly around three key elements: the system-wide decarbonization pathway (when and at what level the sectors decarbonize), the selection and mix of technologies deployed and the overall level of demand reductions needed.

While it was possible to model a working energy system for all the scenarios, the configurations of technologies are challenging and, in some cases, perhaps unrealistic in practice. For instance, in DREAD the power sector lacks diversity, being highly dependent on wind power with high levels of back-up capacity. This reflects the severity of some of the constraints, and the stress that those constraints could impose on the system. In reality, advances in certain technologies (such as storage) and improvements in grid infrastructure and operation could make achieving decarbonization amidst socio-environmental constraints more manageable. Without these other changes, the scenarios highlight the magnitude of the decarbonization challenge; hard choices will need to be made about priorities.

The results suggest that public sensitivities to certain technologies need to be understood and considered early on in planning for decarbonization. If the public is not genuinely engaged, and choices are made that do not take account of public concerns, then the challenge of decarbonization becomes greater. There will be less time to prepare and allow for alternative solutions. Thus, further research, and action, is needed on engagement with society's socio-environmental concerns.

If public acceptance is neglected in UK decarbonization policies and debates, there is a risk of targets not being met, as a public backlash restricts the range of possible system changes. Accommodating these sensitivities also involves a cost: in all of the scenarios presented here, socio-environmental constraints are associated with increased costs to society and the imposition of reductions in demand.

This does not mean that public acceptance should be ignored; rather it suggests that public concerns need more attention, with meaningful public engagement early in the process of decarbonization. This may involve alternative methods of engagement and appropriate levels of engagement in planning processes or changes in the way that the challenges of decarbonization are shared with and discussed across society.

Conclusions

This chapter has explored environmental and social implications of the changing energy system. In the face of climate change, energy and environmental policy focus has tended to shift primarily onto carbon with targets set for decarbonization of society. Along with energy security, carbon emissions have therefore become a key driver for energy system change. However, as this chapter has shown, there are other factors that should be considered both in terms of non-carbon environmental impacts and the role of society in shaping energy system change.

The ecological consequences of the UK energy system include the impact of a number of pollutant emissions (such as carbon monoxide, nitrous oxide and particulate matter) that are variously acidifying, fertilizing and toxic to different organisms. In addition to carbon dioxide, these environmental pressures are also important considerations in the development of the energy system. Driving energy system change purely on the basis of carbon emissions misses out on these important environmental pressures. On the other hand, the environment is more than just a catch-all for pollutants whose impact is negative; it also provides essential resources, only some of which can be exploited sustainably.

The social consequences of energy system change must also be recognized. Energy technologies are not deployed in a social vacuum; people interact with, engage with and respond to technology developments in important ways. The way that the public perceives risks to their landscape (NIMBY), environment (ECO) and health (DREAD) shape the acceptable energy choices that will be made. Yet, decarbonization pathways to 2050 do not often consider how those public socio-environmental sensitivities will alter the options available as well as the cost and achievability of climate change targets.

Thus, this chapter has shown that consideration of energy system change should take account of more than just climate change mitigation and carbon emissions. Other environmental pressures and the public's engagement with energy technologies could and should play an important role in determining the pathways of energy system change.

Note

1   The Natura 2000 series is a Europe-wide network of specially protected sites created as 'Special Areas of Conservation' (SACs) under the EC Habitats Directive (1992) and 'Special Protection Areas' (SPAs) under the Birds Directive (1979).

References

Bell, D., Gray, T. and Haggett, C. (2005) 'The "social gap" in wind farm siting decisions: explanations and policy responses', *Environmental Politics,* vol 14 (4), pp460–477

BERR (Department for Business, Enterprise and Regulatory Reform) (2008) *Digest of United Kingdom Energy Statistics* (DUKES) 2008, The Stationery Office, London

Bickerstaff, K., Lorenzoni, I., Pidgeon, N.F., Poortinga, W. and Simmons, P. (2008) 'Reframing nuclear power in the UK energy debate: nuclear power, climate change mitigation and radioactive waste', *Public Understanding of Science*, vol 17, pp145–169

Bishop, I. and Miller, D. (2007) 'Visual assessment of off-shore wind turbines: The influence of distance, contrast, movement and social variables', *Renewable Energy*, vol 32, pp814–831

Boehlert, G., McMurray, G. and Tortorici, C. (eds) (2007) *Ecological Effects of Wave Energy Development in the Pacific Northwest*, A Scientific Workshop, October 11–12, 2007, NOAA Technical Memorandum NMFS-F/SPO-92

Braunholtz, S. (2003) *Public attitudes to windfarms: a survey of local residents in Scotland*, MORI Scotland, for Scottish Executive Social Research, Edinburgh

BWEA (British Wind Energy Association) (2009) Wind Energy Database, www.bwea.com/ukwed/index.asp, accessed 27 May 2009

Curry, T., Reiner, D., Figueiredo, M. and Herzog, H. (2005) *A Survey of Public Attitudes towards Energy and Environment in Great Britain*, Massachusetts Institute of Technology Laboratory for Energy and Environment, LFEE 2005-001 WP, http://web.mit.edu/mitei/lfee/programs/archive/publications/2005-01-wp.pdf, accessed 27 May 2009

DECC (Department of Energy and Climate Change) (2009) *UK Offshore Energy Strategic Environmental Assessment Environmental Report*, Department of Energy and Climate Change, www.offshore-sea.org.uk/consultations/Offshore_Energy_SEA/OES_Environmental_Report.pdf, accessed 27 May 2009

Devine-Wright, P. (2005) 'Beyond NIMBYism: towards an integrated framework for understanding public perceptions of wind energy', *Wind Energy*, vol 8 (2), pp125–139

Devine-Wright, P. (2009) 'Rethinking Nimbyism: the role of place attachment and place identity in explaining place-protective action', *Journal of Community and Applied Social Psychology*, vol 19 (6), pp426–441

EC (2001) *Directive 2001/80/Ec of the European Parliament and of The Council of 23 October 2001 on the limitation of emissions of certain pollutants into the air from large combustion plants*, eur-lex.europa.eu/LexUriServ/site/en/oj/2001/l_309/l_30920011127en00010021.pdf, accessed 27 May 2009

EEA (European Environment Agency) (1995) *Europe's Environment – the Dobris Assessment*, D. Stanners and P. Bordeau (eds.), EEA, Copenhagen

EEA (European Environment Agency) (2008) *Energy and environment report, 2008*, EEA, Copenhagen

Ek, K. (2005) 'Public and private attitudes towards "green" electricity: the case of Swedish wind power', *Energy Policy*, vol 33, pp1677–1689

Energy Savings Trust (2005) *Potential for Microgeneration Study and Analysis*, Energy Savings Trust, www.berr.gov.uk/files/file27558.pdf, accessed 10 June 2010

Firestone, J. and Kempton, W. (2007) 'Public opinion about large offshore wind power: Underlying factors', *Energy Policy,* vol 35, pp1584–1598

Fowler, D., Cape, J.N., Leith, I.D., Paterson, I.S., Kinnaird, J.W. and Nicholson, I.A. (1982) 'Rainfall acidity in Northern Britain', *Nature,* vol 297, pp383–386

Fowler, D., Muller, J., Smith, R.I., Cape, J.N. and Erisman, J.W. (2005) 'Nonlinearities in source receptor relationships for sulfur and nitrogen compounds', *Ambio,* vol 34, pp41–46

Gregory, R. and Mendelsohn, R. (1993) 'Perceived risk, dread, and benefits', *Risk Analysis,* vol 13, pp259–264

Heiskanen, E., Hodson, M., Mourik, R.M., Raven, R.P.J.M, Feenstra, C.F.J., Alcantud Torrent, A., Brohmann, B., Daniels, A., Di Fiore, M., Farkas, B., Fritsche, U., Fucsko, J., Hünecke, K., Jolivet, E., Maack, M.H., Matschoss, K., Oniszk-Poplawska, A., Poti, B.M., Prasad, G., Schaefer, B. and Willemse, R. (2008) *Factors influencing the societal acceptance of new energy technologies: Meta-analysis of recent European projects,* ECN Policy Studies, www.ecn.nl/publications/default.aspx?nr=ECN-E--07-058, accessed 27 May 2009

Hinman, G., Rosa, E., Kleinhesselink, R. and Lowinger, T. (1993) 'Perception of nuclear and other risks in Japan and the United States', *Risk Analysis,* vol 13 (4), pp449–455

van der Horst, D. (2007) 'NIMBY or not? Exploring the relevance of location and the politics of voiced opinions in renewable energy siting controversies', *Energy Policy,* vol 35, pp2705–2714

Huijts, N., Midden, C. and Meijnders, A. (2007) 'Social acceptance of carbon dioxide storage', *Energy Policy,* vol 35, pp2780–2789

Jay, B., Howard, D., Hughes, N. and Whitaker, J. (2010) *Socio-environmental Sensitivities Scenarios: How public acceptance of energy technologies could affect the UK's low carbon future,* UKERC Working Paper, UKERC, London

Lockyer, D.R. and Jarvis, S.C. (1995) 'The measurement of methane losses from grazing animals', *Environmental Pollution,* vol 90, pp383–390

Matejko, M., Dore, A.J., Hall, J., Dore, C.J., Blas, M., Kryza, M., Smith, R. and Fowler, D. (2009) 'The influence of long term trends in pollutant emissions on deposition of sulphur and nitrogen and exceedance of critical loads in the United Kingdom', *Environmental Science and Policy,* vol 12, pp882–896

McElroy, M.B., Wofsy, S.C. and Yung, Y.L. (1977) 'Nitrogen cycle – perturbations due to man and their impact on atmospheric N_2O and O_3', *Philosophical Transactions of the Royal Society of London Series B: Biological Sciences,* vol 277, pp159–181

Michaud, K., Carlisle, J. and Smith, E. (2008) 'Nimbyism vs. environmentalism in attitudes toward energy development', *Environmental Politics,* vol 17 (1), pp20–39

Millennium Ecosystem Assessment (2005) *Full and Synthesis Reports,* www.maweb.org/en/index.aspx, accessed 21 October 2010

NAEI (National Atmospheric Emissions Inventory) (2006) *Emission Factor Database,* National Atmospheric Emissions Inventory, www.naei.org.uk/, accessed 27 May 2009

Pasqualetti, M. (2001) 'Wind Energy Landscapes: Society and Technology in the California Desert', *Society and Natural Resources,* vol 14, pp689–699

RGAR (Review Group on Acid Rain) (1997) *Acid Deposition in the United Kingdom 1992–94. Fourth report of the Review Group on Acid Rain,* Department of the Environment, Transport and the Regions, London, UK

Ricci, M., Bellaby, P. and Flynn, R. (2008) 'What do we know about public perceptions and acceptance of hydrogen? A critical review and new case study evidence', *International Journal of Hydrogen Energy,* vol 33, pp5868–5880

Shackley, S., Reiner, D., Upham, P., de Coninck, H., Sigurthorsson, G. and Anderson, J. (2008) 'The acceptability of CO_2 capture and storage (CCS) in Europe: An assessment of the key determining factors. Part 2: The social acceptability of CCS and the wider impacts and repercussions of its implementation', *Journal of Greenhouse Gas Control,* vol 3, pp344–356

Singleton, G., Herzog, H. and Ansolabehere, S. (2009) 'Public risk perspectives on the geologic storage of carbon dioxide', *International Journal of Greenhouse Gas Control,* vol 3, pp100–107

Slovic, P. (1987) 'Perception of Risk', *Science,* vol 236, pp280–285

Snyder, B. and Kaiser, M.J. (2009) 'A comparison of offshore wind power development in Europe and the US: Patterns and drivers of development', *Applied Energy,* vol 86, pp1845–1856

Stahl, C.J., McMeekin, R.R., Ruehle, C.J. and Canik, J.J. (1988) 'The medical investigation of airship accidents', *Journal of Forensic Sciences,* vol 33, pp888–898

Sustainable Development Commission (2007) *Turning the Tide: Tidal Power in the UK,* Sustainable Development Commission, London

Thiemens, M.H. and Trogler, W.C. (1991) 'Nylon production – an unknown source of atmospheric nitrous oxide', *Science,* vol 251, pp932–934

UNECE (United Nations Economic Committee for Europe) (2009) *Emission Statistics by UNECE Source Category* www.naei.org.uk/emissions/emissions.php, accessed 21 October 2010

Vorkinn, M. and H. Riese (2001). 'Environmental concern in a local context: The significance of place attachment', *Environment and Behaviour,* vol 33 (2), pp249–263

Warren, C., Lumsden, C., O'Dowd, S. and Birnie, R. (2005) '"Green on Green": Public perceptions of wind power in Scotland and Ireland', *Journal of Environmental Planning and Management,* vol 48 (6), pp853–875

Wolsink, M. (2000) 'Wind power and the NIMBY-myth: Institutional capacity and the limited significance of public support', *Renewable Energy,* vol 21, pp49–64

Wolsink, M. (2007) 'Wind power implementation: The nature of public attitudes: Equity and fairness instead of "backyard motives"', *Renewable and Sustainable Energy Reviews,* vol 11, pp1188–1207

Wong-Parodi, G., Ray, I. and Farrell, A.E. (2008) 'Environmental non-government organizations' perceptions of geologic sequestration', *Environmental Research Letters,* vol 3 (2), p8

11
UK Energy in an Uncertain World

Neil Strachan and Jim Skea

Introduction

In scenario-modelling studies, a range of factors are beyond the control of the user of the study outputs. In the context of long-term UK low carbon energy pathways, the users of scenario exercises such as in this book primarily are UK policy-makers, and a set of *external* drivers are international in nature (Hughes et al, 2009). Critical global uncertainties include the prices of internationally traded fossil fuels (coal, oil, oil products, natural gas), the availability and cost of greenhouse gas (GHG) credits to meet domestic emissions targets, and the availability (and costs) of sustainable biomass imports.

In recent years the application of scenarios in low carbon energy studies has underpinned long-term national energy policy analysis and initiatives (e.g. Berkhout et al, 1999), together with a wide range of country-level studies (e.g. Saddler et al, 2007; van Vuuren et al, 2003) and meta studies (e.g. Silberglitt et al, 2003) that have primarily focused on the implications of scenarios for particular national energy systems. Conventionally, international drivers are treated as exogenous assumptions, which is adequate when analysing national energy policies in isolation but less appropriate when considering long-run carbon emissions reductions as part of a global mitigation strategy. An alternate approach is to utilize global models broken into key regions (e.g. Weyant, 2004), but this approach loses country-level detail.

This chapter undertakes scenario analysis on these key international drivers to investigate the sensitivity of UK long-term stringent carbon dioxide (CO_2) reductions. Therefore the national-level technological, market and policy detail of the UK MARKAL-MED model can be retained, facilitating the comparison with the range of scenarios in other chapters of this book. The core comparator runs are an 80 per cent CO_2 reduction (LC, Chapters 4 and 5) and an 80 per cent CO_2 reduction resilient run (low carbon resilient (LCR), Chapter 6).

Key features of the MARKAL-MED model are described in Chapter 4. A full description is given in Strachan et al (2008a) and Strachan and Kannan

(2008). Detailed national-level modelling assumptions are given in the model documentation (Kannan et al, 2007), with model updates and UKERC Energy 2050 project specific assumptions given in Anandarajah et al (2009). Methodologies for incorporating global drivers in national energy models are discussed via the Japan-UK Low Carbon Societies project (Strachan et al, 2008b).

Assumptions and scenarios

Derivation of assumptions

This chapter focuses on three main global drivers on the UK energy system – fossil fuel prices, sustainable biomass imports, and the availability of CO_2 emission credits. These are key international uncertainties in respect of UK decarbonization pathways in both academic (e.g. Strachan et al, 2008b) and policy (e.g. CCC, 2008; DECC, 2009) publications. Furthermore, although the UK is one of the largest developed economies in the world, it represents a global GDP share of only 3.5 per cent, which is projected to fall to 2.5 per cent in 2050 (EIA, 2007). Hence the UK's direct actions have a very limited impact on these overall drivers.

Fossil fuel prices

In the UK MARKAL-MED model, central global fossil (coal, oil and natural gas) prices are taken from an assessment of the UK government's updated energy projections dating from November 2008 (DECC, 2008a), and from the International Energy Agency's World Energy Outlook (WEO) (IEA, 2007). As the model depicts domestic and imported fossil resources using supply curves, multipliers calibrated from relative historical prices (DUKES, 2005) are used to translate these into prices for both higher priced supply steps as well as imported refined fuels.

In response to substantial rises in global fossil prices through 2007–2008, most forecasts were also raised during this period. In this chapter, assumptions on higher price global resources are derived from WEO 2008 (IEA, 2008) with reference to the DECC (2008a) Hi and Hi Hi scenario runs. All new prices are implemented for oil, oil products, natural gas, LNG, steam coal and coking coal imports and exports. Table 11.1 details the central and high fossil fuel price assumptions.

Note, however, that there is a complex interaction between imported fuel prices and the costs of future domestic resources. These future domestic resource costs (*not* prices) are based on various DECC reports and are broken out into a set of supply curves for the three fossil fuels based on the size and location of the resource base (see the model documentation (Kannan et al, 2007) for details). If the cost of these domestic options is below the (rising) price of imported fuels then the model could choose to moderate the global price increase. Hence to prevent this, in high fossil fuel price runs, we assume no domestic fossil production post-2030.

Table 11.1 *Fossil fuel import prices ($,£ 2009)*

Year	Central			High		
	Oil ($/bbl)	Gas (p/therm)	Coal ($/GJ)	Oil ($/bbl)	Gas p/therm	Coal ($/GJ)
2010	43.6	36.5	2.1	73.0	54.4	2.6
2020	49.1	39.8	2.0	78.5	57.8	2.8
2030	54.5	43.2	2.2	89.4	64.3	3.1
2040	60.0	46.4	2.4	89.4	64.3	3.3
2050	60.0	46.4	2.4	89.4	64.3	3.3

Note: Conversion factors (all GCV): £1 = $1.8, (2000)£1 = (2009)£1.23, 1bbl oil = 6.096GJ, 1therm gas = 9.48GJ, 1tonne coal = 26.8GJ

Carbon dioxide emission credit purchases

For the UK to meet future carbon reduction targets at moderate cost, the purchase of emission credits from mitigation overseas could be critical. As the contribution to radiative forcing from GHG emissions is the same no matter where those emissions are located, if it is cost effective for the UK to purchase emission credits these would be counted under UK emissions reductions. Current UK policy on the long-term scope of potential UK emissions purchases is still evolving, but a general principle is that of supplementarity – that is, the use of the emission credit mechanisms shall be supplemental to domestic action, and domestic action shall thus constitute a significant element of national effort (DECC, 2008b).

The availability and price of any emission credits are driven by other countries (notably large developing countries), in terms of any future emission reductions they themselves legislate, together with the emissions intensity and size of their economies. In light of these and other drivers, marginal abatement cost curves for global CO_2 emissions vary widely and have a very large uncertainty range (Klepper and Peterson, 2006).

In the latest UK government guidance (CCC, 2009), high, central and low CO_2 emission credit availability and price assumptions are only given out to the third budget period (i.e. 2017–2022). In the medium term (2020s) the CCC suggested that if the rest of the world (ROW) increased its ambition, the UK would tighten its target (to a –42 per cent reduction in GHGs from 1990 levels) but allow additional credits to be purchased to meet this more ambitious target. This does not apply in 2050 because one cannot assume how the UK would alter its –80 per cent target in response to action in the ROW. Specifically, the long-term limit in cost effective international emissions credits is highlighted in CCC (2009).

Therefore, for this study we have assumed that only 12.5 per cent of required emission reductions from the 1990 baseline can be met via emissions credits. This ratio is kept in place from 2015–2050 (Table 11.2), and in 2050 this represents $59MtCO_2$. Considering this ceiling in terms of total CO_2 emissions, the UK could emit up to $177MtCO_2$ in 2050 but have one-third of these as purchased credits to meet its −80 per cent target (i.e. 118MtCO$_2$). Table 11.2 gives this upper bound on emission credit purchases together with lower and central prices (CCC, 2009) as used in the scenario definition (Table 11.3).

Sustainable biomass imports

A final international assumption is on the availability of biomass imports that are sustainable with respect to food production, other land use requirements, environmental degradation and biodiversity. The uncertainty in long-term estimates is very large, with UK government reviews (Gallagher, 2008) not going beyond 2020 in any estimation. Other studies (van Vuuren et al, 2010; WWF, 2007) have postulated annual global sustainable biomass with an energy content of between 50 and 250 exajoules (EJ) is feasible by 2050. On a per population basis this range for the UK's share would be 500–2500PJ. The total biomass use under the core low carbon (LC) scenario in Chapters 4 and 5 is around 1100PJ in 2050, reflecting somewhat conservative biomass assumptions already in the UK MARKAL-MED model (Kannan et al, 2007). Instead of further speculating on the availability (and price) of sustainable biomass imports, the relevant sensitivity run has an assumption of zero biomass imports – reflecting indigenous requirements for bio-resources and land use in a low carbon world – leaving the UK with only its domestic biomass resources.

Scenarios

This chapter undertakes scenario analysis on these key international drivers – fossil fuel prices, CO$_2$ emission credits and sustainable biomass imports – on UK long-term stringent CO$_2$ reductions. Note that no international aviation and shipping is contained within the emission budget of these runs – the focus is on domestic CO$_2$ only to facilitate comparability with other chapters in this book. The national-level technological, market and policy detail of the UK MARKAL-MED model is retained.

Eight scenarios are compared (Table 11.3), working off the core comparator runs of an 80 per cent CO$_2$ reduction (low carbon (LC), Chapters 4 and 5) and an 80 per cent CO$_2$ reduction resilient run (low carbon resilient (LCR), Chapter 6) which incorporates additional constraints on fuel shares of primary energy, technology class shares in electricity generation and reductions in final energy

Table 11.2 *International CO$_2$ emissions credits availability and costs*

Metric	Units	2015	2020	2025	2030	2035	2040	2045	2050
Availability upper bound[1]	MtCO$_2$	15.3	21.5	31.5	39.6	46.1	51.5	55.8	59.2
Central cost	(2000)£/tCO$_2$	11.9	12.8	34.6	56.5	83.1	108.9	135.5	161.3
Low cost	(2000)£/tCO$_2$	7.6	8.2	18.2	28.2	41.1	54.8	67.8	80.7

Note: 1) As noted in the text, this is assumed to be 12.5 per cent of the required emission reduction in any given year. Currency conversion factors: (2000)£1 = (2009)£1.23

Table 11.3 *Scenarios undertaken on global energy uncertainties*

	Description	Resilience	Fossil import price	CO_2 credit price	Biomass imports
1 LC	CO_2 –80%	No	Central	–	Yes
2 LCR	Resilient CO_2 –80%	Yes	Central	–	Yes
3 LC-HI	CO_2 –80%, high fossil priced imports	No	High	–	Yes
4 LCR-HI	Resilient CO_2 –80%, high fossil priced imports	Yes	High	–	Yes
5 LC-CC	CO_2 –80%, CCC central cost credit purchases	No	Central	Central	Yes
6 LCR-CC	Resilient CO_2 –80%, CCC central cost credit purchases	Yes	Central	Central	Yes
7 LC-HI-LC	'Best case': CO_2 –80%, high fossil priced imports, CCC low cost credit purchases	No	High	Low	Yes
8 LCR-NB	'Worst case': Resilient CO_2 –80%, no biomass imports	Yes	Central	–	No

demand. Recognizing another set of uncertainties related to the precise timing of emissions reductions and the roles of non-CO_2 GHGs and carbon sinks, a UK 80 per cent emission reduction target – consistent with a global emissions reduction target of around 50 per cent by 2050 – results in a stabilization of GHG concentrations at around 550ppm CO_2e (G8 Communiqué, 2007).

Scenarios 3 and 4 are the LC and LCR –80 per cent CO_2 cases, run with high global fossil fuel prices (Table 11.1). Scenarios 5 and 6 are the LC and LCR –80 per cent CO_2 cases, run with central prices for internationally traded credits, with an availability limited to 12.5 per cent of required UK reductions (Table 11.2). Finally scenarios 7 and 8 are exploratory runs that are intended to represent 'best and worst' cases for the UK to meet domestic emissions reduction targets. A 'best case' for the UK involves higher fossil fuel prices which drive emission reductions, access to low cost emission credit purchases and access to biomass imports. Conversely a 'worst case' for the UK entails central fossil fuel prices, also having to meet resilience constraints, having no access to emissions credits and having no access to (sustainable) biomass imports. Interestingly these 'best and worst' cases for the UK mirror opposite ambition levels at a global level. For example, in the 'best' UK case, this implies that other major economies (especially developing economies), are not substituting away from fossil fuels (hence keeping prices high), are not undertaking stringent emissions reduction (hence the pricing and availability of emissions credits), and are not utilizing their own biomass resources (hence facilitating imports into the UK). It is an open question as to whether in the long term the stringency of UK climate mitigation could or would diverge from global ambition levels.

Results

As summarized in Chapter 6, the core –80 per cent CO_2 cases (LC and LCR) have quite different characteristics. In addition to an 80 per cent emission reduction, the LCR scenario has a maximum share of any fuel of 40 per cent primary energy, has a maximum share of any technology class of 40 per cent electricity generation, and has a 3.2 per cent annual reduction from 2010 in final energy intensity (energy/GDP). As a result the LCR case has much lower final energy (by 2050 this is –40 per cent from 2000 levels in LCR as opposed to –29 per cent in LC), has reduced electricity generation due to the least expensive low carbon options being limited in a portfolio share and has a greater diversity of primary fuels. The combination of both constraints sees a marked increase in costs. By 2050, welfare losses compared with the reference scenario REF (described in Chapter 4) are £36 billion in LC, and are £56 billion in LCR, as the model is forced to use combinations of higher cost fuels and technologies.

Energy system evolution

Table 11.4 details primary energy for the eight scenarios in 2035 and 2050 (along with the base year of 2000). As noted above, the international uncertainties on fossil fuel prices, emissions credit purchases, and sustainable biomass act upon the two quite different core –80 per cent CO_2 cases of LC and LCR.

The impact of higher fossil fuel prices (scenarios 3 and 4) is generally to moderate primary energy consumption. However there is a complex trade-off both between *which* fossil fuel price changes have the most effect and *when* this differential has the most effect. Hence, in both LC and LCR scenarios, the intermediate impacts of higher coal prices are the most severe as they impact early generation CCS plants and other uses. In response nuclear generation is brought forward, together with biomass and renewable electricity. By 2050 in LC-HI, coal use recovers as subsequent generation CCS is still cost effective, and non-fossil sources (notably renewable electricity) are also boosted. In the LCR scenarios, coal and oil are the fossil fuels that see the largest reduction, due to the interplay between CO_2 and resilience constraints (both primary energy share and demand reductions). By 2050 in LCR-HI, the higher cost of fossil fuels has led to extreme efficiency gains and energy service demand reductions together with increases in renewable electricity and biomass use. In the longer term in LCR-HI natural gas use is boosted due to its high efficiency and higher base cost making additional price rises proportionally less effective.

The impact of emissions credit purchases (scenarios 5, 6 and 7) is generally only seen in later years where the price of CO_2 emissions credits is lower than the marginal costs of domestic reduction options (see Figure 11.3). The impact is most pronounced in LCR-CC (owing to high domestic CO_2 marginal prices) and LC-HI-LC (owing to low cost international credits) and in these scenarios this allows the UK energy sector to use 300PJ and 600PJ respectively of additional primary energy. When combined with fuel switching this sees substantial increases in gas, oil and coal use in 2050, alongside reductions in renewable electricity and biomass use.

Table 11.4 *Primary energy (petajoules, PJ)*

	2000	1 LC	2 LCR	3 LC-HI	4 LCR-HI	5 LC-CC	6 LCR-CC	7 LC-HI-LC	8 LCR-NB
					2035				
Renewable electricity	20	178	169	189	195	178	170	191	175
Biomass and waste	121	618	307	673	352	437	305	622	248
Natural gas	3907	2339	1986	2109	1800	2437	1984	2216	2011
Oil	3043	2018	1507	1914	1152	2220	1503	2036	1341
Refined oil	−298	−217	−72	−200	106	−220	−72	−204	−65
Coal	1500	1857	1545	1236	1210	1866	1541	1341	1389
Nuclear electricity	282	240	329	664	561	238	333	640	422
Imported electricity	52	88	88	88	88	88	88	88	88
Total	8628	7121	5859	6673	5464	7244	5852	6929	5608
					2050				
Renewable electricity		283	436	470	506	282	253	270	548
Biomass and waste		1142	723	1144	839	1078	532	1041	615
Natural gas		1173	1053	1114	1288	1262	1283	1409	1119
Oil		385	597	405	197	439	971	852	492
Refined oil		128	146	119	339	101	20	0	224
Coal		1876	685	1797	315	1854	991	2169	514
Nuclear electricity		769	592	641	507	776	505	609	596
Imported electricity		103	103	103	103	103	103	103	103
Total		5859	4367	5794	4093	5896	4659	6452	4248

Finally looking at the 'best case' scenario (7), the availability of substantial emissions purchases means that only a 25 per cent reduction in primary energy is required from 2000–2050 in meeting an 80 per cent CO_2 reduction from the UK energy sector. Furthermore, despite higher fossil energy prices, 69 per cent of all primary energy still comes from natural gas, coal and oil. In a 'worst case' scenario (8), the model is striving to meet CO_2 emission constraints and resilience constraints and does not have access to imported biomass. Faced with this array of constraints, the freedom to make substantial changes is limited although intermediate demand reductions and an overall 100PJ reduction in biomass are seen. Under such conditions, the costs of decarbonizing the energy sector rise still further.

Figure 11.1 expands this discussion of long-term (to 2050) impacts by focusing on final energy. Similar to LC and LCR scenarios, electricity is a key low carbon vector in all scenarios, increasing in absolute size even as the UK final energy demand reduces by 20–40 per cent. Under resilience constraints, final direct heat demands are retained in buildings sectors, although this is reduced under emission purchase of biomass limitations. High fuel price

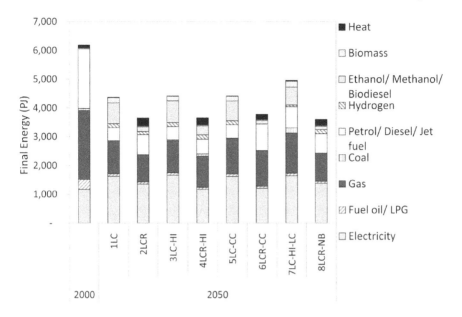

Figure 11.1 *Final energy (petajoules, PJ)*

variants make less difference in final energy (compared to primary energy) although still show modest gains in bio-fuels and hydrogen uses in selected transport modes. The impact of emissions purchases eases the most stringent demand reduction requirements and the most expensive zero carbon electricity generation, whilst allowing more (high efficiency) natural gas under overall portfolios. Finally the impact of no biomass imports, in a worst case scenario (8), sees a 680PJ reduction in bio-related final energy in the transport and buildings sectors and a renewed emphasis on an expanded low carbon electricity sector.

Turning to sectoral CO_2 emissions, Figure 11.2 details end-use sectoral emissions in 2050, with electricity and hydrogen emissions allocated to end-use sectors. The figure of 118MtCO$_2$ represents an 80 per cent reduction from 1990 levels of 592MtCO$_2$. Residual emissions across the scenarios are concentrated in industrial and transport subsectors where emissions are most expensive to reduce and/or few technological options are available in the model. Additional resilience constraints (LCR scenarios) generally place more CO_2 reduction emphasis on the transport sector as zero carbon electricity portfolios are more expensive. Access to international emission purchases facilitates higher emissions across all UK end-use sectors.

Figure 11.3 details the trajectory of domestic CO_2 emissions as well as the magnitude of emissions credit purchases as the UK meets its –80 per cent CO_2 emissions target by 2050. Five scenarios have no access to international credits and their emission trajectory (black dashed line in Figure 11.3) is the extrapolated path from near-term UK carbon budgets through to 2050. Emissions purchases in the remaining three scenarios are generally only a factor in later periods when the marginal costs of domestic action rise steeply. The exception to this is in 2015, when international credits can compensate when existing (and high carbon) energy capital has not yet been replaced. Emissions credits are more important when the costs of domestic action become very high (e.g. in 2050 in LCR-CC) or when emission purchases are very cost effective

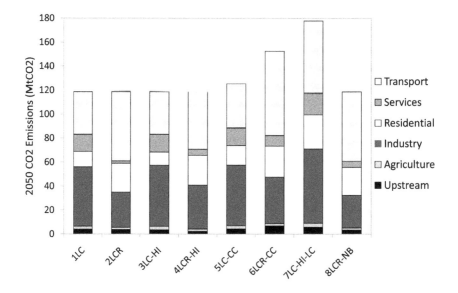

Figure 11.2 *UK sectoral CO₂ emissions (2050)*

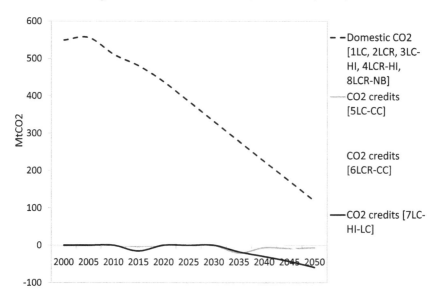

Figure 11.3 *UK CO₂ emissions and use of emission purchases*

(LC-HI-LC). In this latter case, by 2050 the assumed policy limit of 12.5 per cent of required reduction is reached – this is equivalent to one-third of total residual domestic CO_2 emissions, as the combination of domestic emissions minus international purchases meets the 80 per cent reduction target of $118MtCO_2$.

Technology pathways

Electricity generation

Looking at specific low carbon pathways, Table 11.5 details the electricity generation (in TWh) and installed capacity (in GW) for the eight global energy scenarios in the year 2050. A key question is how electricity can fulfil a role as an enabling low carbon pathway, with resultant impact on end-use sectors. Relative to a year 2000 generation of 358TWh, despite substantial reductions in the size of a decarbonized energy system (Figure 11.1), electricity is a low carbon energy vector in the LC scenarios, with electricity in LC growing 56 per cent by 2050. Due to reductions in final energy demand and constraints on electricity generation share by technology, the role of low carbon electricity is less in LCR but still 16 per cent above year 2000 levels.

One surprise is that nuclear does not play a bigger role in higher fossil price scenarios (3 and 4). This is largely due to the fact that nuclear is not cost competitive until later vintages (especially 2030) but the fossil cost penalty on competing fossil based technologies is greatest in intermediate years. A further limit on nuclear is its inflexibility if gas electric plants are too high-cost to be available beyond pure peaking operation. Fossil based electricity technologies do see further declines in LCR-HI, but in all high fossil price cases, the role of wind (cost competitive in the near term) is boosted. In terms of capacity margins, natural gas plants remain a key peaking plant in LC scenarios, with a more balanced installed portfolio in LCR scenarios.

In scenarios with increasing levels of emission credit purchases (5, 6 and 7), the flexibility of emission purchases alleviates some need for low carbon electricity and for more expensive tranches of wind in particular. In the 'worst case' scenario (8), the constraint on biomass imports plays a negligible role in power generation given the predominance of bio resource use in transport (bio-fuels) as well as direct use in buildings.

Transport

Turning to transport, Figure 11.4 details the final energy use across all transport modes in 2035 and in 2050. In the intermediate period, a cost optimal transport future even without a carbon price would see a near-term switch to hybrid vehicles as well as the impact of a modest Renewable Fuel Transport Obligation (RTFO). Thus in 2035 despite increases in the energy service demands (billion vehicle kilometres), transport final energy remains relatively flat under LC-derived scenarios. Under LCR-derived scenarios, the energy intensity constraints see transport final energy reductions even in intermediate periods. Hence the mid-term additional impact of higher oil prices or

Table 11.5 *Electricity generation (TWh) and capacity (GW) in 2050*

[2050]	1 LC		2 LCR		3 LC-HI		4 LCR-HI		5 LC-CC		6 LCR-CC		7 LC-HI-LC		8 LCR-NB	
	TWh	GW	TWh	GW	TWh	GW	TWh	GW	TWh	GW	TWh	GW	TWh	GW	TWh	GW
Coal	–	–	–	8	–	3	–	17	–	–	–	8	–	3	–	11
Coal CCS	225	31	83	16	221	31	26	4	222	31	117	16	233	32	60	10
Gas	–	26	–	8	–	29	–	2	–	24	–	8	–	16	–	8
Gas CCS	–	–	–	–	–	–	–	–	–	–	–	–	–	–	–	–
Nuclear	214	29	164	23	178	24	141	20	216	30	140	19	169	23	166	23
Oil	–	–	–	–	–	–	–	–	–	–	–	–	–	–	–	–
Hydro	9	3	9	3	9	3	9	3	9	3	5	1	9	3	9	3
Wind	52	18	95	31	104	33	114	36	52	18	48	16	49	16	126	40
Bio and waste	11	4	17	3	11	3	16	3	11	4	14	3	11	3	13	3
Imports	29	8	29	11	29	9	29	10	29	9	29	9	29	11	29	12
Marine	18	5	18	5	18	5	18	5	18	5	18	5	18	5	18	5
Storage	–	1	–	1	–	1	–	1	–	1	–	1	–	1	–	1
Total	557	126	414	108	569	143	352	102	555	124	371	87	517	114	420	117

Note: 1TWh = 3.6PJ

In year 2000, total electricity generation was 358TWh and electricity capacity was 84GW

emission purchases is negligible. In the longer term to 2050, a range of transport modes switch to alternate fuels and drive-trains. Cars switch to ethanol and plug-in hybrids, short distance buses move to electric vehicles, with trains also switching completely to electric carriers. The LGV (light goods vehicle) transition is to biodiesel hybrids whilst HGV (heavy goods vehicles) can be biodiesel or hydrogen (benefiting from a limited refuelling network). Under the more stringent LCR runs, (domestic) aviation switches to bio-kerosene, with this more esoteric technology shift illustrating the higher costs of this scenario family.

The impacts of various global uncertainties depend on the model's ability to make changes in an already very optimized sector. For higher fossil prices, the LC runs see changes in other end-use sectors, but under the LCR run, the cost optimal solution for the energy system as a whole is to reduce transport oil use and substitute to ethanol. Similarly under LCR, access to carbon credits allows conventional transport fuels to retain market share. Across carbon credit scenarios, one key tipping point is in HGV hydrogen vehicles, with additional flexibility in meeting a carbon target meaning this new (and hence less cost effective) infrastructure is not required. Finally, in the no-biomass import case (8), higher priced domestic biomass resources ensure a higher retention of fossil fuels, as bio-fuel use drops dramatically.

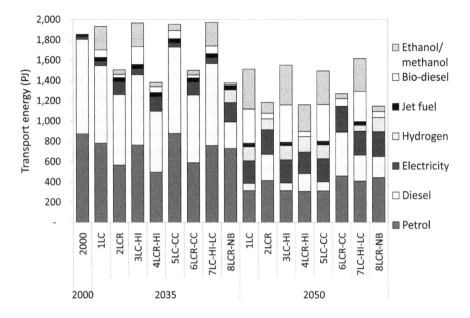

Figure 11.4 *Transport final energy demand (2035, 2050)*

Buildings sectors

Across all eight scenarios, the service (commercial) sector is highly optimized by 2050, through demand reduction and shifts to electricity as a low carbon vector. In this MARKAL-MED model version, the residential sector (which is soft-linked with prior assumptions in the UKDCM model, as discussed in Chapter 4) shows more variation. Under LC scenarios, the residential sector relies on low carbon electricity via electric boiler and heat pumps, while in the LCR scenarios this is supplemented by the retention of district heating as portfolios of low carbon electricity are now more expensive. In all scenarios demand reductions are important but the resilience energy intensity constraint forces an average service demand reduction of around 25 per cent in LCR scenarios compared to only 10 per cent in the LC cases. The variation in individual demand constraints is driven by the price elasticity and by the technology and fuel substitution options. These deep LCR energy service demand reductions are price driven, but are still less than those deriving from lifestyle change in the Chapter 9 scenarios. In terms of other drivers, high resource prices further increase demand reductions, as well as further uptake of technical conservation options. This additional impact is modest as fuel costs play a smaller share in final energy (especially after secondary conversion to electricity, bio-fuels and hydrogen). A similar magnitude but opposite impact is seen in the runs with additional flexibility because of the availability of international emission credits.

Economic impacts

Welfare costs arising from losses in the sum of producer and consumer surplus (Figure 11.5) illustrate the overall costs of meeting the −80 per cent CO_2 target, whilst CO_2 prices (Figure 11.6) are the marginal values, and indicate what was required for the last unit of emissions reduction. All costs in this discussion are in £(2000), with a conversion factor of £(2000) 1 = £(2009) 1.23.

The additional cost of the resilience constraints is considerable, both in increasing the 2050 annual welfare loss from around £36 billion (from a reference scenario (REF) as the baseline) to around £56 billion. In addition, comparable welfare losses are incurred up to 10 years earlier. The highest cost case is the 'worst case' scenario (8) with the additional constraints on resilience, high fossil prices and restrictions on imported biomass imposing welfare costs of around £15 billion as early as 2010–2015, with 2050 costs at £58 billion. Conversely the combination of no resilience constraints, high fossil prices and low cost carbon credits gives the 'best case' scenario (7) a cost of only £26 billion in 2050. Thus the overall cost difference between best and worst cases is greater than a factor of two.

Marginal CO_2 prices show a wide variation (factor of four) between scenarios illustrating the sensitivity of prices to moderate changes in the availability of cost-effective mitigation options. The highest CO_2 prices are £320/tCO_2 in LCR or £365/tCO_2 in LCR-NB (worst case scenario). These values are substantially moderated by high fossil prices and by the removal of resilience constraints. Furthermore, in those cases with international emission credits, this acts as a backstop price and limits the marginal costs of meeting the −80 per cent emission target to £161/tCO_2 in the central price emission credit and to £90/tCO_2 in the low cost emission credit scenarios.

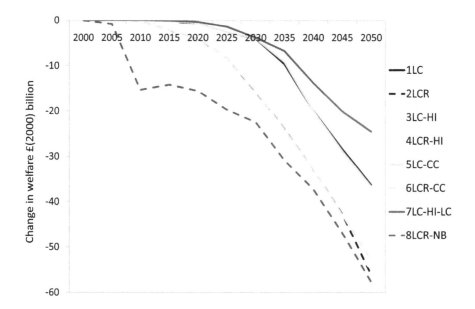

Figure 11.5 *Welfare costs, £(2000) billion*

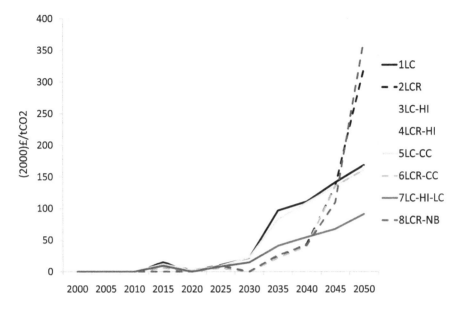

Figure 11.6 *Marginal CO₂ prices, £(2000)/tCO₂*

Conclusions

In the context of long-term UK low carbon energy pathways, the users of scenario exercises such as in this book primarily are UK policy-makers, and a key set of *external* drivers are international in nature. This chapter focuses on three main global drivers on the UK energy system – fossil fuel prices, sustainable biomass imports, and the availability of CO_2 emission credits. These are listed as key international uncertainties on UK decarbonization pathways in a range of academic publications and policy documents.

A set of eight illustrative scenarios focused on global driven uncertainties in meeting 80 per cent CO_2 reduction scenarios, with low carbon (LC) and low carbon resilient (LCR) as core runs. Possible increases in the prices of internationally traded fossil fuels (coal, oil, oil products, natural gas), the availability and cost of greenhouse gas (GHG) credits to meet domestic emissions targets, and the availability of sustainable biomass imports are all highly uncertain in the long term (to 2050). Hence a set of exploratory assumptions investigated these interacting uncertainties. It is particularly important to consider the consistency of scenarios. Notably 'best and worst' cases for the UK can be argued to mirror opposite ambition levels at a global level. It is an open question as to whether in the long term the stringency of UK climate mitigation could or would diverge from global ambition levels.

The international uncertainties on fossil fuel prices, emissions credit purchases, and sustainable biomass act upon the two quite different core –80 per cent CO_2 cases of LC and LCR. The latter's additional resilience constraints see lower final energy demand, less reliance on the key low carbon energy vector of electricity, and a greater reliance of portfolios of energy options in the transport and heat sector. The variation from international drivers is seen most in interactions between decarbonization in the electricity and transport sectors, with residential buildings and heat markets as a secondary impact.

The impact of higher fossil fuel prices (scenarios 3 and 4) is generally to moderate fossil energy use and emissions (especially towards 2050 as the CO_2 constraint tightens). However, there is a complex trade-off both between *which* fossil fuel price changes have the most effect and *when* this differential has the most effect. For example under LC, intermediate coal reductions are most notable, while under LCR long-term oil use sees the largest reductions. One surprising finding is that although nuclear electricity is implemented earlier, it does not play a bigger long-term role in higher fossil price cases. This is due to the fact that the fossil cost penalty on competing fossil based technologies is greatest in intermediate years, and to portfolio balancing of base-load plant in the UK electricity network.

The impact of emissions credit purchases (scenarios 5 and 6) is generally only seen in later years where the price of CO_2 emissions credits is lower than the marginal costs of domestic reduction options. The added flexibility of international emission reductions (up to 600PJ in 2050) allows substantial retention of gas, oil and coal use in 2050, alongside reductions in renewable electricity, less reliance on bio-fuels and hydrogen in transport modes, and less need for energy service demand reductions. In the most optimistic cases, by 2050 the assumed policy limit of 12.5 per cent of required reduction is met via emissions credits – this is equivalent to one-third of total residual domestic CO_2 emissions.

Considering the 'best case' versus 'worst case' (scenarios 7 and 8) gives an overall welfare cost difference of a factor of two – ranging from £26 billion to £58 billion in 2050. The additional constraint on imported biomass imposes welfare costs of around £15 billion as early as 2010–2015, reflecting the added lack of flexibility in near-term emissions mitigation across all sectors. Marginal CO_2 prices illustrate the sensitivity of prices to moderate changes in the quantities of cost-effective mitigation options, and give a CO_2 price variation of a factor of four by 2050 – from £365/tCO_2 to £90/tCO_2. The marginal costs of CO_2 are reduced both by higher fossil prices and by the availability and the effective backstop price of international emission credits.

This exploratory set of scenarios on international drivers – analysed via a national-level energy-economic model – has illustrated the importance of these external assumptions to UK energy system evolution, technology pathways and economic implications. The UK's direct actions have a very limited impact on these overall drivers (notwithstanding the UK's potential role via mitigation negotiations and as an exemplar of a low carbon economy). Therefore it is essential that UK based decision-makers are cognizant of international drivers and iteratively assess energy policy and successive CO_2 emissions budget periods accordingly.

References

Anandarajah G., Strachan N., Ekins P., Kannan, R. and Hughes, N. (2009) *Pathways to a Low Carbon Economy: Energy systems modelling*, UKERC Energy 2050 Research Report 1, UK Energy Research Centre, London, www.ukerc.ac.uk

Berkhout, F., Skea, J. and Eames, M. (1999) *Environmental Futures*, Office of Science and Technology/Foresight Programme, London

CCC (Committee on Climate Change) (2008) *Building a low-carbon economy – the UK's contribution to tackling climate change*, Committee on Climate Change, London, www.theccc.org.uk

CCC (2009) *Meeting Carbon Budgets – the need for a step change*, First annual progress report, Committee on Climate Change, London, www.theccc.org.uk

DECC (2008a) *Updated Energy Projections* (UEP), Department of Energy and Climate Change, London, November 2008, www.decc.gov.uk/en/content/cms/statistics/projections/projections.aspx

DECC (2008b) *Climate Change Act*, Department of Energy and Climate Change, London, November 2008, www.decc.gov.uk/en/content/cms/legislation/cc_act_08/cc_act_08.aspx

DECC (2009) *UK Low Carbon Transition Plan*, Department of Energy and Climate Change, London, www.decc.gov.uk/en/content/cms/publications/lc_trans_plan

DUKES (2005) *Digest of UK Energy Statistics*, Department of Trade and Industry, London

EIA (2007) *International Energy Outlook 2007*, US Energy Information Administration, www.eia.doe.gov/oiaf/ieo/index.html

Gallagher, E. (2008) *The Gallagher Review of the indirect effects of biofuels production*, Renewables Fuels Agency, www.renewablefuelsagency.gov.uk

G8 Communiqué (2007) *Chair's Summary*, G8 Heiligendamm Summit, 8 June 2007, www.g-8.de/Webs/G8/EN/G8Summit/SummitDocuments/summit-documents.html

Hughes, N., Mers, J. and Strachan, N. (2009) *Review and analysis of UK and international low carbon energy scenarios*, EON/EPSRC Transition Pathways working paper, www.lowcarbonpathways.org.uk

Kannan, R., Strachan, N., Pye, S. and Balta-Ozkan, N. (2007) *UK MARKAL model documentation*, www.ukerc.ac.uk

Klepper, G. and Peterson, S. (2006) 'Marginal abatement cost curves in general equilibrium: the influence of world energy prices', *Resource and Energy Economics*, vol 28, pp1–23

IEA (2007) *World Energy Outlook 2007*, International Energy Agency, Paris

IEA (2008) *World Energy Outlook 2008*, International Energy Agency, Paris

Saddler, H., Diesendorf, M. and Denniss, R. (2007) 'Clean energy scenarios for Australia', *Energy Policy*, vol 35 (2), pp1245–1256

Silberglitt, R., Hove, A. and Shulman, P. (2003) 'Analysis of US energy scenarios: Meta-scenarios, pathways, and policy implications', *Technological Forecasting and Social Change*, vol 70 (4), pp297–315

Strachan, N. and Kannan, R. (2008) 'Hybrid modelling of long-term carbon reduction scenarios for the UK', *Energy Economics*, vol 30 (6), pp2947–2963

Strachan, N., Kannan, R. and Pye, S. (2008a) *Scenarios and Sensitivities on Long-term UK Carbon Reductions using the UK MARKAL and MARKAL-Macro Energy System Models*, UKERC Research Report 2, www.ukerc.ac.uk

Strachan, N., Pye, S. and Hughes, N. (2008b) 'The role of international drivers on UK scenarios of a low carbon society', *Climate Policy*, vol 8, ppS125–S139

van Vuuren, D., Fengqi, Z., de Vries, B., Kejun, J., Graveland, C. and Yun, L. (2003) 'Energy and emission scenarios for China in the 21st century – exploration of baseline development and mitigation options', *Energy Policy*, vol 31 (4), pp369–387

van Vuuren, D., Bellevrat, E., Kitous, A. and Isaac, M. (2010) 'Bio-Energy Use and Low Stabilization Scenarios', *The Energy Journal* special issue on *The Economics of Low Stabilization*, pp193–222

Weyant, J. (2004) 'Introduction and Overview: EMF 19 study on technology and climate change policy', *Energy Economics*, vol 26, pp501–515

WWF International (2007) *A first estimate of the global supply potential for bio-energy,* A briefing study commissioned by WWF International, Gland, Switzerland

12
Putting It All Together:
Implications for Policy and Action

Paul Ekins, Mark Winskel and Jim Skea

Introduction

Energy systems are inseparable from the societies that construct them and rely on them, and the production, distribution and use of energy involve a series of interconnected technical and social components. Changing any one of the component parts will have consequences for other parts of the system. Energy policy in the UK and other major industrial countries implies a remaking of the energy system over the next decade and beyond. The prospects for change are both complex and potentially confusing, and energy system change needs to be analysed in ways which recognize and take account of the diversity and interconnectedness of the different components of the system.

This diversity and complexity also mean that no single pathway can be defined that best responds to the policy and other pressures now being exerted on the system. Rather, there are multiple possible responses, and as this book has shown, many possible pathways consistent with meeting policy goals. Fundamentally, then, energy system change is a matter of priority setting though social, political and organizational deliberation and choice.

This book has attempted to explore the multiple dimensions of radical energy system change for one major developed country, the UK. The purpose of this final chapter is to pull together the key themes that emerge from the work, systematically compare the outcomes of the various scenarios described throughout the book and draw, at a fairly high level, some conclusions about policy needs. The chapter begins by setting out the nature of current challenges to the energy system before reflecting on the historic context, back to 1970, within which the UK energy system has operated. Looking back 40 years, a period for which many of the major energy trends and issues were identified in Chapters 2 and 3, gives insight into the nature and possible rates of system change that might be envisaged looking forward to 2050. There is then a systematic comparison of all of the scenarios covered in the book in terms of energy demand and supply, CO_2 emissions and economic variables. The chapter then discusses overall conclusions around a set of key themes: multiple pathways to a low carbon future; energy security and system resilience; supply side technologies and the role of R&D; energy demand and lifestyle change; wider environmental considerations; and the wider international context within which a country like the UK must operate. The chapter concludes with some overall remarks about energy system change and the insights that research can provide.

Current energy system challenges

The challenges now facing the energy system – environmental degradation, security of supply and resource depletion – echo those from previous periods of energy crisis. However, today's primary environmental concern – climate change – is so pervasive and deep-rooted that there can be no single technical or regulatory fix. In addition, the economic, organizational and political backdrop for change has changed profoundly. The forces of liberalization and globalization have remade energy systems; instead of governance by a small number of national and/or corporate actors, energy systems now reflect multiple, fragmented influences, many of them beyond national borders. Governing system change is a much more heterogeneous and multi-faceted problem than ever before, involving combinations of local, national and international actors.

This book identifies and elaborates two powerful drivers of UK energy system change: decarbonization and system resilience. While decarbonization can be defined relatively straightforwardly, energy system resilience is a more complex concept. There is a consequent risk that energy analysis is on shakier ground with respect to resilience and security concerns and, equally, that weaker policy may result.

The *Reference* 'business as usual' scenario described in Chapter 4 demonstrates that, without purposeful policy action, the UK's decarbonization and resilience goals will not be met. The *Reference* scenario describes a future energy system shaped essentially by competitive economic forces resulting in conservative, incremental changes. Energy demand stays at around current levels out to 2050 and coal-fired generation, relying heavily on imports and without CCS technology, dominates the electricity system. However, there are some less expected changes driven by economic forces, notably the rise of electric vehicles. Overall, though, this scenario confirms the disparity between unchecked market-based decision-making, and current policy ambitions for decarbonization and system resilience.

Although this book focuses on the UK, climate change and energy security/resilience are both fundamentally international problems. Adequately addressing them requires international agreements and action. At the same time, national and regional character still matters a great deal in energy systems, and the different local, regional and national resources, infrastructures and demand patterns mean that responses to climate and security concerns will be shaped by local context. The UK situation presents particular challenges: the years of energy self-sufficiency lulled generations of UK energy policy-makers into a false sense of security, and market liberalization was pursued more aggressively in the UK than it was elsewhere. To the extent that policy solutions for the 'market failures' of climate change and energy insecurity require a retreat away from market-based decision-making, in favour of a return to politically-led long-term planning, UK energy policy has further to travel than most. The chapter now moves on to look at the historic context that has helped to shape the energy challenge in the UK.

The historic context

How relevant is the past as a guide to energy system futures? Tracing the historical development of the UK energy system, Chapter 2 reveals a pattern of accumulating trends punctuated by episodes of discontinuity and disruption. Long-running trends provide a degree of predictability for energy system change, but discontinuities and disruptions mean that it is also essential to allow for unpredictability in thinking through energy futures.

The history of the UK energy system provides grounds for both optimistic and pessimistic readings of future prospects. There has been a long-term decline in energy demand and CO_2 emissions, and an even more substantial decline in the carbon intensity of the UK economy. However, a major contributor to the decline in carbon intensity has been de-industrialization and the consequent 'offshoring' of production-related emissions through globalized supply chains.

At the same time, there are also indications of the deep-rooted challenge of system change: long-term rises in energy demand from road transport and aviation, and, until recently, households, an embedded and longstanding reliance on oil-based fuels for transport, and an increasing reliance on natural gas. Even so, there is a high-level message that relationships between economy, energy and emissions are malleable over relatively short timescales. This provides the basis for some optimism. In addition, the last few years have shown some encouraging changes, such as declining household energy demand and a now rapidly rising share of UK power generation from renewables.

Until recently, the UK enjoyed an enviable position in terms of energy supply resource security and relative self-sufficiency. Reflecting, this, the UK energy system has often been seen as a largely independent system with only minor links to the wider European energy system. However, this overlooks the international trade flows which have become increasingly important over the past 20 years, and can only grow more important over time. The likelihood of increasing dependence on intermittent renewables also suggests that greater electrical interconnection with continental Europe and other North Sea countries may be desirable. This means shifting the perspective from a largely independent transmission and distribution infrastructure, to a perspective that sees the UK energy system within the context of wider European systems and system change.

Chapters 2 and 3 showed how energy systems and energy policies co-evolve. For example, increasing import dependence has recently provoked policy interest in fuel storage. More generally, the emergence of supply security and especially climate change concerns have driven the re-emergence of energy policy and planning. In terms of policy style, these have led to a change toward more interventionist and prescriptive policy-making. A market-oriented energy policy, with minimal policy intervention, is no longer widely believed to be able to deliver change in the required direction and at a sufficient scale. The last decade has seen multiple policy and regulatory initiatives and market interventions across generation, transmission and supply. Rather than any radical remaking of the energy system, multiple layers of reforms have been imposed in a piecemeal, incremental fashion.

These changes, and their associated costs, have been rather inconspicuous to the energy consumer. As the magnitude of interventions and their costs have increased, this style of 'stealth' intervention is becoming less tenable. As the scale of policy ambitions and interventions increase further in the future, and extend more widely across society and lifestyles, there may be a need for a more explicit 'social contract' for energy system change between policy-makers and the public.

Comparing the scenarios

Before exploring the key themes emerging from the book, this section systematically compares, out to 2050, the major scenarios summarized in Table 4.2 and reported in earlier chapters. Table 12.1 compares energy and carbon outcomes. Table 12.2 focuses on economic variables.

In respect of CO_2 reduction (column 3 of Table 12.1), the *Reference* scenario, which seeks to take account of UK Government carbon reduction policies up to 2007, produces only 2 per cent reduction from the 1990 level. Of the other non-carbon constrained scenarios, the *Reference Lifestyle* (REF-LS) and *Resilient* (R) scenarios produce 31 per cent and 52 per cent CO_2 emission reductions respectively, indicating the carbon-reducing power of lifestyle change and making energy system resilience a policy priority. However, this is still well short of the 80 per cent reduction target, so that stringent additional carbon-reducing policies are clearly necessary if the target is to be achieved. An 80 per cent carbon reduction will not be achieved as a by-product of other goals. It must be explicitly pursued.

In respect of cumulative CO_2 emissions 2000–2050 (column 4), the *Least Cost* (LC-LC) and *Socially Optimal Least Cost* (LC-SO) scenarios show that the same cumulative emissions ($19GtCO_2$) can be achieved over different time-paths. However, the *Least Cost* scenario, which delays carbon reduction, has to achieve 89 per cent decarbonization by 2050 in order to achieve this.

In respect of energy demand, columns 5 and 6 show for each scenario the difference in 2050 from the energy demand in 2000. Thus in the *Reference* scenario in 2050 primary energy demand is 4 per cent below, and final energy demand is 4 percent above, the year 2000 level. It can be seen that all of the scenarios which achieve an 80 per cent reduction or more by 2050 reduce primary energy demand by 2050 by 31–51 per cent below the year 2000 level (27–47 per cent below the *Reference* scenario level). Final energy demand is reduced by 25–42 per cent below the year 2000 level (29–46 per cent below the *Reference* scenario level). These reductions occur through the combined impact of the carbon price, and efficiency and conservation measures. In the *Low Carbon Lifestyle* scenario non-price related behaviour change also plays a role. At present, the policies in place neither establish the necessary carbon price across the economy, nor stimulate the necessary uptake of efficiency and conservation measures.

Achieving the 80 per cent target is also associated with either considerable increases in electricity consumption (typically around 50 per cent more than in the year 2000), considerable reductions in residential energy demand (typically around 50 per cent less than in the year 2000), or both (column 6). *Low Carbon Resilient*, *DREAD* and *Low Carbon Lifestyle* have the smallest increases in electricity demand, for different reasons. In *Low Carbon Resilient*, decarbonization is largely attributable to reducing energy demand because of the energy intensity constraint. *DREAD* uses less electricity in transport. *Low Carbon Lifestyle* has among the highest reductions in residential energy demand (57 per cent). It also uses less transport fuel and, like *DREAD*, uses more bio-fuels. The *Super-ambition* scenario (*LC-90*), *LC Acctech*, and *ECO* all have increases in electricity demand of greater than 70 per cent compared to the *Reference* scenario: in *LC-90* this is because of the high level of decarbonization required (it has the lowest cumulative CO_2 emissions of all the major carbon reduction scenarios); in *LC Acctech* because technology acceleration makes low carbon electricity technologies relatively cheap; and in *ECO* because there are no limits on nuclear and coal CCS.

Of the diversity indicators (column 7) only *Reference Lifestyle* and *Socially Optimal Least Cost Path* breach the 40 per cent maximum share of any primary fuel, with both using too much coal for power generation. Far more scenarios breach the 40 per cent maximum share in generation, with the *Reference, Reference Lifestyle* and *High Prices Cheap Credits* scenarios having too much coal-fired generation, and *DREAD* too much wind because of social concerns about the other major low carbon electricity sources, nuclear and CCS.

Column 8 shows the extent of decarbonization of the electricity system by 2050. The CO_2 intensity of the least decarbonized of the major carbon reduction scenarios (*Socially Optimal Least Cost Path*) is only 7 per cent that of the *Reference* scenario by 2050. *DREAD, ECO, NIMBY* and *Super-ambition (LC-90)* are almost completely decarbonized. Most of those with relatively little power sector decarbonization tend to use more bio-fuels (*Early Action, Least Cost Path* and *Low Carbon Lifestyle*), allowing the power sector to take more of the carbon budget. The *Socially Optimal Least Cost Path* scenario uniquely goes for a mixture of hydrogen in transport and demand reduction.

The final two columns relate to EU targets for 2020. In respect of the first, it can be seen from column 9 that in no scenario does the UK achieve the mandatory 15 per cent of renewables in final energy demand by 2020. The maximum achieved is 7 per cent, when all seven decarbonization technologies are accelerated, but CCS is delayed. Column 10 shows that the maximum reduction in primary energy demand by 2020 is 11 per cent in the *Socially Optimal Least Cost Path* scenario, 10–11 per cent in the two core resilience scenarios,

Table 12.1 Comparative table of Energy 2050 scenarios: carbon and energy

Scenario	Scenario name/Major characteristics	Carbon reduction targets/2050 reductions if no targets (from 1990 level)	Cumulative emissions GTCO$_2$ (2000–2050)/2050 emissions MTCO$_2$	Primary energy demand/Final energy demand %	Total electricity demand/Residential (all fuels) demand %	Max primary share/Max genmix % at 2050	CO$_2$ intensity of power g/kWh at 2050	Share of renewables in final energy demand at 2020 %	Primary energy demand reduction at 2020 (% from 2020 REF)
(1)	(2)	(3)	(4)	(5)	(6)	(7)	(8)	(9)	(10)
CORE SCENARIOS				Difference from 2000 baseline at year 2050					
REF	Reference	2%	30/583.0	−4/+4	+24/−2	38/81	592	5	0
LC	Low Carbon	26% by 2020 80% by 2050	20/118.5	−32/−29	+56/−55	32/41	31	5	1.3
R	Resilient	52%	23/285.3	−44/−38	+1/−50	31/40	352	5	10.6
LCR	Low Carbon Resilient	26% by 2020 80% by 2050	20/118.5	−49/−40	+17/−50	24/40	15	5	10.2
CARBON REDUCTION									
LC-40	Faint-heart	15% by 2020 40% by 2050	26/355.4	−5/0	+25/−6	39/74	110	5	1.6
LC-60	Medium carbon	26% by 2020 60% by 2050	22/236.9	−19/−14	+33/−19	34/57	44	5	3.5
LC-90	Super-ambition	32% by 2020 90% by 2050	18/59.0	−41/−28	+73/−58	34/45	8	5	6.1
LC-EA	Early action	32% by 2020 80% by 2050	19/118.5	−32/−30	+56/−56	33/43	31	5	3.8
LC-LC	Least-cost path	Same cumulative as LC-EA 89%	19/67.1	−37/−30	+60/−58	30/49	22	5	3.3
LC-SO	Socially optimal least-cost path	Same cumulative as LC-EA 70%	19/178.6	−29/−31	+59/−58	43/54	41	6	11.3

Scenario	Scenario name/ Major characteristics	Carbon reduction targets/2050 reductions if no targets (from 1990 level)	Cumulative emissions GTCO2 (2000– 2050)/2050 emissions MTCO2	Primary energy demand/ Final energy demand %	Total electricity demand/ Residential (all fuels) demand %	Max primary share/Max genmix % at 2050	CO2 intensity of power g/kWh at 2050	Share of renewables in final energy demand at 2020 %	Primary energy demand reduction at 2020 (% from 2020 REF)
TECHNOLOGY ACCELERATION									
LC Renew	All four renewables	26% by 2020 80% by 2050	20/118.5	−36/−26	+60/−48	24/41	15	6	0.3
LC Acctech	All seven technologies	26% by 2020 80% by 2050	20/118.5	−31/−25	+89/−28	26/34	20	6	−0.2
LC Acctech (late CCS)	All seven technologies, CCS only available from 2030	26% by 2020 80% by 2050	20/118.5	−34/−25	+82/−30	29/38	14	7	6
ENVIRONMENTAL SENSITIVITIES									
DREAD	No new nuclear, CCS, hydrogen	26% by 2020 80% by 2050	20/118.5	−44/−27	+17/−47	32/84	4	6	7.0
ECO	Constraints on wind and marine power, imported biomass, higher fossil fuel prices	26% by 2020 80% by 2050	20/118.5	−45/−34	+78/−58	23/48	8	4	8.2
NIMBY	Constraints on new renewables, wind and CCS	26% by 2020 80% by 2050	20/118.5	−44/−32	+51/−57	25/41	9	5	1.5
ENERGY LIFESTYLES									
REF-LS	Reference lifestyle	31%	23/407.1	−33/−29	+2/−52	46/77	571	6	21
LC-LS	Low carbon lifestyle	26% by 2020 80% by 2050	20/118.5	−44/−39	+15/−57	31/43	32	6	21
GLOBAL UNCERTAINTIES									
LC-HI	High fossil prices	26% by 2020 80% by 2050	20/118.5	−33/−29	+42/−55	31/39	28	6	5.9
LC-HI-LC	High prices, cheap credits	26% by 2020 80% by 2050	20/118.5	−25/−20	+40/−45	34/45	33	6	5.6
LCR-NB	No imported biomass	26% by 2020 80% by 2050	20/118.5	−51/−42	+18/−50	26/39	11	6	14.5

Table 12.2 *Comparative table of Energy 2050 scenarios: economic variables*

Scenario	Scenario name	Carbon reduction targets/2050 reductions if no targets (from 1990 level)	Marginal cost of carbon abatement (carbon price) at 2050 £(2000)/tCO$_2$	Energy system costs £bn (change from REF at 2050)[1]	Welfare costs £bn (change from REF at 2050)[1]
(1)	(2)	(3)	(4)	(5)	(6)
CORE SCENARIOS					
REF	Reference	2%	0	0	0
LC	Low Carbon	26% by 2020 80% by 2050	169	17	38
R	Resilient	52%	0	−11	49
LCR	Low Carbon Resilient	26% by 2020 80% by 2050	20	2	59
CARBON REDUCTION					
LC-40	Faint-heart	15% by 2020 40% by 2050	20	3	5
LC-60	Low carbon	26% by 2020 60% by 2050	85	8	20
LC-90	Super ambition	32% by 2020 90% by 2050	299	30	52
LC-EA	Early action	32% by 2020 80% by 2050	173	17	37
LC-LC	Least-cost path	Same cumulative as LC-EA 89%	360	23	48
LC-SO	Socially optimal least-cost path	Same cumulative as LC-EA 70%	66	2	7
TECHNOLOGY ACCELERATION					
LC Renew	All four renewables	26% by 2020 80% by 2050	148	19	34
LC Acctech	All seven technologies	26% by 2020 80% by 2050	131	18	28
LC Acctech (late CCS)	All seven technologies, CCS only available from 2030	26% by 2020 80% by 2050	127	18	28
ENVIRONMENTAL SENSITIVITIES					
DREAD	No new nuclear, CCS, hydrogen	26% by 2020 80% by 2050	190	20	43
ECO	Constraints on wind and marine power, imported biomass, higher fossil fuel prices	26% by 2020 80% by 2050	234	28	53
NIMBY	Constraints on new renewables, wind and CCS	26% by 2020 80% by 2050	187	20	42

Scenario	Scenario name	Carbon reduction targets/2050 reductions if no targets (from 1990 level)	Marginal cost of carbon abatement (carbon price) at 2050 £(2000)/tCO₂	Energy system costs £bn (change from REF at 2050)[1]	Welfare costs £bn (change from REF at 2050)[1]
ENERGY LIFESTYLES					
REF-LS	Reference lifestyle	31%	0	−89[2]	–[3]
LC-LS	Low carbon lifestyle	26% by 2020 80% by 2050	163	−94[2]	–[3]
GLOBAL UNCERTAINTIES					
LC-HI	High fossil prices	26% by 2020 80% by 2050	166	31	30
LC-HI-LC	High prices, cheap credits	26% by 2020 80% by 2050	91	31	26
LCR-NB	No imported biomass	26% by 2020 80% by 2050	364	1	59

Notes: 1) undiscounted costs are reported in order to compare scenarios with different discount rates; 2) because of differences in the modelling, these numbers are not strictly comparable with those in the rest of the column; 3) because underlying preferences have changed in these scenarios, no meaningful welfare comparison can be drawn with REF or LC

and more than 20 per cent in the two lifestyle scenarios. This compares with the EU aspiration to reduce primary energy demand by 20 per cent below the 'business-as-usual' level by that date, which therefore looks to be difficult to attain without lifestyle change.

Turning to economic indicators (Table 12.2), the highest CO_2 marginal abatement costs occur in those scenarios that have the highest carbon reduction targets (*Super-ambition*), delayed carbon reduction (*Least-Cost Path*), or rely for decarbonization on both high levels of power generation and demand reduction (*ECO*). For the rest, meeting the 80 per cent carbon reduction target can be achieved with a carbon price of below £200/tCO₂. For comparison, if the current rate of fuel duty in the UK (about 50p/litre) were considered to reflect a carbon price, this would equate to about £200/tCO₂ (see Chapter 5).

With regard to energy system costs, as discussed in Chapter 4, there are two opposing tendencies when a carbon or resilience constraint is applied in the MARKAL-MED model. The use of more expensive low carbon or other technologies tends to increase the energy system cost, while reductions in energy demand (from the higher energy price) tend to reduce it. It can be seen that only in the *Resilient* and the two *Lifestyle* scenarios does the demand effect dominate the impact of more expensive supply technology. In the *Resilient* scenario, the energy system cost falls by £11bn. In all other scenarios, system costs increase. In the high carbon reduction scenarios and high fossil fuel price scenario the ranges from £18bn to £31bn. As a point of reference, in 2050 the total energy system cost in 2050 across the scenarios is £250–300bn. The increase in energy system costs due to decarbonization is therefore around 10 per cent in most cases.

The energy system cost calculation in the two lifestyle scenarios is not strictly comparable with that in the other scenarios because of modelling differences. However, the intuition behind the result seems sound: people demand less energy because their preferences have changed, rather than because energy prices have been increased through the deployment of more expensive technologies. The energy system is therefore smaller. The reduced size (about 30 per cent) is approximately in line with the reduction in energy demand.

Constraining the MARKAL-MED model always reduces social welfare because its unconstrained run is by definition optimal. The reductions in social welfare in most of the 80 per cent or more decarbonization scenarios amount to £28–59bn. The reduction in *Socially Optimal Least-Cost Path* is much lower because the lower discount rate involves earlier decarbonization and a lower carbon price. The losses of social welfare in the *Low Carbon Resilient* scenario are at the top end of the range. As noted in Chapter 6, this is an upper bound, because of the pessimistic assumption that energy price increases cause reductions in demand rather than an increase in uptake of relatively low-cost conservation measures. There is no obvious comparator to this reduction in social welfare in MARKAL-MED. However, to put its size in some kind of perspective, a £50 billion welfare loss as a proportion of UK current GDP (about £1.5 trillion) is about 3 per cent, but would be less than half that in 2050 assuming that UK GDP grows at 2 per cent annually in line with historic rates.

Multiple pathways to a low carbon economy

Chapter 5 presented a set of scenarios for UK energy system evolution from now to 2050 responding to the decarbonization imperative. The different pathways described feature different levels of carbon emission reductions by 2050, different interim reductions to 2020 and different weightings for early or later action. A key conclusion from across the scenarios is that achieving a low carbon energy system in the UK is technically feasible at an affordable cost.

Overall, three opportunities for decarbonization emerge across many of the scenarios: much greater emphasis on energy efficiency and demand reduction; deep decarbonization of electricity supply; and the use of low carbon electricity for decarbonizing transport and heat. Different parts of the energy system are decarbonized at different times from now to 2050. The early emphasis is on electricity decarbonization, followed by transport and heat sectors. Within this, there is some variation in the technologies and fuels deployed with, for example, different types of low carbon vehicles (e.g. electric vehicles, or biomass or hydrogen powered), and different levels of biomass or electricity use for domestic heating. Given the technical and social uncertainties involved here, this implies a role for policy in keeping options open. A key trade-off across the energy system is the speed of reduction in energy demand versus decarbonization of energy supply.

The emphasis on decarbonization of electricity supply, and then electrification of transport and heating, has become a new orthodoxy in research on UK decarbonization pathways. It is important to note, then, that the emphasis on different parts of the energy system at different times is sensitive to a number of input assumptions. For example, early action brings forward the date of transport sector decarbonization and the use of wind power as opposed to nuclear power in power sector decarbonization.

The use of a 'social' discount rate, which is lower than the market rate and gives greater relative weight to costs and benefits in later periods, inevitably results in earlier action than is the case with higher 'investor' discount rates, because early investment costs become relatively less expensive than later capital and running costs. Social discount rates, unsurprisingly, suggest lower overall costs for decarbonization. The message that emerges is the need for a society-wide response to the decarbonization challenge, reiterating the desirability of a wide ranging social contract for change.

Building in resilience

Chapter 6 presented a detailed analysis of the other major driver of energy system change identified in this book: maintaining and promoting system resilience. The key elements of resilience are: promoting reductions in energy demand and import dependence through efficiency; encouraging diversity of supply; and ensuring adequate investment in capacity and infrastructure. This led to the identification of a set of resilience indicators covering primary energy supply, energy infrastructure and end use. The 'macro' indicators address primary energy supply diversity, power generation diversity and the level of final energy demand (as a proxy for imports). There were also sector-specific reliability indicators for gas supply, gas infrastructure and electricity deriving from 'value of lost load' considerations. A key conclusion is that reducing energy demand is the key to a resilient energy system. This will reduce the UK's exposure to energy price shocks and could help the UK to ride out major disruptions to infrastructure

The *Resilient* scenario of UK energy system change that emerged from the analysis differed from the *Reference* and *Low Carbon* scenarios in important ways. Compared to the *Reference* scenario, for example, the *Resilient* scenario features significantly lower overall energy demand and CO_2 emissions, and a greater role for nuclear power rather than coal. Ensuring resilience entails some welfare losses under normal energy operations. However, reduced energy demand and carbon emissions result in lower energy system costs, suggesting a partial overlap or alignment between decarbonization and energy security drivers.

Chapter 6 also investigated the implications of major disruptions to gas supply taking place in the mid-2020s. The energy systems described by the *Resilient* and *Low Carbon Resilient* scenarios were more able to ride through the disruptions, mainly because of much lower levels of residential gas demand.

Considerable investment in gas storage, new interconnectors and LNG import facilities has already taken place on the basis of market incentives, and further such investment is likely in the future. However, the level of market investment could still leave the UK vulnerable in the event of catastrophic loss of infrastructure for periods of days or weeks. Chapter 6 assessed the value of additional investment in gas infrastructure in mitigating the impact of hypothesized disruptions. Using an insurance analogy, the costs of the mitigating investments were compared to the benefits that would arise in the event that disruptions took place. A key conclusion was that it would be hard to undertake additional infrastructure investment at an 'investor' rate of return but that, at a long-term 'social' discount rate, investment could be justified. Public policy would need to play a role in bringing forward such investments. The cost of 'strategic' investment is relatively modest compared to aggressive energy efficiency measures or back-up capacity for renewables. The degree to which strategic storage is justified is also dependent on the level of effort in bringing down final energy demand, especially in the residential sector. The analysis here suggests that the gas and/or electricity system could 'ride out' all but the worst infrastructure outages if demand levels fall.

Ensuring adequate capacity and infrastructure is largely down to market design and regulation. The current system has achieved this since electricity privatization, but the consequence of having large volumes of intermittent renewable energy on the system and the loss of indigenous natural gas supplies needs consideration. Current market arrangements may not induce sufficient investment in back-up capacity to ensure system reliability. Options such as capacity payments or allowing grid operators to earn a regulated return on back-up capacity need to be considered. Changes will be needed to the design and regulation of energy markets to facilitate the move to a resilient low carbon energy system, with, for example, much stronger incentives for transformational investments in supply and transmission infrastructures.

However, the key imperative in terms of resilience is to promote reductions in energy demand, either through the take-up of cost-effective energy conservation measures or through lifestyle changes that reduce demand for energy services. If supply-side measures that decarbonize electricity supply are not available, because of delays in deploying technology such as renewables or CCS for example, then greater levels of energy demand reduction will be required to stay on course for an 80 per cent reduction in CO_2 emissions by 2050. Seeking out cost effective efficiency is vital because there could be major welfare losses associated with forcing down energy demand through the price mechanism. A step change in policy delivery is required and the residential sector will be critical. Measures such as a new Supplier Obligation beyond 2012 and a careful look at the business models for delivering energy efficiency, particularly the role of the utilities vis-à-vis local authorities and others, is needed.

Accelerating technological change

Chapter 7 presented a series of 'accelerated technology development' scenarios which highlighted the potential role of technological innovation in underpinning long-term decarbonization pathways. By opening up new and diverse sources of low carbon energy supply, technological innovation has the potential to reduce significantly the overall cost to society of deep decarbonization between now and 2050. The more ambitious are the CO_2 reduction targets, the greater the long-term pay-off from investing in innovation. A key conclusion was that deploying new and improved technologies on the supply side will require a substantially increased long-term commitment to RD&D, the strengthening of financial incentives and the dismantling of regulatory and market barriers.

One concern raised by this chapter is the possible longer-term implications of ambitious short-term policy targets for decarbonization and renewable energy deployment. The accelerated technology development scenarios suggest that the required responses to meet 2020 targets – the setting up of large technology deployment programmes – should avoid locking out emerging supply technologies, such as advanced photovoltaics and marine energy, which have the potential to play a significant role in UK decarbonization pathways over the longer term.

There are multiple low carbon energy supply technologies now under development, many of which could potentially contribute to future energy mixes in the UK. Rather than attempting to draw up an exclusive list of technology 'winners' in the face of uncertainty about their progress, the recommendation is for a close alignment between innovation policy and energy policy, with an emphasis on maintaining diversity and option creation in early-stage innovation. Again, there are risk mitigation advantages here, as a broad portfolio of emerging options insures against the failure of particular technologies to develop.

Low carbon technology acceleration is essentially a global challenge and opportunity, but the UK has an important role to play where it has particular technological strengths. There are high-level implications for UK energy and innovation policy: the need to recognize the powerful role of innovation over time in generating fundamental system change, the need for sustained investment in innovation to secure substantial long-term benefit, the need to consider carefully the interrelatedness of system change over different timescales, and the path dependence of such change, and the possible longer-term consequences of short-term actions.

Decarbonizing electricity supply

A major theme that emerges from nearly all the scenarios is the importance of decarbonizing electricity. The three big centralized electricity options that are most important are large-scale renewables (mainly wind, but also biomass and marine renewables in the longer term), new nuclear and fossil fuels with CCS. The modelling indicates that, provided good levels of technological development are achieved over time, the costs entailed in the large-scale deployment of these technologies, while still uncertain, are likely to be affordable.

Renewables

Broadly, for renewables it is clear that time is running out if the UK's Renewable Energy Strategy target of achieving a 15 per cent share for renewable energy in final energy demand is to be achieved by 2020, as required by the EU Renewables Directive. This could require around 3000 extra 5MW offshore wind turbines, a deployment rate of about 5 per week, from a base of 0.8GW at the end of 2008. In addition, seven 3MW onshore wind turbines machines per week would be required, when at the end of 2008 the total operational capacity was only 1.7GW. The realization of these numbers strains credibility in the absence of a completely transformed policy landscape that clearly addresses all the issues relating to both financial incentives and non-financial constraints. In spite of these challenges associated with the existing target, one of the early actions of the UK Coalition Government formed in 2010 was to invite the Committee on Climate Change to assess the feasibility of increasing the ambition of the renewables targets, although the Committee's subsequent advice was against such an increase.

Nuclear

Looking further into the future, the Government is keen to ensure that there is a favourable policy framework for a new generation of nuclear power stations, with a view to new nuclear plant coming onstream by 2020 at the latest. The 2008 Nuclear White Paper set out a range of issues needing to be addressed include planning, site assessment, assessment of potential health impacts, design assessment and licensing, and review of the regulatory regime in general.

On the economics of nuclear, and therefore the potential need for public subsidy of new nuclear build, the Government had a remarkable change of mind in the early years of this century, from thinking in 2003 that it was 'an unattractive option', to believing in 2007 'that nuclear power stations would yield economic benefits to the UK in terms of reduced carbon emissions and security of supply benefits'. There is a presumption, which the new UK Government has clearly articulated, that new nuclear build would neither need nor get public subsidy. The Government has noted that:

> the Treasury and HMRC are, however, exploring the possibility that the timing of nuclear decommissioning could create a potential tax disadvantage for nuclear operators and, if so, whether it may be appropriate to take action to ensure a level fiscal playing field between nuclear power and other forms of electricity generation.

This may open the door to some public subsidy of decommissioning costs at least.

The issue is important, because if nuclear power is an important contributor to UK energy system resilience and decarbonization – as is suggested in some of the scenarios presented here – and if private companies decide that it is not in fact financially viable without public subsidy (as has been the case in the past), then new nuclear stations will not be built and energy system resilience and/or decarbonization will not be delivered.

The new UK Government's proposals for a minimum carbon price are clearly relevant here. This is more a reflection of the unaccounted environmental costs of fossil fuel combustion than a subsidy for low carbon energy, but its effect of improving the economic position of low carbon energy is much the same. Clearly much hangs on the level at which the minimum carbon price is set.

Carbon capture and storage

Compared to nuclear and large-scale wind, CCS is not a mature technology. Each component of CCS has been demonstrated, but a complete CCS system at the scale of a major power station has never been proven. The twin policy challenges associated with CCS are to demonstrate the technology, technically and economically, and to put in place a regulatory regime to encourage commercial deployment at an appropriate future time. The current UK policy intention is that no coal-fired power stations should operate without CCS by the mid-2020s.

As of 2010, the UK was running a domestic CCS competition covering post-combustion CCS (which can be fitted to existing plant), with only one project, Longannet, left in contention. A 900MW coal project at Hatfield was one of six successful bids for the EU CCS competition announced in 2009. In addition, the UK is prepared to support between one and three additional demonstration projects to be financed through a levy on electricity consumers provided for in the 2010 Energy Act.

Current policy is that no new coal plant will be permitted unless CCS is fitted to at least 300MW of capacity. The Coalition Government has announced that it intends to introduce an Emissions Performance Standard (EPS) which would effectively mandate CCS at coal-fired power stations. Fearing a further dash for unabated gas, the Committee on Climate Change has recommended that gas-fired CCGTs be brought under any future EPS and that at least one of the future demonstration projects be associated with gas-fired plant.

Micro-generation

Although the book does not consider any high-level system scenarios of high micro-generation futures, Chapter 8 considers the potentially important role of distributed generation – especially micro-generation – in the UK.

Three prospective types of micro-generation were identified and reviewed: low carbon heating (biomass boilers and heat pumps), micro-renewables (solar PV and solar thermal, micro-wind) and micro-CHP. Because 80 per cent of domestic energy demand is for heating, low carbon heating technologies, or the electrification of heating using low carbon electricity technologies, has a key role in high micro-generation pathways.

The chapter stressed the need to take a comprehensive view of micro-generation as ensembles of humans, building and technologies. The way domestic technologies are installed and used, and their interaction with behaviour, is key to their role in reducing demand and emissions. Institutions beyond the home also make a difference here: the training of suppliers, levels of manufacturing capacity (including the presence of enough large manufacturers to create competition in supply), and the key role of independent providers of information and advice. Although technology cost is an important issue and barrier, there are also many non-cost issues.

Micro-generation could challenge, quite fundamentally, current patterns of supply and use, and their supporting institutions. These have been optimized for large-scale supply, but this 'lock-in' is inimical to system innovation toward decentralized futures. More appropriate models of technology diffusion and business and/or enterprise are needed here, at community as well as house-holder level, with a focus on energy services, rather than sales. Without taking seriously this wider and more disruptive or radical agenda, it is argued, micro-generation could exacerbate lock-in around current relationships between energy suppliers and users, with little to show by way of demand-led system change.

The prospects for micro-generation and/or CHP in the medium term depend on the speed with which centralized electricity generation is decarbonized. The slower electricity decarbonization scenarios considered in the book, such as *Low Carbon Resilient* and *Low Carbon Lifestyles*, will leave the window for cost-effective deployment open for a much longer period.

Finally, on policy, Chapter 8 recognized a rapidly changing scene, with the introduction of feed-in tariffs for micro-generation, the renewable heat incentive, and the introduction of smart meters and smart networks. All of these promise a more dynamic relationship between supply and demand. At the same time, there is a need for policy design to avoid perverse incentives (such as excess supply or wasteful use being created by feed-in tariffs), and also, to take account of the wider system context for householder-level changes, such as the carbon intensity of grid electricity. These wider dependencies and linkages will to a large extent define the impact of micro-generation on policy objectives for decarbonization and security.

Policies for electricity decarbonization

Despite abundant technical potential at a seemingly affordable cost, renewables are not being deployed at anything like the required rate to meet the EU's 2020 target; the economics of new nuclear are still uncertain; CCS is still commercially unproven at the requisite scale; the next generation of renewable technologies are, as far as the UK is concerned, being developed and deployed very slowly; and non-financial barriers to new energy technologies of all kinds still seem to be pervasive and are difficult to remove. Despite the broadly optimistic conclusions that can be derived from the scenarios described in earlier chapters, there are serious risks that neither carbon reduction targets nor the necessary diversity in the energy system for resilience will be achieved.

Policy in this area must therefore focus on the following.

- Resolving outstanding technical uncertainties about these options, both for those technologies which are close to deployment (large-scale wind, new nuclear and CCS) and those which are currently further away on technical or cost grounds (e.g. marine renewables, photovoltaics, hydrogen fuel cells) but which could have an important role to play in the longer term. Substantially increasing UK commitment to RD&D for these technologies, as part of expanded wider international efforts, is of primary importance.

- Putting in place the necessary financial incentives, such as, for example, a substantial minimum carbon price, to cause the private sector to start making the investments in deployment at the necessary scale. There is also a need to bring forward important renewable technologies that the Renewables Obligation has so far marginalized by, for example, changing the banding of the RO to provide greater support for technologies further from full commercial deployment. Longer-term investments to develop more emergent options will require a greater share of public investment, alongside some private investment.

- Removing the non-financial barriers to large-scale deployment of the technologies. These seem particularly problematic in the UK context. For renewables, some of the relevant issues are the effectiveness of financial incentives, planning issues, grid issues, supply chain issues, information issues, network issues, and market structure.

- Reforming the governance arrangements for both electricity and gas. This was not a focus for the Energy 2050 project, and its discussion in the chapters above has therefore been limited. However, the views of both Ofgem and the UK Government on this issue, and their agreement on the need for reform, were noted in Chapter 3. It seems unlikely, in line with their analysis, that moves to a low carbon and secure energy system in the UK will take place at the required speed, or even at all, if the current arrangements stay in place.

Energy demand

A reduction in demand for energy through improved energy efficiency could play an important role both in achieving a more resilient energy system and in reducing the costs of CO_2 abatement. The results would be cost reductions deriving from a smaller energy system and reduced welfare costs. A step change is needed in the rate of improvement of energy efficiency.

In the *Reference* scenario, primary energy demand in 2025 is 7 per cent below that in 2000, and, in the *Resilient* scenario, because of the imposed 3.2 per cent annual reduction in energy intensity, is 14 per cent below the *Reference* case. This is broadly consistent with the 2006 EU Energy Efficiency Action Plan and with government estimates of planned energy efficiency measures having a high impact out to 2020. Beyond 2020, the same annual rate of improvement is assumed to be maintained. The *Resilient* and *Low Carbon Resilient* scenarios achieve greater percentage reductions in primary energy than all of the decarbonization scenarios (see Table 12.1, final column), other than the *Lifestyle* scenarios. Decarbonization strategies need to promote behaviour change as well as the uptake of low carbon technologies if the UK is to be in line with EU energy efficiency aspirations.

In principle, the means of reducing the economy's energy intensity seem clear: a combination of rising energy prices and measures to enhance the development and deployment of energy efficiency measures in the various end-use sectors of industrial processes, electric motors, buildings, vehicles and appliances, and to induce more energy-conserving behaviours. Although many energy efficiency measures are cost effective, achieving a 3.2 per cent annual reduction in energy intensity will be challenging.

Delivering improvements in energy intensity through measures that are economically attractive at prevailing prices has not often been achieved. Partly this has been because governments have been unwilling to increase energy prices above market rates to the extent that would be required. It has also been because of now well documented barriers to the take up of energy efficiency measures. People are unaware of the prices they pay for energy, the scale of their energy consumption and size of their energy bills. Even with recent energy price increases, energy bills are a low proportion of most people's expenditure. Energy efficiency technologies have a low profile and are of little interest to energy consumers. Consumers do not trust the expertise of the installers of energy efficiency technology or energy suppliers. People (both consumers and installers) are unaware or sceptical of the technologies that would help save energy. The installation of some energy efficiency technologies causes disruption to the household.

A range of policy instruments is therefore needed, including market-based instruments, information and advice and regulation. Governments in the UK have traditionally not favoured raising energy prices for fear of exacerbating fuel poverty, which adds to the priority of raising energy efficiency in existing housing to address this policy tension. A combination of advice, product regulation and increasingly large energy efficiency programmes (such as those mandated under CERT, the Carbon Emission Reduction Target) have been successful in raising the uptake of low-cost measures such as insulation, condensing boilers and efficient lights and appliances. In the future, more intrusive and expensive measures such as solid wall insulation may be needed.

The last UK Government's Heat and Energy Saving Strategy and the Coalition Government's proposed Green Deal recognize that the continuation of the current policy framework will be insufficient to deliver long-term goals. Deployment of higher-cost energy efficiency and micro-generation measures to further reduce demand will face all the familiar barriers as well as larger financing constraints. The need for a major shift in energy sector investment towards efficiency to address carbon goals is international, but because of the age and condition of the housing stock, the issue is particularly acute in the UK.

The required policies will need to be stronger than historically and different from those for low carbon energy supply, because households are less responsive to energy price incentives than energy businesses. Measures such as a new Supplier Obligation beyond 2012, availability of finance on attractive terms, and a careful look at the business models for delivering energy efficiency, particularly the role of the utilities vis-à-vis local authorities and others, are needed.

Lifestyle change

Chapter 9 took forward the analysis of energy demand by attempting to incorporate the prospect of radical changes in lifestyles and behaviours in system level scenarios. The basic question addressed was: how can lifestyle change drive reductions in energy use and carbon emissions?

Lifestyle changes that reduce energy demand would enhance energy system resilience and reduce the costs of CO_2 reduction. Compared to the *Low Carbon* scenario, overall energy system costs are much reduced, with savings of almost £100bn annually by 2050.

Explicit political intervention in lifestyle choices has not been on the agenda in the UK for many years. However, lifestyles have actually changed quite markedly since the 1980s, with the creation and legitimation of consumption to a far more significant degree than previously. In the energy sector, this was associated with rising expectations of abundant energy for households and for transport.

Such expectations can change. There is clearly potential for lifestyle change to contribute to decarbonization. Realizing this potential calls for a range of policy interventions that go beyond technology development, and that would seek to encourage and legitimize a different set of cultural, social, lifestyle and behavioural futures. Public policy of this kind seems likely to co-develop alongside, and seek to support, cultural change, rather than to lead or follow it.

The lifestyle scenarios described in Chapter 9 were based on the recognition that lifestyle change is already implicit in the socio-economic drivers associated with the more conventional carbon reduction scenarios. The lifestyle scenarios made it explicit by developing narratives and associated data assumptions about the social dynamics of energy service demands.

The lifestyle scenarios were developed using three different modelling tools, covering residential buildings, transport and the energy system overall. A historic analysis of changing lifestyles and social norms informed narratives of future changes, such as the rise of localism and the decline of long-distance transport. The system level scenarios that emerged from this detailed analysis are striking, with demand-related lifestyle changes providing major reductions in overall energy demand of 30 per cent, and reductions in the household sector of 50 per cent by 2050. These are comparable to the changes envisaged under 80 per cent decarbonization scenarios, but here without any explicit carbon emission constraint.

The lifestyle scenarios are also characterized by much less growth in the power supply sector, despite the electrification of the heating and transport sectors. Power sector CO_2 emissions fall by 31 per cent relative to the *Reference* scenario, despite the absence of any carbon emission constraint. This suggests that very substantial reductions in energy demand and emissions are possible from lifestyle changes, and that these can have an earlier impact than many supply-side changes. With lifestyle change contributing significantly to overall decarbonization pathways, much less fundamental changes to supply portfolios and infrastructures would be required between now and 2050. This implies that a policy approach with a successful early emphasis on behaviour change could apply less pressure for disruptive supply technology changes.

Historically there has been little emphasis in energy policy on how life-styles might be induced to change to reduce energy demand, or even whether this is something that policy can influence, except through prices. Yet making high-energy lifestyles much more expensive is likely to be all but infeasible politically unless these lifestyles come increasingly to be perceived (perhaps like smoking) as socially damaging. Addressing this issue effectively points to extending the reach and targets of energy policy to include the infrastruc-tures and institutions that provide the context within which energy choices are made, notably in housing and transport; and to education, both to ensure citi-zens have a clear understanding of climate and energy challenges and to reskill the huge numbers of professionals and tradespeople who will be employed in the required refashioning of the energy system. This in turn points to a greater emphasis in energy policy on the role of local government, where many of the relevant decisions on housing, transport and education are made.

Policy-makers could perhaps start by making high-energy lifestyles less necessary, for example by focusing more systematically, especially in towns and cities, on ensuring ready and safe access to basic services and amenities without the use of private vehicles, i.e. by walking, cycling or public transport. It is possible then that social norms would develop that would increasingly allow private vehicle access to be restricted, so that the low carbon modes could be expanded and used even more effectively. There is already evidence of this beginning to happen in some cities, notably London where car ownership is relatively low. However, the policy priority that has been given for so long to car ownership and use is still prevalent. Its abandonment would make a low carbon energy system much easier and less expensive to achieve.

Environmental concerns

In the face of climate change, the main environmental driver for energy policy has become CO_2 and other greenhouse gases. Chapter 10 considered wider environmental and socio-environmental aspects of decarbonization and resil-ience pathways. There were two distinctive parts to this analysis.

The first part dealt with environmental pressures in the core scenarios. This considered the possible environmental consequences associated with decarbonization and resilience pathways arising from energy system change, in terms of the emissions of several non-greenhouse gas pollutants. In addition a preliminary assessment of the consequences for water demand, land take and upstream carbon emissions was undertaken.

The second concern in this chapter was the possible impact of public atti-tudes towards new technologies associated with decarbonization pathways. Three distinctive *socio-environmental sensitivities* scenarios were devised, to represent different social aspects of socio-environmental concern: local environ-mental impacts (NIMBY), wider environmental concerns (ECO) and a concern for the environmental impacts of technology risk and failure (DREAD). The scenarios involved imposing particular constraints on different supply tech-nologies or fuels which reflected the underlying causes of concern.

Environmental pressures in the core scenarios

The overall outcome of this analysis was encouraging, in that reducing CO_2 will broadly lead to improvements in other environmental areas. However, regulatory attention may be needed in some areas (air quality, water stress) where there are potentially adverse effects.

Irrespective of the strategy, policies that manage the consumption of fossil fuels to deliver low carbon energy and secure supply tend also to reduce emissions of non-CO_2 pollutants. After 2020, the *Low Carbon Resilient* (LCR) scenario has significantly lower total pollution emissions than either the *Low Carbon* or *Resilient* scenarios. This suggests that while the pursuit of both low carbon and resilience goals benefits the environment, the greatest improvement is seen when they operate in combination. Each driver delivers distinctive environmental benefits.

However, there are some exceptions, with, for example, higher radioactive discharges in the *Low Carbon* scenario compared to the *Reference* scenario, associated with the use of nuclear power in meeting decarbonization targets. There are also likely to be additional pressures on water and land use resources associated with decarbonization and resilience, reflecting the much greater use of bio-energy crops and cooling water at thermal (fossil fuel and nuclear) power stations. The seriousness of these wider environmental consequences will vary greatly with location.

Environmental sensitivities

The consequences for decarbonization pathways arising from the concerns embodied in the *NIMBY*, *ECO* and *DREAD* scenarios are potentially very significant, with major restrictions on the availability or affordability of supply technologies and fuel resources. The impact of imposing specific constraints on one or more supply technologies, to reflect possible public opposition, has consequences across the energy system. For example, where supply-side constraints on technologies and fuels were particularly acute (in the *ECO* scenario), there is a much greater emphasis on demand reduction. The assumptions underlying the environmental sensitivity scenarios also result in much higher costs being incurred in reaching CO_2 emission reduction goals.

On the supply side, public opposition to low carbon technologies may have negative consequences for supply diversity. For example, heavy reliance on offshore wind power is suggested if other (non-renewable) low carbon supply options are greatly restricted, as in the *DREAD* scenario. Such scenarios would have significant impacts on energy infrastructure and the cost of energy with, for example, a much greater need for energy storage or back-up plant. At the same time, the constraints on large-scale supply technologies in the *DREAD* scenario are associated with a much greater uptake of micro-generation technologies.

Overall, the socio-environmental scenarios suggest that public concerns could significantly restrict the major changes in energy supply technologies and infrastructures needed to achieve an 80 per cent CO_2 reduction by 2050. The scenarios also indicate that there are potentially large additional costs involved in accommodating public opposition to supply-side change. For policy, this suggests the need to take public attitudes and opinion into account in the overall policy process, with more thorough-going and early processes of deliberation and engagement than seen to date. It also suggests that policy needs to be developed for other options, such as demand reduction and micro-generation, in case supply-side expansion is restricted by public concerns.

The international context

There are many uncertainties at the international level that will affect the evolution of the UK energy system. Chapter 11 focused on three of these: fossil fuel prices, biomass imports and the use of international CO_2 emission credits. The first two are in large part outside the control of UK policy, but each has potentially important consequences for the cost and feasibility of achieving a low carbon resilient energy system.

For example, with high fossil fuel prices, coal-fired generation using CCS has a significantly diminished role in achieving decarbonization, especially when combined with the pursuit of energy system resilience. In such scenarios, renewables and nuclear power have more significant roles. There are also consequences for energy system costs and demand. The scenarios also suggest that emission credits offer a useful source of long-term flexibility, as credit prices are anticipated to be lower than the cost of domestic action in the longer term.

Comparing 'best' and 'worst' case combinations of the three international uncertainties confirms the sensitivity of UK decarbonization pathways to international forces. The 'worst' case involved applying a complete restriction on biomass imports to the *Low Carbon Resilient* scenario. The 'best' case involved a combination of high fossil fuel prices and the availability of low-cost international credits. Paradoxically, high fossil fuel prices are associated with the 'best' case because they do some of the work that would otherwise need to be undertaken by explicit climate policies (although they could also substantially reduce GDP, which is not considered here). In the 'worst' case, there is a doubling of welfare costs by 2050, and a fourfold increase in the marginal cost of CO_2 abatement. The overall policy message is that the path towards UK decarbonization will be heavily influenced by international drivers. There is a consequent need to mitigate the associated vulnerabilities where possible.

In conclusion

A key role of research on energy systems – such as the work described in this book – is to help inform political and societal decision-making. In this way, the setting of priorities, targets and measures can be made in awareness of the full range of options and contingencies. In this spirit, the scenarios presented in this book explore possible pathways for UK energy system change from now to 2050. Each scenario prioritizes one or more of the different pressures being exerted on the system – e.g. economic, environmental or security – and each places a different emphasis on different aspects of system change such as electricity decarbonization, demand reduction or the deployment of bio-energy.

The basic message is optimistic, in that multiple possible pathways have been identified for an affordable transition to a resilient and low carbon UK energy system by 2050. This said, the practical task of translating this potential into reality faces a number of challenges. These are not easily represented in scenario exercises. Political, technological, economic and societal responses are needed that reflect the scale and extent of policy ambition.

Policy orientation

As noted in the first chapter of this book, private capital plays a key role in the energy system. This creates an imperative to establish stable and predictable conditions for large and small-scale project investment. At the same time, there is a need for flexibility of policy responses to changing conditions and priorities, and technological and social innovations. One key principle is that retrospective policy actions which discourage investment should be avoided.

Nevertheless, there is an inescapable trade-off between stability and flexibility. Energy technology history demonstrates the danger of inflexibly pursuing long-term programmes. On the other hand, there is the danger that 'keeping options open' will be interpreted as an excuse for incrementalism, prevarication and a shying away from necessary decisions.

A blend of stability and flexibility is required. In considering the balance, the fact that different incentives and actions have an effect over different time-scales needs to be considered. This theme is repeated throughout the book. At the same time, it should also be noted that 'contextual instability' – unexpected external events which actors cannot control and are unable to anticipate – can also be a catalyst for rapid system change. The business and financial sectors can sometimes respond unexpectedly and quickly to changing contexts, with dramatic – and sometimes positively innovative – consequences. The dash for gas, the largest change to the make-up of the UK energy supply over the last 20–30 years, was in large part the result of such an institutional change – privatization and market liberalization – that had not sought to promote it.

Policy will also need to promote the right mix of market dynamism and competition and state guidance and regulation. Precisely what this mix is will need to be determined, and the appropriate new governance arrangements put in place, in the near future if the required investment for a low carbon and secure energy system is to be unlocked. This is a major ongoing research focus for UKERC, and is certain to be a major source of public debate over 2011 and 2012, by which time decisions will need to be made.

The role of research

Energy systems are assemblies of connected parts. This interconnectedness needs to be understood, or policy may result in perverse incentives and unintended consequences. System level research and research tools allow structured exploration of the complex interdependencies involved. Research also needs to respond to the prospect of more radical system change, such as highly distributed systems of energy production and use.

Energy system change operates at different social and technological scales from the level of individual behaviours and lifestyles through to households, communities, public and private organizations, cities and regions, nations and international groups and fora. This book has focused mainly on one level – the UK energy system. However, the micro-level dynamics (such as those associated with lifestyles and households), and some of the UK–international linkages and dependencies involved, have also been considered.

Analysing long-term system change requires a thinking through of distinctive enablers and barriers to change which can be conceptualized as a series of technological, economic, institutional and societal/behavioural 'lock-ins'. Lock-in permits system optimization and incremental innovation, but represents a barrier to more radical system innovation.

Increasingly, there is a need for different research disciplines, perspectives and tools to be brought together to respond properly to the breadth and heterogeneity of the challenges involved in energy system change. This means difficulties, in terms of co-ordinating and reconciling different inputs, but there are also many rewards.

The breadth and complexity of system change also means that no single analytical perspective, or research discipline, can fully address it in its entirety. This book has taken a deliberately eclectic approach to scenario building, involving a combination of trend analysis, back-casting and more speculative exploration of technological, social and environmental drivers. This approach was taken in response to the underlying complexity, providing space for contributions from different disciplines. This underlines the value of a cross-disciplinary, whole-systems approach to energy systems research. There is a need to draw on different disciplinary insights and tools, and to iterate between a system level view and the detailed but inevitably partial insights deriving from individual disciplines.

The price of policy failure

This book has set out many possible pathways towards a low carbon and more resilient energy system for the UK. One thing that is certain is that none of these pathways will come about without strong and sustained policy intervention. Just as there is no single pathway to be predicted and preferred, so there is no unique policy package that will achieve it. All the packages will need some combination of the three core policy elements – carbon and energy pricing, technology support, and lifestyle and behaviour change – but the scenarios clearly show that different mixes of these elements are possible, and can achieve climate and energy security objectives at affordable cost, provided that the conditions for their successful implementation exist. The political challenge is both to help to create those conditions and articulate a policy approach that is sufficiently ambitious and can command adequate public support.

This will not be an easy challenge to address successfully, so it is worth remembering that failure to do so does not promise an easy political life either. The lack of energy system resilience in the face of the burgeoning energy uncertainties facing the UK over the next few decades will expose the UK to serious risks both to its economy and the way of life of its citizens. Policies that reduce these risks now, in a measured and deliberate way, seem on the balance of evidence likely to be very cost-effective insurance against the considerable risks of economic and social disruption that energy insecurity may otherwise bring about.

The same is even more true for climate change. Climate science now places a high probability on enormous disruption to human societies, some of which is already occurring, if carbon emissions are not stringently controlled. The UK can make little impact on global emissions by itself. But its endowment of low carbon resources and technological capacities and skills means that it is as well placed as any other country to reduce its emissions by a large margin. There is an urgent need for a large developed country to show that a low carbon transition is consistent with economic competitiveness, energy security and a high quality of life. The vast majority of expert analysis, including that in this book, consistently shows that this is possible. Future generations seeking to cope with runaway climate change will not judge kindly the failure of this generation to pursue vigorously some or many of the many low carbon opportunities and policies that this book has discussed.

Index

Accelerated Technology Development
(ATD): decarbonization pathways
198–208
  effects 203–6
  electricity demand 345
  electricity generation 204f7.6
  clcctricity supply 199–203
  hydrogen fuel cells 214
  innovation 352–3
  long term benefits 206–8
  marine renewables 202f7.4
  risks 213
  scenario analysis 195t7.1, 200t7.2,
  201t7.2
  technologies 192–6
  welfare 208f7.9
accidents 146, 148t6.2
affordability 1, 198, 206, 215
air pollution 33–5
assumptions: Accelerated Technology
Development (ATD) scenarios 193,
214–15
  behavioural factors 287
  core scenarios 91t4.4
  decarbonization 196
  demand 182
  energy usage 262
  environmental sensitivities 361
  gas supplies 176
  macro level resilience indicators 153–4
  MARKAL-MED (Markal elastic demand
  model) 93
  optimization model 134
  prices 92t4.5
  scenario analysis 325–8
  transport 137
awareness raising 50, 139, 222, 241–3,
259, 282

Bacton 180, 181t6.17
barriers 93, 237, 259, 307
barriers removal 50, 53, 61, 197, 357
behavioural factors: consumption 258
  lifestyle changes 275, 281, 282–4, 357
  low carbon 345
  micro-generation technologies 235,
  237–8, 241, 243, 252
  solar thermal water heating 228
  transport 286

BERR (Department for Business, Enterprise
and Regulatory Reform) 54, 58–9
bio-energy: Accelerated Technology
Development (ATD) scenarios 193
  constraints 315
  electricity generation 25, 212, 213
  emerging technologies 199
  land use 305f10.8
  NIMBY scenario 309
bio-fuels: bio-ethanol 18
  buses 101
  demand reduction 158
  energy security 279–80
  final energy demand 331
  goods vehicles 139
  LC-SO scenario 137
  sectoral energy demand 127f5.23
  transport 18, 98, 116, 129
  vehicles 136, 162
biomass: constraints 338
  decarbonization 51
  demand 113
  demand reduction 278
  ECO scenario 311
  electricity generation 22, 205f7.7
  energy supplies 17
  enhanced use 126
  flexibility 204
  heating 224
  imports 112, 114f5.9, 325, 327, 331
  modelling tools 94
  natural gas 116–17
  primary energy demand 156, 329
  renewables 141
  residential heating 35
  residential sector 223f8.3, 270
  service sector 116, 136
  small particulate matter (PM-10s) 301
  vehicles 334
buses 98, 98f4.12, 101, 130

capacity: back up 131
  capacity mix 99f4.13
  coal 166–7
  development 214
  electricity generation 23, 85–6, 130f5.27,
  334t11.5
  gas supplies 101, 176f6.19
  installed capacity 132f5.28, 315f10.11

intermittent generation 94*t4.7*
loss-of-load-expectation (LOLE)
175*f6.18*
margins 174*f6.17*
modelling tools 79
regulation 352
storage 152
transmission 89
vulnerability 150
car technologies 95*t4.8*
carbon ambition scenarios 110*f5.4*, 124
carbon budgets 75
carbon capture and storage (CCS):
Accelerated Technology Development
(ATD) scenarios 193, 199
assumptions 93–4
availability 212
coal 117–18
decarbonization 112, 131, 134, 197
demand reduction 164, 167, 278
development acceleration 187
DREAD scenario 312
electricity generation 209, 213, 355
emissions 140
gas supplies 30
low carbon 135
NIMBY scenario 309
non-availability 205
primary energy demand 157
subsidies 52
technologies 26–7
carbon dioxide (CO$_2$): cumulative emissions
108*f5.2*
electricity generation 170*f6.14*
emissions 32, 33*f2.17*, 69, 168,
314*f10.9*, 343
emissions by sector 100*f4.16*
emissions reduction 6, 48, 50–4, 75
emissions reduction policies 49–57
fuel cells 231–2
gas supplies 135
marginal costs 120–2, 318*f10.14*,
337*/11.6*, 338
pollutants 297
reduction 47
scenario results 107–8
sectoral emissions 109*f5.3*
transport 19
carbon dioxide (CO$_2$) emissions: combined
heat and power (CHP) 251*f8.9*
constraints 134–6
core scenarios 169*f6.13*
credit purchases 332*f11.3*
heat pumps 225
increases 162
marginal costs 121*f5.17*
marginal price 120*f5.16*
power sector 109–11

scenario analysis 264
sectoral emissions 332*f11.2*
carbon dioxide (CO$_2$) emissions credits:
availability 325
international availability 327*t11.2*
purchases 326–7
carbon dioxide (CO$_2$) emissions reduction:
Accelerated Technology Development
(ATD) scenarios 195, 198
coal 101
constraints 108*f5.1*, 131
energy demand 54–7
final energy demand 118
fuel switching 115
Large Combustion Plant Directive
(LCPD, European Union) 296
lifestyle changes 261, 359
low carbon marginal costs 121*f5.18*
marginal costs 137–8, 168, 170*f6.15*,
207, 281, 349
nuclear power 54
performance assessment 250
power sector 202
reference 344
residential sector 242
sectoral emissions 111*f5.5*
social welfare 124
targets 279
Carbon Emissions Reduction Target
(CERT) 46, 55, 56, 57, 247, 284
carbon monoxide (CO) 297, 301, 302*f10.5*
carbon pathways 106*t5.1*
carbon pricing 50, 54, 282, 285
cars: carbon dioxide (CO$_2$) emissions reduction 136
core scenarios 162*f.67*
demand 122
electricity generation 158
nitrogen oxides (NOx) 302
occupancy rates 128
ownership 276
passenger distance travelled 97*f4.11*,
273
challenges 208–14, 219, 245–52, 342–3
changes: energy systems 188–92, 245–6,
295–8
locking-in 214
NIMBY scenario 307
potential 242–4
residential sector 219–22
climate change: behavioural factors 243
concerns 11
emissions 2
energy 1, 342
environmental consequences 294
fossil fuels 69
greenhouse gases (GHGs) 45
mitigation 51

policy agenda 61–2, 344
technologies 189
UN Framework Convention on Climate
Change (UNFCC) 47
vulnerability 70–1
water 36
Climate Change Act (2008): carbon budgets
75
emissions reduction 4, 195–6
emissions reduction targets 105
greenhouse gases (GHGs) 49
policy agenda 45, 49
climate change levy 55, 56, 138
coal: capacity 166–7
carbon capture and storage (CCS) 199
demand 14
electricity generation 117, 158, 164, 191,
342
energy supplies 12, 13
environmental concerns 31
final energy demand 101
power sector 113
power stations 21–2, 131
primary energy demand 20, 156
residential sector 220
sulphur dioxide (SO$_2$) 299
coefficient of performance (COP) 225, 227
Combined Cycle Gas Turbine (CCGT)
21–2, 23, 45, 131, 199
Combined Gas and Electricity (CGEN)
model 79, 87, 88$f4.5$, 89–91, 145, 175
combined heat and power (CHP) 17, 222,
229, 249, 251$f8.9$, 355
Committee on Climate Change (CCC) 27,
49, 105
Community Energy Savings Programme
(CESP) 55, 247–8, 253
conservation measures 134, 138, 183, 191,
270
constraints: Accelerated Technology
Development (ATD) scenarios 195–6
biomass 338
carbon dioxide (CO$_2$) emissions 134–6
carbon dioxide (CO$_2$) emissions reduction 108, 131
cumulative emissions 137
decarbonization 110
demand 122
demand reduction 154, 315
diversity 155
final energy demand 154$t6.8$, 260,
330–1, 335
lifestyle scenarios 263
low carbon resilient 172
micro-wind turbines 228–9
mitigation 116
NIMBY scenario 307
resilience 155–6, 168, 182

social welfare 350
trade-offs 130
consumers 74, 238–41
consumption 189, 219, 222, 260–1, 280,
345
core scenarios: assumptions 91$t4.4$
capacity 167$f6.12$
carbon dioxide (CO$_2$) emissions
169$f6.13$
carbon monoxide (CO) 302$f10.5$
cars 162$f.67$
demand reduction 172$t6.9$
electricity demand 165$f6.10$
emissions 296
environmental pressures 360, 361
final energy demand 159$f6.3$
gas supplies 176$f6.19$
hypothetical shocks 178$t6.14$
lifestyle scenarios 262$t9.1$
pollutant emissions 298–306
primary energy demand 156
primary energy supplies 157$f6.2$
radioactive releases 304$t10.7$
residential sector 163$f6.8$
sectoral energy demand 160$f6.4$
social change 258–9
transport fuel demand 161$f6.5$
costs: Accelerated Technology Development
(ATD) scenarios 194$f7.3$
carbon dioxide (CO$_2$) 120–2
Combined Gas and Electricity (CGEN)
model 89
condensing boilers 224
decarbonization 138, 212, 215, 353
emissions credits 362
emissions reduction 361
energy systems 350
fuel cells 232
greenhouse gases (GHGs) 324
heat pumps 227
increases 206
internal combustion engines 229
investment 351
lifestyle changes 280–1
locking-in 189
micro-generation technologies 236–7,
240
reliability 175$t6.11$
research, development and demonstration
208–9
resilience 181, 182$t6.18$, 183
resource supply curves 81–2
resources 325
shocks 145–6
socio-environmental sensitivities 317
solar PV (solar photovoltaic) 227
solar thermal water heating 228
Stirling engines 231

technologies 306
transport 272–3
uncertainties 133
credit purchases 329–30, 332, 332*f11.3*, 337–8
cultural attitudes 282
cumulative emissions 108*f5.2*, 133, 137, 344
cycling 276

decarbonization: Accelerated Technology Development (ATD) scenarios 192, 198–208, 214
acceptability 317
affordability 206
behavioural factors 357
co-ordination 251
constraints 110
costs 138, 212, 215, 289, 353
development acceleration 187
diversity 157, 187
efficiency 115
electricity generation 135, 162–3, 345, 353–7
electricity sector 19, 139
emissions reduction 50–1, 221, 350
energy sector 39
energy supplies 112
energy systems 69–73, 342
flexibility 204
fossil fuels 117
gas supplies 158
global uncertainties 337
history 190–1
innovation 188
LC-SO scenario 107
lifestyle changes 359
micro-generation 222
pathways 105, 134, 313
policy agenda 191
policy instruments 51
power generation 118, 349
power sector 109, 131, 202
resilience 73
responses 196
short term 213
socio-environmental sensitivities 319, 360
targets 191–2
technologies 125, 195
transport 116, 129, 136
welfare costs 183, 207
demand: biomass 113
elasticity 125
electricity 11–12, 118, 316*f10.12*
electricity generation 120
electricity sector 28
energy 2, 7

energy sector 342
energy services 123*f5.19*, 259
final energy demand 14
hydrogen 206
management 191
natural gas 28*f2.14*
oil 27*f2.13*
price response 161*f6.6*
primary energy 12*f2.1*
reduction 32, 122–3
residential sector 269–70
transport 116
variability 172
demand reduction: bio-fuels 158
carbon capture and storage (CCS) 167
carbon monoxide (CO) 302
constraints 154, 315
core scenarios 172*t6.9*
decarbonization 350
energy 184
final energy demand 168, 335
hypothetical shocks 178
LC-LC scenario 137
price induced 169
prices 136, 160
development: acceleration 187
capacity 214
infrastructure 283–4
locking-in 190
micro-generation technologies 245
nuclear power 26
policy agenda 47–9
power sector 203
technologies 19
vehicles 142
Devine-Wright, P. 307
diffusion of innovation theory 238*f8.6*
distributed energy resources 219, 245–6
disturbance 70, 145, 150, 150*f6.1*, 279, 351
diversity: decarbonization 187
decline 101
DREAD scenario 313
electricity supply 164
energy security 58
energy supplies 69, 214
energy systems 341
gas supplies 158
low carbon 199
primary energy supplies 153, 157
reference 345
residential sector 220
socio-environmental sensitivities 319
technologies 130
vulnerability 150
DPSIR model 295, 296*f10.1*
DREAD scenario 311–13, 317, 345, 360, 361

ECO scenario 310–11, 313, 360, 361
economic implications 280–1
economic instruments 52, 55
economics 53, 354
Economics of Climate Change: The Stern Review, The 50
economy: contraction 32
  energy sector 1, 39
  energy security 2
  energy systems 111–14
  impacts 136, 336
  long term benefits 206
  transport 272
efficiency: appliances 285
  biomass heating 224
  condensing boilers 223
  decarbonization 115, 135, 350
  decline 191
  demand reduction 357
  electricity use 125, 268
  emissions reduction 69
  energy 2, 33
  energy conversion processes 111
  Energy Efficiency Action Plan 154
  energy generation 51
  energy policies 16
  energy sector 188
  energy usage 260–1
  engineering resilience 70
  fuel cells 232
  gas consumption 222
  heat pumps 225
  internal combustion engines 229
  micro-generation technologies 234, 241–2
  solar PV (solar photovoltaic) 227
  Stirling engines 231
  technologies 139
  transport 136, 272, 286
  vehicles 128, 274
elasticity 92, 93*t4.6*, 122, 206, 335
electricity: condensing boilers 224
  demand 14, 269
  demand reduction 205, 316*f10.12*
  final energy demand 115
  generation trends 21–3
  low carbon 330
  management 43–4
  transport 19
electricity demand 165*f6.10*, 345
electricity generation 99*f4.14*
  Accelerated Technology Development (ATD) 204*f7.6*
  capacity 130*f5.27*, 167*f6.12*, 334*t11.5*
  capacity mix 99*f4.13*
  carbon dioxide (CO_2) 171*f6.14*
  carbon dioxide (CO_2) emissions 134
  cars 158

changes 23*f2.11*
coal 342
decarbonization 135, 350, 353–5, 356–7
diversity 153, 155
energy supplies 11–12
environmental concerns 31–2
existing capacity 24*f2.12*
feed-in tariffs 52
final energy demand 331
fuel mix 118*f5.13*, 119*f5.14*
generation mix 164, 166*f6.11*, 314*f10.10*
infrastructure 28
investment 146
low carbon 333, 335
micro-generation 222
modelling tools 79
reliability 182, 183
renewables 22*f2.10*, 141, 189, 205–6, 209, 212
residential sector 205*f7.7*
scenario analysis 117–20, 306
technologies 130–3
technology mix 313
water 305
Wien Automatic System Planning (WASP) model 85–7
wind 120
electricity networks 87
electricity sector 19, 21*f2.9*, 29*f2.15*, 42, 139
electricity supply: Accelerated Technology Development (ATD) scenarios 199–203
  accidents 146
  decarbonization 221
  major disturbances 147*t6.1*
  reliability 173–4
  reliability indicators 173*t6.10*
  resilience 164–8
electricity use 135, 267–8, 277–8
emerging technologies 197, 198–9, 214, 216
emissions: carbon capture and storage (CCS) 27
  carbon dioxide (CO_2) 32, 33*f2.17*, 69, 168, 314*f10.9*
  carbon monoxide (CO) 302*f10.5*
  core scenarios 296
  credit purchases 329
  electricity generation 86
  energy sector 14
  fuel cells 232
  greenhouse gases (GHGs) 2
  hydrogen 331
  importance 297
  internal combustion engines 229
  nitrogen oxides (NO_x) 35*f2.19*
  nitrous oxide (N_2O) 303*f10.6*

power sector 164
radioactive waste 37–8
reduction 241–2, 328
residual 140
small particulate matter (PM-10s)
36*f2.20*, 301*f10.4*
sulphur dioxide (SO2) 34*f2.18*, 300*f10.3*
emissions by sector 100*f4.16*
emissions credits 334, 335, 338, 362
emissions reduction 6, 18, 49, 219
emissions reduction targets 105
end use sectors 125, 131, 134, 169, 172
energy 14–17, 184, 241–2
energy demand: flexibility 252
  goods vehicles 128
  households 16*f2.5*
  industry 18*f2.7*
  MARKAL-MED (Markal elastic demand)
  model 83–4, 137–9
  policy agenda 54–7
  reduction 357–8
  residential sector 263–5, 267
  sectoral demands 125*f5.21*
  technologies 17–19
  transport 17*f2.6*, 265–7, 271–2
  welfare costs 168–72
energy flows 230*f8.5*, 264
energy infrastructure 150, 151*t6.5*
Energy Market Assessment 60–1
energy markets 155
Energy Markets Outlook 58–9
energy mix 13, 220*f8.1*, 330*t11.4*
Energy Performance Certificates 285
energy policy 45–6, 47–9
energy production 36–7, 191
energy rating 57*f3.2*
energy sector: Carbon Emissions Reduction
  Target (CERT) 56
  competition 59
  consumption 189
  decarbonization 50–1
  economy 39
  environmental concerns 31–2
  installed capacity 315*f10.11*
  pollutants 306
  privatization 58, 191
  reference 344
  reforms 60
  role 1
energy security: climate change 342
  conceptualizing 71*f4.1*
  economy 2
  fears 146
  implications 279–80
  policy agenda 57–61, 62
  resilience 69–72
energy services: delivery 68
  demand 92, 115, 122, 259, 263

demand reduction 123*f5.19*, 136
lifestyle changes 359
micro-generation technologies 243
energy supplies: changes 306
  decarbonization 112
  diversity 155, 214
  end use 89
  GDP (gross domestic product) 13–14
  investment 172
  policies 139–42
  reliability 351
  Renewable Energy Strategy 51
  sources 152
  technologies 193
  trends 39
energy systems: Accelerated Technology
  Development (ATD) scenarios 203–6
  challenges 342–3
  changes 188–92, 245–6
  economy 111–14
  history 343–4
  impact of events 72*t4.1*
  investment 196
  MARKAL model 80–3
  methane (CH4) 300
  policy agenda 209, 341
  resilience 69–73, 145
  shocks 146–50
  social attitudes 320
  vulnerability 73
  welfare costs 171*f6.16*
Energy Technology and Systems Analysis
  Program (ETSAP) 78
energy usage 150, 260–1, 268*f9.2*
energy users 152*t6.6*, 172
Energy White Paper (2003) 53
Energy White Paper (2007) 53–4, 58–9, 69
*Energy White Paper: Meeting the Energy
  Challenge* 49
environment 1, 9, 31–8
environmental concerns 240, 258, 269–70,
  294, 342, 360–2
environmental pressures 294, 295–304,
  305–6, 360
EU Climate Policy 154–5
EU energy labelling 56
EU Integrated Product Policy 57
EU Renewables Directive 59, 117, 189,
  209, 354
European Union 2, 3, 48, 58
existing capacity 24*f2.12*
expansion 13, 85–6, 89, 91
expenditure 189, 190*f7.1*, 207–8

facilities lost 177*t6.13*
feed-in tariffs 52, 141–2, 237, 249, 356
final energy consumption 220–1, 230*f8.5*,
  258, 260

final energy demand: constraints 154*t6.8*,
172, 260
  core scenarios 159*f6.3*
  demand reduction 168
  different scenarios 115*f5.10*
  emissions reduction 107, 110
  end use sectors 124
  energy mix 331*f11.1*
  by fuel 15*f2.4*, 96*f4.9*
  fuel 116*f5.11*, 117*f5.12*
  gas supplies 175–6
  households 16*f2.5*, 243, 267, 343
  levels 152
  lifestyle changes 258, 288
  primary energy demand 19–20
  reduction 130, 135, 138, 154, 276–7,
338
  reference scenario 344–5
  residential heating 163*f6.9*
  residential sector 101, 125, 163*f6.8*
  resilience 158, 352
  scenario analysis 115–17
  by sector 96*f4.8*
  sectoral 15*f2.3*
  sectoral mix 14
  transport 17*f2.6*, 162, 333–4, 335*f11.4*
  value 173
flexibility: bio-energy 315
  biomass 204
  distributed energy resources 245
  emissions credits 333, 335, 338, 362
  energy demand 252
Flowers, Sir Brian: *Nuclear Power and the
Environment* (Flowers Report) 37
fossil fuel prices 325, 326*t11.1*
fossil fuels: consumption 298
  ECO 311, 313
  electricity generation 117
  energy production 191
  energy sources 69
  international trade 324
  low carbon 333
  NIMBY scenario 309
  prices 337–8, 349
  primary energy demand 20, 111–12, 113
fuel 20*f2.8*, 96*f4.9*, 97*f4.10*, 117*f5.12*,
130*f5.27*, 286
fuel cells 231–2
fuel demand 128–30
fuel flexibility 94–5
fuel mix 118*f5.13*, 119*f5.14*
fuel poverty 46, 47*f3.1*
fuel switching: bio-fuels 98
  carbon dioxide (CO_2) emissions reduction 115
  emissions 168
  emissions reduction 107
  final energy demand 116*f5.11*, 158

goods vehicles 128
greenhouse gases (GHGs) 32
  residential sector 221, 277
  technology switching 125
  transport 136, 334
fuel types 159*f6.3*, 163*f6.9*, 206*f7.8*
future 25, 269–71

gas capacity 59*f3.3*
gas networks 43, 87, 91, 149
gas projects 179*t6.16*
gas storage 30, 176, 351
gas supplies: accidents 148*t6.2*
  capacity 166, 315
  carbon dioxide (CO_2) emissions 116,
251
  diversity 158
  electricity generation 164
  Europe 1
  final energy demand 101, 115
  hypothetical shocks 177–9
  interruptions 179
  investment 176*t6.12*, 184
  management 44
  power generation 202
  primary energy demand 156
  privatization 43
  reduced 59
  reliability 174, 175*t6.11*, 176–7, 351
  reliability indicators 173*t6.10*
  reserve capacity 135
  residential sector 163, 220, 316*f10.13*
  resilience 179
  supply interruptions 146
  United Kingdom (UK) 30, 31*f2.16*
GDP (gross domestic product) 2, 13*f2.2*
generation mix: capacity 167*f6.12*
  electricity generation 155, 164,
166*f6.11*, 314*f10.10*
  electricity sector 21*f2.9*
  power generation 197, 203*f7.5*
  power sector 278
  transformed 199
global energy uncertainties 328*t11.3*
global production 149*t6.3*
globalization 2, 3, 193, 215, 342
goods vehicles 128, 130, 139
governance 3, 283, 342, 357
greenhouse gases (GHGs): carbon dioxide
(CO_2) 32
  climate change 45
  Climate Change Act (2008) 49
  costs 324
  emissions 2, 328
  emissions reduction 3–4, 48
  energy policy 360
  environmental concerns 31–2
  environmental pressures 295

final energy consumption 220–1
Kyoto Protocol 195
residential sector 221*f8.2*
growth 14, 25, 101, 115, 269, 273

heat flows 226*f8.4*
heat pumps: combined heat and power
(CHP) 251*f8.9*
decarbonization 251–2
electricity demand 127*f5.24*
emissions 168
installation 139
micro-generation technologies 355
residential sector 162–3, 225–7, 270,
277
heating 222, 231, 264, 267–9, 270, 300
Helm, D. 41
Holling, C. 70
households: appropriate technologies 236
biomass heating 224
condensing boilers 223–4
emissions 33
energy demand 14, 16*f2.5*
energy management 55
energy usage 268–9
final energy demand 16, 243, 260, 267,
277
fuel cell based micro-CHP 230*f8.5*
heat pumps 228
heating 16
internet 274
micro-generation 243
micro-wind turbines 228–9
property size 234–5, 272
solar PV (solar photovoltaic) 227
solar thermal water heating 228
hurdle rates 93, 106, 142
hybrid vehicles 19, 162, 333
hydrogen 19, 95, 113, 130, 206, 231
hydrogen fuel cells: Accelerated Technology
Development (ATD) scenarios 193, 199
development acceleration 187
DREAD scenario 312–13
electricity generation 203
transport 204–5
vehicles 213–14, 334
hydrogen vehicles 116
hypothetical shocks 179–81
core scenarios 178*t6.14*
facilities lost 177*t6.13*
impacts 179*t6.15*

IEA (international Energy Agency) 51–2, 78
impacts: Accelerated Technology
Development (ATD) scenarios 203–4
development acceleration 187
DREAD scenario 312–13
ECO scenario 310–11

hypothetical shocks 178, 179*t6.15*
NIMBY scenario 308–9
research, development and demonstration
208
socio-environmental sensitivities 319
welfare costs 336*f11.5*, 338
implications 184, 208–14, 279–80, 281–7
imports: biomass 112, 114*f5.9*, 327
decreased 222
dependence 12, 39, 45, 343, 351
electricity generation 88, 164
energy 3, 49, 101
energy security 58, 279
fossil fuel prices 326*t11.1*
gas supplies 174–5, 176–7
vulnerability 150
industry: decarbonization 109, 135
demand reduction 122, 136, 164
emissions 34–5
energy demand 18*f2.7*
final energy demand 14, 260
natural gas 116
process heat 17
sulphur dioxide (SO₂) 34
information: consumers 239
lack of 242
micro-generation technologies 237, 244,
355
policy instruments 50, 54, 56
smart metering 250
information technology 274, 275–6
infrastructure: bio-fuels 18
Combined Gas and Electricity (CGEN)
model 87–91
development 283–4
electric vehicles 19
energy sector 1, 4, 28–30
energy systems 342
gas projects 179*t6.16*
investment 153, 352
network expansion 90*f4.7*
NIMBY scenario 307
offshore 30
innovation: Accelerated Technology
Development (ATD) scenarios 352–3
chain 192*f7.2*
globalization 193
investment 208
mature technologies 197
reduction 282
research priorities 210*t7.3*, 211*t7.3*
risks 198
role 196
social attitudes 286
technologies 187, 188–92
installation 224, 225, 228, 233–4, 235–44
installed capacity 315*f10.11*
interconnection 30, 44, 175, 343

intermittent generation 94*t4.7*, 145
internal combustion engines 229–31
International Atomic Energy Agency (IAEA) 78–9, 85
International Energy Agency (IEA) 5, 149–50, 207–8, 325
international energy systems 324, 362
investment: costs 75, 351
   decline 191
   demand reduction 278
   distributed energy resources 245
   electricity generation 146
   energy sector 3, 74
   energy security 57, 58
   energy supplies 45, 172
   energy systems 196
   gas infrastructure 176*t6.12*
   gas storage 30, 351
   gas supplies 175
   hypothetical shocks 179
   infrastructure 153
   lack of 53
   locking-in 197
   low carbon technologies 189
   marginal costs 134
   micro-generation technologies 243
   mitigation 146, 180, 181*t6.17*
   nuclear power 54, 140
   oil 27–8
   private sector 51
   reliability 145
   renewables 184
   Renewables Obligation (RO) 52
   research, development and demonstration 188
   risks 21
   self-regulation 50
   technologies 284
   wind 167, 212

Kannan, R. 80, 324
key characteristics 80*t4.3*
Kyoto Protocol 47, 195

labour market 274, 276
land use 272, 275, 305*f10.8*, 306
Large Combustion Plant Directive (LCPD, European Union) 23, 34, 35, 59, 296, 299
LC-EA scenario: demand reduction 126
   electricity generation mix 119*f5.15*
   emissions reduction 108
   final energy demand 117*f5.12*
   installed capacity 132*f5.29*
   primary energy demand 113, 114*f5.8*
   sectoral energy demand 126*f5.22*
   transport sectoral energy demand 129*f5.26*

LC-LC: biomass scenario 142
   cumulative emissions 344
   demand reduction 126
   electricity generation mix 119*f5.15*
   final energy demand 117*f5.12*
   installed capacity 132*f5.29*
   least-cost scenario 106
   MARKAL-MED (Markal elastic demand) model 107
   mitigation 137
   primary energy demand 113, 114*f5.8*
   sectoral energy demand 126*f5.22*
   transport sectoral energy demand 129*f5.26*
LC-SO scenario: bio-fuels 137
   costs 122
   cumulative emissions 344
   decarbonization 130
   demand reduction 123, 126
   electricity generation mix 119*f5.15*
   emissions reduction 108
   final energy demand 117*f5.12*
   hurdle rates 142
   installed capacity 132*f5.29*
   MARKAL-MED (Markal elastic demand) model 107
   primary energy demand 113, 114*f5.8*
   sectoral energy demand 126*f5.22*
   social discount rate 106
   transport sectoral energy demand 129*f5.26*
liberalization 2–3, 21, 58, 191, 342
lifestyle 9, 261–7, 278*f9.3*
lifestyle changes: analysis 287
   behavioural factors 281
   consumption 189
   drivers 259–60
   economic implications 280–1
   energy consumption 258
   energy security 279
   evolution 74
   final energy demand 222, 358–60
   future trends 269–71
   low carbon 276–81
   low energy 274–6
   micro-generation technologies 235
   policy agenda 289, 360
   residential sector 267–71
   scenario analysis 73
   transport 266
locking-in: Accelerated Technology Development (ATD) scenarios 353
   changes 214
   consumption 260
   emerging technologies 216
   micro-generation technologies 356
   residential sector 243
   technologies 190, 197

long term benefits 206–8
loss-of-load expectation (LOLE): capacity
   margins 175*f6.18*
   reliability 86
   renewables 174
   resilience 153
   shocks 146
   vulnerability 150
   Wien Automatic System Planning (WASP)
      model 79
loss-of-load probability (LOLP) 172–3
low carbon: Accelerated Technology
   Development (ATD) scenarios 194
   assumptions 75
   behavioural factors 345
   constraints 172
   demand reduction 126
   diversity 199
   electricity 330
   electricity generation 333
   electricity generation mix 119*f5.15*
   end use sectors 118
   energy pathways 337
   final energy demand 117*f5.12*, 158
   installed capacity 132*f5.29*
   lifestyle changes 276–81
   marginal costs 121*f5.18*
   nuclear power 168
   pathways 350–1
   pollutants 298–9
   power generation 131
   primary energy demand 114*f5.8*, 156
   residential sector 162
   scenario analysis 6, 8, 74, 205, 306
   sectoral energy demand 126*f5.22*
   technologies 51, 189, 214, 349
   transport sectoral energy demand
      129*f5.26*
low carbon heating 222, 355
low carbon resilient: assumptions 75
   conservation measures 183
   constraints 172
   electricity generation 164
   final energy demand 158
   MARKAL-MED (Markal elastic demand)
      model 145
   pollutants 298–9
   primary energy demand 156
   scenario analysis 6, 8, 74
Low Carbon Transition Plan 141, 213
LS-REF scenario 262, 276, 277*t9.2*

macro level 153*t6.7*, 155
major disturbances 147*t6.1*
marginal costs: carbon dioxide (CO₂)
   318*f10.14*, 336, 337*f11.6*, 338
   carbon dioxide (CO₂) emissions reduc-
      tion 168, 170*f6.15*, 207, 281, 349

investment 134
   socio-environmental sensitivities 317
margins 174*f6.17*
marine renewables: Accelerated Technology
   Development (ATD) scenario 193, 199,
   202, 202*f7.4*
   costs 131
   electricity generation 24–5
   targets 141
   zero-carbon source 135
MARKAL-MED (Markal elastic demand)
   model: Accelerated Technology
   Development (ATD) scenarios 194
   assumptions 93
   buildings sectors 335
   car technologies 95*t4.8*
   core scenarios 296
   decarbonization 131
   demand reduction 183
   diversity 155
   elasticity 206
   energy demand 137–9
   energy systems 349
   international energy systems 324
   least-cost scenario 120
   lifestyle changes 276
   modelling tools 79, 83–4, 195, 262,
   264–5
   reliability 172
   resilience 145
   results 105
   scenario analysis 141, 196, 306
   sensitivity tests 153, 154
   supply and demand 85*f4.4*
   Wien Automatic System Planning (WASP)
      model 87
markets 41, 45, 49, 93
methane (CH₄) 297, 299, 300
micro-generation 219, 222, 223*f8.3*,
   228–9, 232, 233
micro-generation diffusion 239*f8.7*,
   244*f8.8*
micro-generation technologies: comparison
   233*t8.1*
   consumers 238–41
   costs 240
   distributed energy resources 246
   electricity generation 355–6
   equipment size 234
   feed-in tariffs 249
   installation 358
   installers 236–8
   net zero carbon 247
   operation 235
   policy agenda 245–52
   renewables 271
micro-wind turbines 228–9, 233–4, 317
mitigation: carbon dioxide (CO₂) emissions

reduction 115
constraints 116
decarbonization 136
demand reduction 123
hypothetical shocks 179–81
investment 181t6.17
policy agenda 279
shocks 146
mobility 265–7, 271–3, 286
modelling process 263f9.1
modelling studies 258, 262
modelling tools: Accelerated Technology
Development (ATD) scenarios 188
choices 265
DPSIR model 295, 296f10.1
energy systems 203
gas networks 87
innovation 192
key characteristics 80t4.3
lifestyle changes 288
scenario analysis 78–91
tandem operation 81f4.3
technologies 190
monitoring 304

national energy systems 67–8, 324
natural gas: demand 14
demand reduction 277
dependence 17
depletion 45
energy supplies 13
hydrogen 231
industry 116–17
internal combustion engines 229
power stations 21
primary energy demand 20
production 28f2.14
sulphur dioxide (SO₂) 34
network expansion 90f4.7
NIMBY scenario 307–9, 360, 361
nitrogen oxides (NOₓ) 33, 34, 35f2.19,
297, 301, 302–3
nitrous oxide (N₂O) 303f10.6
nuclear power: Accelerated Technology
Development (ATD) scenarios 193, 199,
202
capacity 166
decarbonization 51
development acceleration 187
DREAD scenario 312, 313
electricity generation 26, 164, 191, 213,
354–5
energy security 279–80
energy supplies 13
environmental concerns 32
final energy demand 338
France 36
investment 184

low carbon 168, 333
marginal costs 134
NIMBY scenario 309
power generation 197
power stations 53, 131, 140
primary energy demand 112, 329
resilience 351
transport 19
zero-carbon source 135
Nuclear Power and the Environment
(Flowers Report) 37

Office of Gas and Electricity Markets
(OFGEM) 42, 43, 45, 60, 146
offshore wind 30, 133f5.30
oil: crises 13
demand 14
demand reduction 277
dependence 18
depletion 1, 45
ECO 311
energy security 59
global production 149t6.3
primary energy demand 20, 156
resource availability 279
supply disruptions 149, 150f6.1
supply interruptions 146
transport 16
UK production 27f2.13
optimization model 88f4.5, 134

passenger distance travelled 97f4.11,
98f4.12, 128, 273
pathways: decarbonization 105, 134, 313
locking-in 189
low carbon 337, 350–1
technologies 333–5
perceptions 294, 312, 313
performance and suitability 222–3, 232,
235, 246, 249
policy agenda: awareness raising 139
carbon dioxide (CO₂) emissions reduction 49
challenges 245–52
climate change 45, 47–9
decarbonization 134, 191
efficiency 16
electricity generation 356–7
electricity supply 44
emissions reduction 50–4, 107
energy demand 54–7
energy policy 45–7
energy resilience 68
energy sector 1, 2, 3
energy security 57–61
energy systems 363
environmental concerns 258
failure 364–5

goals 12
greenhouse gases (GHGs) 32
implications 105, 184, 281–7
incertitude 72
intervention 344
lifestyle changes 289
longer term 213–14
low carbon 324
micro-generation technologies 243
mitigation 279
public institutions 41
Renewable Energy Strategy (RES) 141
residential sector 244
role 259
scenario analysis 138
short term 209–13
policy aspirations 101, 189
policy framework 7, 252–3, 354
policy instruments 49, 246, 249, 282, 358
policy mechanisms 246–50
policy support 250–2
politics 149, 364–5
pollutants 298f10.2, 299–306, 320, 360
power generation: decarbonization 118, 349
expansion 85
generation mix 197, 203f7.5
low carbon options 131
technologies 89
power sector: carbon dioxide (CO_2) emissions 109–11
coal 113
decarbonization 134, 135, 199, 202, 313
development 203
diversity 164
DREAD scenario 313
emissions 32–3
generation mix 278
lifestyle changes 359
power stations: coal 131, 342
Combined Cycle Gas Turbine (CCGT) 21–2, 45
emissions 34
Large Combustion Plant Directive (LCPD, European Union) 59
locations 89
nuclear power 53, 140
radioactive waste 38
sulphur dioxide (SO_2) 299
Pressure-State-Response (PSR) model 295
prices: assumptions 92t4.5
carbon dioxide (CO_2) emissions credits 326–7
demand 164
demand reduction 136, 160, 183
elasticity 93t4.6, 122, 206, 335
energy demand 83–4
final energy demand 138, 269, 358

fossil fuel imports 325
fossil fuels 329, 349
internal combustion engines 229
micro-generation technologies 237
solar PV (solar photovoltaic) 227
volatility 72
primary energy demand 100f4.15
changes 19–20
electricity generation 13
energy mix 330t11.4
energy security 279
fuel 20f2.8
fuel types 206f7.8
GDP (gross domestic product) 13f2.2
international uncertainties 329
reduction 48, 349
reference 344, 357
resilience 156–7
scenario analysis 111, 112f5.6, 113f5.7, 114f5.8
trends 14
primary energy supplies 73, 88, 150, 151t6.4, 153, 157f6.2
priorities 139, 232
privatization 42, 58, 191
production 28f2.14
Pye, S. 80

quantification 261–7

radioactive releases 304t10.7
radioactive waste 37, 38t2.1
Radioactive Waste Management: *Nuclear Power and the Environment* (Flowers Report) 37–8
radioactivity 297, 303–4, 361
real fuel prices 47f3.1
reduction: demand 122–3
emissions 328
final energy demand 138, 276–7, 282, 338, 345
primary energy demand 349
Reference scenario: baseline scenario 75
carbon dioxide (CO_2) emissions reduction 344
demand reduction 172t6.9
electricity generation 117, 130
electricity supply 164
Energy White Paper (2007) 74
final energy demand 158
final energy use 278f9.3
MARKAL-MED (Markal elastic demand) model 145
policy aspirations 101
power sector 109
primary energy demand 156
scenario analysis 5, 6, 8, 68–9, 342
scenario results 95–101

technologies 134
transport 158
welfare costs 171*f6.16*
reforms 60, 142, 344
regulation: capacity 352
  energy sector 3, 42, 59, 283
  micro-generation technologies 244
  new buildings 247, 253, 284–5
  nuclear power 140
  Office of Gas and Electricity Markets
  (OFGEM) 46
  policy instruments 49, 52, 54
  radioactive waste 304
  transport 273
  white certificates 247–8
reliability: biomass heating 224
  capacity 174*f6.17*, 175*t6.11*
  electricity generation 153
  electricity supply 173–4
  energy supplies 69, 280, 351
  energy systems 145
  fuel cells 232
  gas supplies 174–7
  indicators 173*t6.10*
  loss-of-load-expectation (LOLE) 86
  micro-wind turbines 229
  network industries 172–7
  solar PV (solar photovoltaic) 227
Renewable Energy Strategy (RES) 51,
  52–3, 141, 212, 353
Renewable Heat Incentive 52, 141, 249
renewables: Accelerated Technology
  Development (ATD) scenarios 193
  assumptions 94
  decarbonization 51
  deployment 140–1, 209, 213
  development acceleration 187
  electricity generation 22*f2.10*, 164, 191,
  203, 205–6, 353–4
  energy 2
  energy supplies 13
  environmental concerns 32
  EU climate and energy package (2008) 48
  final energy demand 14, 101
  interconnection 343
  intermittent generation 145
  investment 184, 352
  marine renewables 24–5
  micro-generation technologies 271
  primary energy demand 329
  reliability 173
  solar PV (solar photovoltaic) 222
  targets 192
  technologies 189
  transmission 30
  transport 19
Renewables Obligation (RO) 46, 52, 109,
  111, 141, 191

Renewables Transport Fuel Obligation
  (RTFO) 53, 158
research, development and demonstration (RD&D): Accelerated Technology
  Development (ATD) 192, 353
  energy sector 3
  energy systems 362
  expenditure 187–8, 189, 190*f7.1*, 207–8
  globalization 215
  investment 51
  priorities 232
  role 363–4
  technologies 8
research priorities 208–9, 210*t7.3*, 211*t7.3*
residential heating 35, 163*f6.9*
residential sector *see also* households:
  biomass 112, 205*f7.7*
  carbon dioxide (CO$_2$) emissions reduction 242
  challenges 219–22
  conservation measures 183
  consumption 252
  decarbonization 110–11, 126
  demand reduction 122–3, 136
  distributed energy resources 245
  DREAD scenario 313
  electricity use 135
  emissions 168, 331, 333
  energy demand 263–5
  energy mix 220*f8.1*
  final energy demand 101, 125, 163*f6.8*,
  316*f10.12*, 316*f10.13*, 343
  greenhouse gases (GHGs) 221*f8.2*
  heat flows 226*f8.4*
  heat pumps 205, 225–7
  lifestyle changes 267–71
  low carbon 335
  micro-generation 223*f8.3*
  micro-generation diffusion 239*f8.7*,
  244*f8.8*
  micro-wind turbines 228–9
  natural gas 116
  policy mechanisms 246–50
  pollutants 299
  primary energy demand 156
  refurbishment 270
  resilience 162–3
  scenario analysis 329–33
  small particulate matter (PM-10s) 301
  solar PV (solar photovoltaic) 227
  Stirling engines 231
resilience: assumptions 75
  constraints 168
  costs 75, 146, 181, 182*t6.18*, 183, 336
  DREAD scenario 317
  energy infrastructure 151*t6.5*
  energy sector 8, 365
  energy security 69–72

energy systems 69–73, 342, 351–2
energy users 152*t6.6*
final energy demand 158, 330–1
goals 183
imports 279
indicators 150–2, 153*t6.7*
primary energy demand 156–7
primary energy supplies 151*t6.4*
residential sector 162–3, 335
scenario analysis 6, 8, 73, 74, 154
shocks 72–3, 145
supply and demand 68
transport 158–62
resources 1, 81–2, 88, 198, 204, 219
results 105, 313–18
Riese, H. 307
risks 72, 198, 213, 294
Royal Commission on Environmental
Pollution 6
Nuclear Power and the Environment
(Flowers Report) 37

sankey diagram 226*f8.4*, 230*f8.5*
scenario analysis: Accelerated Technology
Development (ATD) scenarios 188,
192–3, 194*f7.3*, 195*t7.1*, 200*t7.2*,
201*t7.2*
bio-fuels 18
carbon dioxide (CO$_2$) emissions 264
comparison 344–5, 346*t12.1*, 347*t12.1*,
348–50
cumulative emissions 129
design 105–6
different scenarios 115*f5.10*
DREAD scenario 311–13
economic variables 348*t12.2*, 349*t12.2*
electricity generation 119*f5.14*
energy sector 4–5
energy supply policies 139–42
environmental pressures 297
final energy use 278*f9.3*
framework 73–8
global energy uncertainties 328*t11.3*
heat pumps 127*f5.24*
installed capacity 132*f5.28*
international energy systems 324, 327–8
lifestyle changes 281
low carbon 205, 306
MARKAL-MED (Markal elastic demand)
model 141
modelling tools 78–91
parameters 137
pathways 136
primary energy demand 111, 112*f5.6*,
113*f5.7*, 114*f5.8*
reference 342
research, development and demonstration
188

resilience 154, 181
results 95–101, 313–18, 329–33
scenarios and variants 76*t4.2*, 77*t4.2*
sectoral energy demand 125*f5.21*
shocks 145
social welfare 124*f5.20*
systems modelling 196–8
technologies 189
scenario results 107, 108*f5.1*
sectoral emissions 109*f5.3*, 110*f5.4*,
111*f5.5*, 314*f10.9*, 332*f11.2*
sectoral energy demand: bio-fuels 127*f5.23*
carbon dioxide (CO$_2$) emissions
169*f6.13*
core scenarios 160*f6.4*
electricity demand 165*f6.10*
energy mix 14
scenario analysis 125*f5.21*, 126*f5.22*
technologies 124–8
transport 128*f5.25*, 129*f5.26*
sectoral modelling 263, 266
sectors 96*f4.8*
security 2, 4, 11, 155
selective catalytic reduction (SCR) 35
self-sufficiency 27–8, 174, 343
sensitivity tests 153, 154, 183
service sector 116, 126, 135–6, 138
Severn tidal barrage 309, 311
shocks 69, 72–3, 145, 146–50, 179, 181
skill base 284
small particulate matter (PM-10s) 35,
36*f2.20*, 297, 299, 300, 301*f10.4*
smart metering 249–50, 270, 285
social attitudes: carbon pricing 282
decarbonization 307
environmental concerns 360
lifestyle changes 271
NIMBY scenario 308
technologies 361–2
transport 275, 287
social change 258–9, 282–4
social discount rate 106, 351
social welfare: Accelerated Technology
Development (ATD) scenarios 195
constraints 350
decarbonization 196
energy surplus 318*f10.15*
lifestyle changes 280
loss 123–4
scenario analysis 124*f5.20*
socio-environmental sensitivities scenarios
294, 306, 319, 360
solar PV (solar photovoltaic): Accelerated
Technology Development (ATD)
scenarios 193, 194*f7.3*, 199
electricity generation 25–6
installers 236
locations 233

renewables 222
residential sector 223*f8.3*, 227, 271
solar thermal water heating 228, 242
sources 152, 155, 298*f10.2*, 303–4
Stern, N. 50, 281
storage 37, 44, 152, 180–1, 196, 224
Stirling, A. 71
Strachan, N. 80, 95, 324
subsidies 54
sulphur dioxide (SO₂) 33, 34*f2.18*, 297, 299, 300*f10.3*
summary results 277*t9.2*
supply and demand 41, 59*f3.3*, 68, 74, 85*f4.4*, 176*f6.19*
supply system 29*f2.15*
sustainability: bio-energy 19
bio-fuels 18
biomass 204, 311, 327
consumption 258
Office of Gas and Electricity Markets (OFGEM) 45–6
Sustainable Energies Act (2003) 45
systems modelling 196–8

tandem operation 81*f4.3*
targets: carbon dioxide (CO₂) emissions reduction 279, 326
Carbon Emissions Reduction Target (CERT) 247
decarbonization 188
Renewable Transport Fuel Obligation 53
renewables 142
Renewables Obligation (RO) 141
United Kingdom (UK) 189
taxation 55, 138, 283, 285
technologies *see also* emerging technologies: accelerated development scenarios 192–6
acceptability 317
buildings sectors 335
car technologies 95*t4.8*
carbon capture and storage (CCS) 355
carbon dioxide (CO₂) emissions reduction 69
Carbon Trust 214
choices 142
costs 208–9
decarbonization 126, 195
development 19, 140
development acceleration 187
diffusion 356
diversity 130
DREAD scenario 311
ECO scenario 310
efficiency 139, 358
electricity generation 23–7, 130–3, 213
emissions reduction 107
energy sector 7, 39

energy supplies 193
expenditure 189
future 17–19
hurdle rates 106, 232
impacts 80
innovation 188–92
interaction 259
investment 284
lifestyle changes 288
locking-in 190, 197
low carbon 349
low energy 269–70
micro-generation 219, 222
monitoring 235
NIMBY scenario 307
nitrogen oxides (NOₓ) 303
parameters 133
pathways 333–5
performance and suitability 222–3
policy agenda 356
policy instruments 282
power generation 89
price support 51
reference 134
Renewable Heat Incentive 52
renewables 189
residential sector 285–6
resilience 155
resistance 313–18
sectoral energy demand 124–8
social attitudes 278, 361–2
social status 239
socio-environmental sensitivities 319
storage 196
supply and demand 68, 74
transport 136, 160
vehicles 128, 158, 162, 287, 334
technology switching 129, 135
tidal power 310–11
toxicity 300–1, 302, 303
trade 2, 3, 12, 18, 27–8, 324
trade-offs 124–8, 130, 137
transport: bio-fuels 112, 116
carbon monoxide (CO) 301
cycling 276
decarbonization 51, 110, 126, 135, 136, 350
demand 343
demand price response 161*f6.6*
demand reduction 125, 277
electricity 19
emissions 33
energy demand 265–7
final energy demand 130, 159*f6.3*, 161*f6.5*, 172, 333–4, 335*f11.4*
fuel demand 97*f4.10*, 128–30
growth 14
hydrogen 113

hydrogen fuel cells 204–5, 331
lifestyle changes 271–6
nitrogen oxides (NO$_x$) 34–5
policy agenda 286–7
pollutants 299
resilience 158–62
sectoral energy demand 128f5.25
small particulate matter (PM-10s) 35, 301
taxation 283
technologies 17–18, 160
trends 7, 12–14, 39, 269–71, 273–4

UK Domestic Carbon Model (UKDCM) 92, 139, 182, 262, 264
UK Energy Research Centre (UKERC) 6–7, 74f4.2, 188, 212
UK fridge freezers 57f3.2
UK Low-Carbon Transition Plan: National strategy for climate and energy 48–9
UK Transport Carbon Model (UKTCM) 262, 266–7
UN Framework Convention on Climate Change (UNFCC) 47
United Kingdom (UK): carbon dioxide (CO$_2$) emissions 48
    Emissions Trading Scheme (ETS) 52
    gas supplies 31f2.16
    GDP 2
    liberalization 3
    targets 189
    water 36

value of lost load (VOLL) 79, 153, 173
vehicles: bio-fuels 116, 136
    decarbonization 350
    development 139, 142
    efficiency 274
    electric 119, 213

hybrid vehicles 162
low carbon 287
MARKAL-MED (Markal elastic demand) model 95t4.8
technologies 128, 158, 286, 334, 342
voluntary agreements 50, 54
vulnerability 70–1, 72–3, 150, 352

water 16, 32, 36, 37f2.21, 305
welfare 69, 79, 123–4, 208f7.9
welfare costs: cumulative costs 137
    decarbonization 136, 207
    energy demand 168–72
    energy systems 171f6.16
    impacts 183, 336f11.5, 338
    investment 181t6.17
    resilience 351
    value of lost load (VOLL) 173
white certificates 247–8
Wien Automatic System Planning (WASP) model 78, 85–7, 173
wind: Accelerated Technology Development (ATD) 199, 213
    assumptions 94
    development acceleration 187
    DREAD scenario 313
    ECO scenario 310
    electricity generation 24, 120, 131, 164, 212
    installed capacity 133f5.30
    investment 167
    land use 305
    NIMBY scenario 308
    planning 142
    renewables 141
Winskel, M. 93, 209
World Energy Outlook 5, 325

zero-carbon source 135

Milton Keynes UK
Ingram Content Group UK Ltd.
UKHW031140141024
449569UK00024B/1188